ADVANCES IN

CHROMATOGRAPHY

Volume 25

ADVANCES IN

CHROMATOGRAPHY

Volume 25

Edited by

J. CALVIN GIDDINGS

EXECUTIVE EDITOR

DEPARTMENT OF CHEMISTRY
UNIVERSITY OF UTAH
SALT LAKE CITY, UTAH

ELI GRUSHKA

GAS CHROMATOGRAPHY AND LIQUID CHROMATOGRAPHY

DEPARTMENT OF INORGANIC AND ANALYTICAL CHEMISTRY
THE HEBREW UNIVERSITY OF JERUSALEM
JERUSALEM, ISRAEL

JACK CAZES

MACROMOLECULAR AND INDUSTRIAL CHROMATOGRAPHY

SILVER SPRING, MARYLAND

PHYLLIS R. BROWN

BIOCHEMICAL CHROMATOGRAPHY

DEPARTMENT OF CHEMISTRY
UNIVERSITY OF RHODE ISLAND
KINGSTON, RHODE ISLAND

MARCEL DEKKER, Inc.   New York and Basel

Library of Congress Cataloging in Publication Data
Main entry under title:

Advances in chromatography. v. 1-
  1965-
  New York, M. Dekker
    v. illus. 24 cm.
    Editors: v.1- J.C. Giddings and R.A. Keller.
    1. Chromatographic analysis-Addresses, essays, lectures.
I. Giddings, John Calvin, [date] ed. II. Keller, Roy A., [date]
ed.
QD271.A23 544.92 65-27435
ISBN 0-8247-7546-5

MARCEL DEKKER, INC.
270 Madison Avenue, New York, New York 10016

Current printing (last digit):
10 9 8 7 6 5 4 3 2 1

PRINTED IN THE UNITED STATES OF AMERICA

Contributors to Volume 25

Hugo A. H. Billiet   Analytical Chemistry Department, Delft University of Technology, Delft, The Netherlands

Christine M. Conroy   Department of Chemistry, University of California, Riverside, California

Leo de Galan   Analytical Chemistry Department, Delft University of Technology, Delft, The Netherlands

Peter R. Griffiths   Department of Chemistry, University of California, Riverside, California

Theo L. Hafkenscheid*   Physical Pharmacy Group, Subfaculty of Pharmacy, University of Amsterdam, Amsterdam, The Netherlands

Toshihiko Hanai   Research Institute, Gasukuro Kogyo Inc., Iruma, Japan

Firoze B. Jungalwala   Biochemistry Department, Eunice Kennedy Shriver Center, Waltham, Massachusetts

---

Current affiliation: Chemistry Department, Directorate-General of Labour, Voorburg, The Netherlands

Robert H. McCluer   Biochemistry Department, Eunice Kennedy Shriver Center, Waltham, Massachusetts

Katsuyuki Nakano   PL Medical Data Center, Tondabayashi, Osaka, Japan

Roswitha S. Ramsey   Analytical Chemistry Division, Oak Ridge National Laboratory, Oak Ridge, Tennessee

Lane C. Sander   Organic Analytical Research Division, Center for Analytical Chemistry, National Bureau of Standards, Gaithersburg, Maryland

Eric Tomlinson*   Physical Pharmacy Group, Subfaculty of Pharmacy, University of Amsterdam, Amsterdam, The Netherlands

M. David Ullman   Research Service, Edith Nourse Rogers Memorial Veterans Hospital, Bedford, Massachusetts

Stephen A. Wise   Organic Analytical Research Division, Center for Analytical Chemistry, National Bureau of Standards, Gaithersburg, Maryland

Current affiliation: Ciba-Geigy Pharmaceuticals, Horsham, West Sussex, United Kingdom

Contents

Contents of Other Volumes

ADVANCES IN

CHROMATOGRAPHY

Volume 25

1

Estimation of Physicochemical Properties of Organic Solutes Using HPLC Retention Parameters

Theo L. Hafkenscheid* and Eric Tomlinson†   *University of Amsterdam, Amsterdam, The Netherlands*

Current affiliation: Directorate-General of Labour, Voorburg, The Netherlands

†*Current affiliation*: Ciba-Geigy Pharmaceuticals, Horsham, West Sussex, United Kingdom

I. PERSPECTIVE

Pharmaceutical, biological, and environmental sciences require knowledge regarding the physicochemistry of solutes of interest. For example, most mathematical models used in drug design and pollutant bioaccumulation studies are based on quantitative relationships between physicochemical (structural) parameters and solute (biological) behavior. Because the thermodynamic activity of a solute in its aqueous solution can be related both to its chromatographic behavior as well as its (biological) activity, it is not surprising that numerous attempts have been (and are being) made to relate solute chromatographic migration to (*inter alia*) solute biological activity and bioaccumulation [1]. Concimitant with such studies have been the numerous theoretical and experimental attempts to appreciate the phenomena of chromatographic retention on the basis of quantitative structure-retention relationships. This present contribution aims to identify those solute physicochemical parameters that may either be readily obtained or estimated from chromatographic retention data, or that can be used in the above-described (predictive) models; many examples are taken from our own studies in this area. Attention will be focused on the use of silica-based stationary-phase materials, as used in high-performance liquid chromatography (HPLC). Intended for chromatographers and other scientists involved in the above-mentioned correlation studies, this work covers the estimation of solute hydrophobic-lipophilic balance, aqueous solubility, and various complex-formation parameters (including electronic parameters). It is intended to be comprehensive in identifying published studies. In addition, this article examines related studies on the prediction of solute liquid chromatographic retention using known physicochemical parameters.

For related reviews on the use of thin-layer chromatography (TLC) for obtaining physicochemical information for use in drug design models, for example, and for articles on structure-retention relationships mainly in HPLC, the reader is directed to Refs. 1 and 2 and 3–13, respectively. Also of use to the reader is the work of Conder and Young on the application of gas chromatography (GC) to the evaluation of solute physicochemical properties [14], and the application of size-exclusion chromatography (SEC) to the study of effective sizes of (bio)polymers [15,16].

II. CHROMATOGRAPHIC RETENTION

If one considers chromatographic retention to be a dynamic equilibrium process with an equilibrium constant K_r, and to have n process variables (i.e., temperature, pressure, phase ratio, phase compositions), then fixing n − 1 variables results in a mutual dependence between

K_r and the remaining variable. In thermodynamic terms, chromatographic retention may be regarded as a process occurring between two "immiscible" phases, as described by a Gibbs free-energy term, namely:

$$\Delta G_r = -RT \ln K_r \tag{1}$$

where ΔG_r, R, and T are the free energy of retention, the gas constant, and absolute temperature, respectively, and where K_r is related to the solute capacity factor, k, by

$$k = K_r \cdot (V_{st} \cdot V_{mob}^{-1}) = K_r \cdot \Phi \tag{2}$$

where V_{st} and V_{mob} are the volumes occupied by the stationary and mobile phases, respectively, the ratio of which, Φ, is difficult to quantify because of problems in distinguishing a definite boundary between stationary and mobile phases. The capacity factor, k, of a solute, i, is calculated as the normalized retention time (volume) according the Eq. (3):

$$k_i = [t_{r,i} \cdot t_o^{-1} - 1] = [V_{r,i} \cdot V_o^{-1} - 1] \tag{3}$$

where t_r (V_r) is the solute retention time (volume) and t_o (V_o) is the mobile phase holdup time (volume), which in size exclusion chromatography is equivalent to the exclusion time (volume)

A. Determination of t_o (V_o)

For estimation of physicochemical solute parameters via k values, it is critical that t_o (V_o) be correctly evaluated. Equation (3) shows this to be particularly so for solutes that are hardly retained. In HPLC, t_o (V_o) is generally given as equal to the retention time (volume) of a "nonretained" marker; ideally such a marker should be identical with the mobile phase. Because in practice this is impossible to achieve, a number of methods have been suggested to obtain a good approximation of the mobile-phase holdup parameters. In normal-phase liquid chromatography (NPLC), apolar solvents are generally used as markers. Results are considered to be reliable when apolar mobile phases are used, but to be unreliable when relatively polar mobile phases are used because of partial exclusion from the pores of the stationary phase [17]. A similar phenomenon is encountered in reversed-phase liquid chromatography (RPLC), when anionic solutes are used as markers with mobile phases of low to zero ionic strength [17,18]. In bonded-phase RPLC, the problem of correctly evaluating t_o (V_o) is compounded by the potential for mobile-phase components to influence the orientation and the composition of the stationary phase; this results in a still greater uncertainty in any attempts to establish a definite boundary between stationary and mobile phases.

Frequently employed methods for estimating t_0 (V_0) include (a) the use of pure or diluted mobile-phase components, (b) the use of polar organic solutes, (c) the use of radiolabeled mobile-phase components, (d) differential weighing of the column, (e) linearization of the retention of homologous solute series, and (f) the use of ultraviolet (UV)-absorbing inorganic anions (such as nitrate, nitrite, and dichromate). In general, methods (a) through (d) lead to an overestimation of mobile-phase holdup values. (Even the use of pure water may lead to t_0 (V_0) values that are considerably higher than those obtained using a dilute solution of water in mobile phase, or via linearization of retention of n-alcohols [18,19].) Method (f) is applicable only when particular salt solutions are used [18], or when mobile phases of high ionic strength are employed. Also, the results obtained with method (e) depend very much on the type and the "size" of the homologous solute series used. For practical purposes, we have found that the use of a dilute solution of water in the mobile phase is highly appropriate, giving good approximations of t_0.

B. Advantages of HPLC

Measured physicochemical parameters should be readily determined with simple, reliable, and flexible techniques. In these respects, the availability of high-performance stationary-phase materials of small particle-diameter, manufactured with a high batch-to-batch reproducibility is an important contribution to the use of column-liquid chromatography for nonanalytical purposes. In addition, HPLC methods have a number of advantages over conventional techniques used for measurement of physicochemical solute properties. These include (a) their relative rapidity, (b) their low solute consumption, and (c) their applicability to impure and/or unstable compounds.

III. PHYSICOCHEMICAL PROPERTIES

There are four broad classes of physicochemical parameters that may be estimated by HPLC techniques. These are (a) hydrophobic-lipophilic solute parameters, (b) electronic solute parameters, (c) complex-formation constants, and (d) miscellaneous parameters, a group that includes activity coefficients in organic solvents.

A. Hydrophobic-Lipophilic Parameters

The hydrophobic effect (i.e., the expulsion of a structure from an aqueous environment) often dominates various phenomena of biological, pharmaceutical, and general physicochemical interest, including (a) oil/water distribution, (b) aqueous solubility, (c) micelle formation in aqueous solutions, and (d) bioavailability (bioaccumulation) of drugs and environmental pollutants. The hydrophobic effect often occurs in parallel with a lipophilic effect, such as specific solvation in an oil phase in liquid/liquid distribution. The presence of common effects such as hydrophobicity and lipophilicity has implied that many relationships have been found between these various phenomena. For example, Meyer [20] and Overton [21] related narcotic activities of inhalation anesthetics to their abilities to distribute between oils and water, and as a result a large number of subsequent studies have examined the relations between biological or pharmacological activities and organic solvent/water distribution coefficients (K_d) (e.g., see Refs. 1,17,21–26). After Hansch and Dunn [24] suggested 1-octanol/water as the reference solvent pair for such correlations, distribution coefficients measured in this system [27] have been widely used as indicators of (variously) hydrophobicity, lipophilicity, and hydrophobic-lipophilic balance. Clearly, solvation in the octanol phase and the high solubility of water in 1-octanol (27% on a mole fraction scale at 25°C) suggests that such distribution coefficients should not be used as true descriptors of solute hydrophobicity, and that only by use of, for example, alkane/water solvent systems can distribution coefficients be regarded as such [28].

The aqueous solubility of organic solutes can also be related to hydrophobic-lipophilic parameters such as the 1-octanol/water distribution coefficient. It has been shown that a linear relationship exists between the logarithm of aqueous solubility (log S_w on a molar basis, with the subscript w referring to water as the solvent) and log K_d in the octanol/water system for a large number of nonionized organic liquids [29,30]. For solids it has been shown by Yalkowsky and co-workers [31,32] that the aqueous solubility of nonelectrolytes may be related to the 1-octanol/water distribution as long as the crystalline state of the compound is taken into account. These workers have assumed that the activity coefficient of a nonionized organic solute in a saturated aqueous solution may be quantatively described by its octanol-water distribution coefficient, because of the considerable "similarity" between the intramolecular forces existing in most organic solutes and those in 1-octanol.

Both octanol/water distribution coefficients and aqueous solubilities are used as solute parameters describing solute transport through membranes [33]; they are also frequently used in relation to solute uptake, distribution, and accumulation in biological systems [34].

However, conventional (shake-flask) methods used for measuring K_d or S_w have a number of practical disadvantages, particularly when applied to highly apolar compounds. These methods (a) are time consuming, due to slow equilibration; and (b) require relatively large amounts of pure, stable compounds.

Problems in obtaining accurately measured values of octanol/water distribution coefficients have stimulated the development of methods for systematic estimation of such parameter values. Such methods are generally based on the finding of Fujita et al. [35] that, in the absence of special steric or electronic effects a solute's octanol/water log K_d may be described by the sum of the separate contributions of the functionalities constituting the solute molecule (the group contribution, or π concept). In particular, Hansch and Leo's π substituent constants [27] and Rekker's hydrophobic fragmental f constants [36] are well known in this respect.

The so-called methylene group effect, i.e., the linear dependence of the magnitude of solute hydrophobic (or, in general, solvophobic) properties on the number of $-CH_2-$ groups within homologous solute series may be considered as a special group contribution effect. Various other geometric solute parameters may also be categorized as solvophobic properties. These include substituent alkyl chain length, carbon number, molecular volumes and surface areas, and molecular connectivity indexes.

As a consequence, the hydrophobic-lipophilic parameter section will deal with (a) liquid/liquid distribution coefficients, (b) geometric properties, and (c) aqueous solubilities.

1. Liquid/Liquid Distribution Coefficients

Early studies intended to elucidate the mechanism of retention in liquid/liquid high-pressure chromatographic systems led Huber and his co-workers to propose that such assemblies could be used for measuring solute distribution coefficients [37,38]. Their dynamic method was used to determine the distribution coefficients of some benzenoids, pesticides, and steroids in six equilibrium phase systems consisting of water, ethanol, and isooctane. Depending upon the solute studied, the stationary supports to be coated were either a hydrophilic phase (Kieselguhr) or a hydrophobic phase (Hyflo Supercel). The technique required calibration of the liquid chromatographic (LC) systems with selected standard solute whose K_d values had been measured via static methods.

A similar approach was taken by Haggerty and Murrill [39], who used a chemically bonded stationary phase (Bondapak C-18) for the specific estimation of octanol/water K_d values ($K_{d,oct}$, with the subscript "oct" referring to 1-octanol as the organic solvent). By assuming that "under the conditions used for assay, separations were effected by liquid/liquid partitioning only," these workers used the retention times of substituted nitrosureas obtained with an aqueous acetonitrile mobile phase to calculate log $K_{d,oct}$ values. This approach gave estimated values that agreed closely with those measured when static techniques were used. Yet, it is to be noted that both of the above methods use non-time-normalized retention data, a feature requiring that [38] "the retention time measurements have to be carried out under exactly the same conditions."

These tentative steps toward modern LC for estimating liquid/liquid distribution coefficients were soon to be amplified by studies of various other groups. Carlson et al. [40] apparently were the first to attempt to relate log k values to measured log $K_{d,oct}$ values. When Bondapak C-18 was used as stationary-phase material and various acetone-water mobile phases were employed, reasonable correlations were found for phenols and for anilines when examined separately as solute groups; whereas for both groups combined, much poorer correlations were found between log k and log $K_{d,oct}$ values. Their work included the first attempts to obtain group contribution values from bonded-phase RPLC retention data to be used in structure-activity relationships. A priori interactions occurring in an octadecylsilica (ODS)/acetone-water system would be expected to be somewhat different from those occurring in the 1-octanol/water system. It is not surprising, therefore, that until such mobile phases as methanol-water were used, no general correlations of good statistical significance could be obtained between log k and log $K_{d,oct}$ for various types of solutes.

Another phenomenon that has to be taken into account is the influence of silica hydroxyl groups (silanol groups) on the RPLC retention of, in particular, basic compounds. This was emphasized by McCall [41], who found that a mobile phase consisting of 1% (v/v) triethylamine in water, used in combination with a vigorously silanized stationary phase (Corasil C-18), yielded favorable results with basic solutes, because of a presumed competitive adsorption of the amine for the residual silanol groups. This observation led McCall to use mobile phases consisting of methanol, triethylamine, and water for successful estimation of log $K_{d,oct}$ values of a series of triaminopyrimidine-3-oxides, although scaling factors were needed to account for the different mobile-phase compositions used. On a practical note, it is doubtful whether such mobile phases of pH∿9 should be used because of their potential for inducing the deteriorating of (bonded-phase) silica materials.

Studies reported thus far, were concerned with obtaining octanol/water distribution coefficients from a measured solute parameter intrinsically different from the solute parameter to be evaluated. Mirrlees et al. [42], emphasizing that "the only true model for octanol is octanol itself," used water-saturated 1-octanol impregnated on Kieselguhr as the stationary phase in liquid/liquid RPLC to determine log $K_{d,oct}$ values. Employing octanol-saturated aqueous buffers as mobile phases, these workers found an excellent linear relationship between the logarithm of normalized solute retention time $(t_r - t_o)$ and log $K_{d,oct}$ for solutes of varying physicochemical character. Problems arise, however, when octanol is used as a stationary phase in a liquid/liquid RPLC system. First, complete mutual presaturation of both phases is necessary, with precise thermostatting thereafter;

second, mobile phases are restricted to octanol-saturated aqueous solutions, leading to a limitation in the range of log $K_{d,oct}$ values that can be measured; and third, poorly water-soluble compounds tend to give anomalously low retention times [42], due, probably, to partial demixing of octanol in the mobile phase.

Arguing that 1-octanol is relatively hydrophobic and should therefore bond strongly to an octadecylsilica stationary phase, Unger et al. [43] used octanol-coated silanized Corasil C-18 as stationary phase with octanol-saturated aqueous mobile phases. Log k values determined with this liquid/liquid RPLC system for a series of reference compounds were found to be highly correlated with the corresponding log $K_{d,oct}$ values (see Fig. 1).

When this LC system was used, log k values obtained for tuberin derivatives could be well correlated with their antibacterial activities [44]. Unger et al. further argued [43] that under the assumption that the ionized form of the solute does not distribute between octanol and water, solute retention obtained at two sufficiently different mobile-phase pH values can be used to calculate the pK_a of the solute.

This latter approach had already been used by Johansson and Wahlund [45], who, for a series of sulfonamides and barbiturates, estimated both the retention of the un-ionized solute and its pK_a from

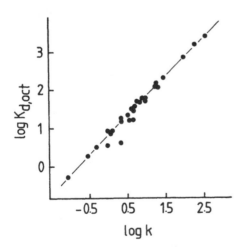

Figure 1   Comparison of log $K_{d,oct}$ from shake-flask experiments, with log k determined by LC on an octanol-coated octadecylsilica stationary phase, and octanol-saturated 0.01 mol dm^{-3} phosphate buffer (pH 7.00) as mobile phase. The solid line represents the linear regression line between log $K_{d,oct}$ and log k. (Modified from Ref. 43 with permission of the copyright owner.)

solute retention as a function of mobile-phase pH. Their RPLC system consisted of 1-pentanol coated onto LiChrosorb RP-2 as stationary phase, with pentanol-saturated phosphate or citrate buffers as mobile phases. It was found that both log K_d and pK_a values obtained with this LC system were in good agreement with values obtained by conventional methods. Moreover, the authors found a reasonable agreement between dynamic 1-pentanol/buffer K_d values and literature values for solute octanol/water distribution coefficients, suggesting that retention in their liquid/liquid RPLC system could be predicted from $K_{d,oct}$ values.

Similarly, Hulshoff and Perrin [46] studied relations between log K_d values for a series of benzodiazepines, measured in either oleyl alcohol/water or octanol/water solvent systems, and retention data obtained using reversed-phase column-liquid as well as reversed-phase thin-layer chromatography (RPTLC). The chromatographic systems that were employed consisted of: (a) Kieselguhr impregnated with oleyl alcohol (TLC), (b) silanized silica coated with oleyl alcohol, or (c) untreated or silanized Corasil C-18 (LC), with aqueous buffer (pH$\sim$9)-methanol mobile phases of varying compositions. For all systems, the logarithm of normalized solute retention (log k, or R_M, its TLC "equivalent") was found to be linearly related to ϕ_m, the volume fraction of organic modifier in the mobile phase. These relations were used to calculate values of log k_w and $R_{M,w}$, i.e. the logarithm of hypothetical normalized solute retention for totally aqueous mobile phases. The resulting $R_{M,w}$ values correlated well with both oleyl alcohol/water and octanol/water log K_d values, and were, in this respect, found to compare favorably with log k_w values. However, the latter values were found to be determined with greater precision.

Clearly a primary condition for comparing dynamic and static parameters that describe a solute's hydrophobic-lipophilic properties, is that both types of parameters be obtained with the solute in a comparable physicochemical state. For ionizable solutes, the un-ionized form is generally chosen as the reference state. However, one of the major disadvantages of using bonded-phase silica stationary-phase materials is the limited operating pH range (1.5—8.0). This implies that only a limited group of solutes may be chromatographed in their un-ionized state when such phases are used: i.e., neutrals, acids with a pK_a greater than $\sim$3, or bases with a pK_a less than $\sim$6.5. Because many compounds of, for example, pharmaceutical interest do not satisfy any of these conditions, a number of complications will arise when RPLC is used to determine their hydrophobic-lipophilic balance. These include (a) effects of organic modifiers on the pK_a values of both solutes and buffer components, (b) solutes being (partly) chromatographed as ion pairs with buffer components, and (c) silanol groups exerting a (not necessarily unambigous) influence on retention. These latter two complications have been discussed by Unger and Chiang [47] in a study on the relationships between RPLC retention and octanol/buffer distribution coefficients for a variety of basic drugs. These workers were able to eliminate the influence of silanol groups on the retention of basic solutes by the addition of a sufficient concentration of N,N-dimethylaminooctane to the mobile phase. However, when they attempted to eliminate ion-pairing effects by using the same buffer for both static and dynamic experiments, these workers observed that the effects of different anion concentrations on the results of the individual methods differed considerably, indicating a difficulty in unambiguously correcting for ion-pair formation.

Certainly the problem of correcting for (partial) ionization and ion-pairing effects also plays an important role in static measurements of both liquid/liquid distribution coefficients and aqueous solubilities of very hydrophobic compounds. Determination of K_d or S_w at pH values close to the solute pK_a, combined with the proper corrections, offers the possibility of approximating K_d or S_w of the un-ionized solute.

Another way to circumvent the problems enumerated above is to use ion-pair reversed-phase liquid chromatography (IP-RPLC), in which the retention of a partly ionized solute is enhanced (and controlled) upon interaction with a (hydrophobic) pairing ion added to the mobile phase. Under mobile-phase conditions in which the solutes are totally ionized, and the ionic strength and pairing-ion concentration are fixed, IP-RPLC data can be used as indicators of solute hydrophobic-lipophilic properties. This can be done, however, only within a series of structurally related compounds.

An example of this use of IP-RPLC is given in the study of Riley et al. [48], who examined functional-group behavior in IP-RPLC, using both anionic and cationic solutes and pairing ions. Although no general correlation could be found between solute retention and distribution coefficients, functional-group contributions to retention, τ_X, where $\tau_X = \log k_{RX} - \log k_{RH}$, with solutes RX and RH differing by a functional group X, were found to be well correlated with corresponding octanol/water π values, as is illustrated by Fig. 2. Here τ_X values for substituted benzoic acids, azapurines, and triazines, obtained using a Spherisorb ODS stationary phase and aqueous methanol mobile phases containing either alkylbenzyldimethylammonium, or dodecylsulfate pairing ions, are plotted against literature π values. It is seen that observed relations between τ and π are indistinguishable for the three solute series. Furthermore, τ values of triazines were shown to be linearly related to solute biological activity (i.e., their minimum inhibitory concentrations against *Staphylococcus aureus*).

Via nonlinear regression analysis, these workers also fitted the chromatographic data to a phenomenological model describing solute retention in ion-pair RPLC, enabling the evaluation of the "dynamic" distribution coefficients of anion-ammonium ion pairs. The calculated coefficients were linearly related to the corresponding substituent

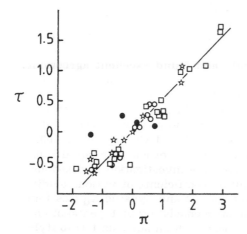

Figure 2 Relationship between group contribution values, π, and IP-RPLC functional group values τ, obtained using a Spherisorb ODS stationary phase and aqueous methanol ($\phi_m = 0.50$) mobile phases. (□) Triazines, using a mobile phase of pH 2.2, containing 0.26 mmol dm^{-3} sodium dodecylsulfate. (○, ●) Benzoic acids, using a mobile phase of pH 7.5, containing 0.40 mmol dm^{-3} tridecylbenzyldimethylammonium chloride; (○) m- and p-substituted; (●) o-substituted benzoic acids. (✰) Azapurines, using a mobile phase of pH 7.5, containing 0.10 mmol dm^{-3} undecylbenzyldimethylammonium chloride. (Modified from Ref. 48.)

π values (except for the hydroxyl and amino compounds). In a subsequent study [49], Riley et al. found similar relationships for substituted benzoic acids between τ_w values (τ values for totally aqueous mobile phases obtained by linear extrapolation of τ as a function of mobile-phase surface tension) and the corresponding π values. The best correlations were observed when the τ_w values used were calculated from retention data obtained with methanol or 2-propanol as organic modifiers; whereas a τ_w versus π slope coefficient close to unity was observed only when "methanolic" τ_w values were used. The lack of correlation observed with τ_w values from retention data measured with acetonitrile or tetrahydrofuran as organic modifiers suggests that these proton acceptor solvents are less suitable for these types of studies.

Studies related to the work of Riley et al. were performed in order to elucidate contributions of liquid/liquid distribution of aqueous ion-pair complexes to their retention in ion-pair liquid chromatography. In these studies, static distribution coefficients of ion-pair complexes were used to calculate k values for the corresponding ion-pair liquid/liquid chromatographic system, and these were subsequently compared to experimental k values. Karger et al. [50] compared static K_d values of catecholamines in the phase system 0.2 mol dm^{-3} HClO$_4$-0.8 mol dm^{-3} NaClO4/1-butanol-methylenechloride (2:3) to liquid chromatographic K_d values measured with an NPLC system (with the aqueous phase coated on Merckosorb SI-100), and found excellent agreement. Further, Su et al. [51] found good agreement between static and dynamic K_d values of some sulfonamides for the phase systems 0.1 mol dm^{-3} tetrabutylammonium hydrogen sulfate in pH 7.4 or 8.4 buffer/1-butanol-1-heptane (1:3). LC experiments were again carried out in the NPLC mode, with aqueous phases coated on LiChrospher SI-100. Fransson et al. [52] compared different modes of ion-pair liquid/liquid chromatography. Their study included the investigation of relations between static and dynamic distribution coefficients of various acidic solutes. The best agreement between static and dynamic K_d was found for a reversed-phase ion-pair LC system consisting of 1-pentanol on LiChrosorb RP-2 as the stationary phase with an aqueous tetrabutylammonium hydrogen sulfate mobile phase, and for normal-phase LC systems with a high stationary-phase loading.

Other studies on relationships between liquid/liquid distribution parameters and LC retention data obtained using silica stationary phases are presented in Table 1 (see Refs. 62—120). As can be observed in Table 1, liquid chromatographic retention parameters, to be compared with static octanol/water distribution coefficients, were determined almost exclusively with alkyl-bonded RPLC stationary phases, although Mirrlees et al. [42] and Unger et al. [43] argued that such phases behave more like alkanes than like alcohols. This argument has been superceded by more recent observations that reversed-phase LC retention parameters and alkane/water distribution coefficients are correlated only for apolar solutes [53—58], whereas RPLC log k values and log $K_{d,oct}$ are found to be linearly related for various types of solute groups [53,56,59,60]; see, for examples, Figs. 3 and 4. In the latter figure, it may be observed that deviations in the log k versus alkane/water log K_d (log $K_{d,alk}$, with subscript "alk" referring to an alkane as the organic solvent) relationship can be adequately accounted for by incorporation of hydrogen-bonding correction terms [61].

2. Geometric Properties

As discussed previously, the term *geometric properties* is used here to indicate those parameters that directly affect a solute's hydrophobicity, including both volume and surface area properties, and parameters related to the hydrocarbon part of a molecule. A number of these geometric properties can be identified, i.e., (a) total carbon number C_n [122—126], (b) substituent alkyl chain length n_C [11,85,127—141], (c) molecular volume or surface area [8,60,104,119,120,133,142—144], (d) molecular connectivity [10,91,119—121,133,137,145,146], and (e) length-to-breadth ratio [121] (see Fig. 7). The molecular connectivity

Table 1 Studies on Relationships Between Liquid Chromatographic Retention and Solute Liquid/Liquid Distribution Parameters (Presented in Chronological Order)

Compound(s)	Stationary phase(s)	Mobile phase(s)	Temperature(s)	Observations	Refs.
Various acidic and basic drugs	μ-Bondapak C-18	Various combinations of methanol and aqueous buffers	Not stated	A quantitative correlation was observed between V_r and log K_d,oct for both acidic and basic drugs.	62
Sulfonamides, barbiturates	Corasil C-18; 1% octanol on Corasil II; 1% squalene on Corasil II	Aqueous buffers of pH 4.0, 5.0 or 6.5; Acetonitrile-buffer pH 5.0 ($\phi_m = 0.10$)	Not stated	Relations were studied between log V_r, and log K_d,oct or solute biological activities. Highest correlations were found when Corasil C-18 was used as stationary phase; the authors, however, emphasize not to have optimized the loading percentages of the liquid/liquid LC systems.	63
Carboxylic acids	0.1 mol dm^{-3} tetrabutylammonium hydrogen sulfate in a pH 7.4 buffer on Partisil 10	Methylene chloride - n-hexane - 1-butanol (16:3:1 by vol.)	Not stated	Reasonable agreement was found between measured k values and those calculated from shake-flask K_d values for the corresponding liquid/liquid phase system.	64
Aliphatic amines	0.01 mol dm^{-3} naphthalene sulfonate in pH 2.2 phosphate buffer on LiChrospher SI-100	Chloroform-1-pentanol ($\phi_m = 0.05$, 0.10)	25°C	Good agreement was observed between experimental k values and those calculated from corresponding static K_d values for short-chain amines ($n_C \leqslant 6$). Above $n_C = 6$, deviations occurred.	65

Table 1 (continued)

Compound(s)	Stationary phase(s)	Mobile phase(s)	Temperature(s)	Observations	Refs.
Amino acids, peptides	LiChrosorb RP-8; LiChrosorb RP-18	Various aqueous buffers	70°C	Relationships between log k values, and log K_d,oct, side-chain hydrophobicities, and geometrical properties were studied. In general, linear relations were found.	66
Estradiol, diethylstilbestrol, and iodo derivatives	μ-Porasil	Chloroform	Not stated	NPLC retention was related to calculated log K_d,oct values and measured chloroform/water log K_d values. Through introduction of R_L (a nondimensionless functional group retention parameter) log K_d values could be accurately estimated.	67
Benzenoid carboxylic acids	1-pentanol on LiChrosorb RP-8	Phosphate buffers (pH 2.08, 5.44)	24° or 25°C	LC distribution coefficients could be calculated from the solutes' log K_d,oct values. Hence the RPLC system used was suggested to be suitable for estimation of log K_d,oct values.	68
Neutral and acidic benzenoids	Octanol on Corasil I	Octanol-saturated aqueous HCl-KCl (pH 2)	Not stated	"Bulk-impregnation" of Corasil I with octanol was compared with the in situ coating	69

14

Compound	Stationary phase	Temperature	Mobile phase	Comments	Ref.
Benzenoids	Silanized ODS-Hypersil	30°C	Methanol-water ($\phi_m = 0.50$) Acetonitrile-water ($\phi_m = 0.30$) THF-water ($\phi_m = 0.25$)	procedure of Ref. 42. Bulk-impregnation was found to give a higher column stability. Excellent correlations were observed between log k and log $K_{d,oct}$ values. The effects of different organic modifiers on RPLC retention were examined. Mobile phases were "isoeluotropic" for the methylene group. Whereas log k and log $K_{d,oct}$ were linearly related when methanol was used as modifier, significant deviations were found when acetonitrile or THF was used.	70
Xanthone-2-carboxylic acid, tiopinac	Octanol on silanized Corasil C-18	Ambient	Octanol-saturated aqueous buffers (ionic strength, 0.15)	Log K_d values for different octanol/buffer systems were directly measured. These values were subsequently used to calculate solute pK_a values.	71
Propranolol and derivatives, anthranilic acids, barbiturates	μ-Bondapak C-18	Not stated	Methanol-phosphate buffer, pH 7.0 ($\phi_m = 0.40$)	RPLC retention indexes (with 2-alkanones as reference solutes) were related to biological activities. Correlations were found in only one case (anthranilic acids) to be less than those found when log $K_{d,oct}$ was used. Yet retention indexes and log $K_{d,oct}$ were found to be linearly related.	72

Table 1 (continued)

Compound(s)	Stationary phase(s)	Mobile phase(s)	Temperature(s)	Observations	Refs.
Barbiturates	μ-Bondapak C-18	Methanol-phosphate buffer, pH 7.0 ($\phi_m = 0.40$)	Not stated	A linear relationship was observed between log K_d,oct calculated using π values, and retention times relative to that of phenacetin, despite partial ionization of the barbiturates under mobile-phase conditions.	73,74
Benzene, (poly)methyl-benzenes	RP-5, RP-6, RP-7, RP-8, RP-10, RP-12, RP-18, and RP-20, plain or coated with the corresponding alkane	Methanol-water ($\phi_m = 0.50$)	25°C	RPLC retention values relative to benzene were found to be correlated to the substituent "K_d values" in the corresponding alkane/methanol-water liquid/liquid systems. The authors therefore proposed a distribution retention mechanism to occur in bonded-phase RPLC.	55
Mono- and polychloro-benzenes, -toluenes, and -anilines	LiChrosorb RP-18	Methanol-water ($\phi_m = 0.70$)	35°C	Log k values were related to log K_d,oct values that were either measured or calculated using Rekker's f values. A comparison was made between estimations of log K_d,oct, using log k and f values.	59

16

Compound	Column	Mobile phase	Temperature	Remarks	Ref.
Organic pollutants	Micropak CH-10	Methanol-water (ϕ_m = 0.22 to 0.75: linear gradient) / Methanol-water (ϕ_m = 0.85)	50°C / Ambient	A nonlinear relationship was found between log $K_{d,oct}$ and retention relative to phenol when the gradient elution system was used. When isocratic RPLC was used, a linear relationship was observed between log t_r and log $K_{d,oct}$, which could be used to estimate unknown K_d values.	75
Peptides	Hypersil ODS	0.1 mol dm^{-3} phosphate buffer (pH 2.1)-acetonitrile (ϕ_m = 0 to 0.60: linear gradient)	Not stated	For small peptides, retention could be reasonably well described using the sum of f values of their "hydrophobic residues." For larger peptides, anomalies were observed when this approach was used.	76
Narcotic analgesics	Permaphase ODS	Methanol-aqueous buffer	Not stated	Relationships between log k, log $K_{d,oct}$, and analgesic activities were studied. Activities were found to be best described when log k parameters were used.	77
Cinnamic acid derivatives	Permaphase ODS	Aqueous acetic acid or aqueous methanol	Not stated	Good agreement was found between RPLC log k values and substituent π-values of 16 cinnamic acid derivatives.	78

Table 1 (continued)

Compound(s)	Stationary phase(s)	Mobile phase(s)	Temperature(s)	Observations	Refs.
Phenols	Micropak CH-10	Methanol-water ($\phi_m = 0.20$); Dioxane-water ($\phi_m = 0.20$)	Not stated	Linear relations were found between log k and substituent π values. Solute biological activities could be described by multiple linear combinations of log k for the methanol-water RPLC system, or π values, and solute pK_a values. Dioxane-water log k values were found to be less suitable here.	79
Benzene derivatives	Whatman ODS	Methanol-water ($\phi_m = 0.50$)	Not stated	Two different linear relationships were found between RPLC τ values and π values for either polar, or apolar benzene derivatives.	80
Iodoamino acids; iodo-thyrocarboxylic acids	μ-Bondapak C-18	Various combinations of methanol and aqueous phosphoric acid (pH 3.0)	About 20°C	Linear relationships were found between log k and log $K_{d,oct}$ calculated using Rekker's f values. Two different relations were observed for normal and reversed iodothyronines.	81
Pyridazinone herbicides	LiChrosorb RP-18	Various combinations of methanol and water	Ambient	Relations were studied between both isocratic and extrapolated retention parameters (log k,	82

18

Compounds	Stationary phase	Mobile phase	Temperature	Remarks	Ref.
				log k_w) and log $K_{d,oct}$. A good correlation was found using log k determined at $\phi_m = 0.55$. Log $K_{d,oct}$ and log k_w were excellently correlated.	
Alkyl-benzenes, (poly)methyl-benzenes	Hypersil SAS	Methanol-water (ϕ_m = 0.40, 0.50)	Not stated	Relationships were studied between log k, and geometrical, or liquid/liquid distribution parameters. Log k was found to be linearly related to solute-cavity surface area, Bondi volume, molecular connectivity index, and both octanol/water and heptane/water log K_d values.	83
	Hypersil ODS	Methanol-water (ϕ_m = 0.70)			
	Magnusil C-22	Methanol-water (ϕ_m = 0.50)			
Barbiturates; phenylureas	Micropak CH-10	Methanol-water (ϕ_m = 0.20, 0.30)	Not stated	Linear relationships were found between log k and log k_d, oct, or substituent π values.	84
		Dioxane-water (ϕ_m = 0.20)			
n-Alkyl-benzoates and derivatives	LiChrosorb RP-8	Methanol-water (ϕ_m = 0.60, 0.70, 0.80)	20°C	Structural effects on retention were studied. Only for higher homologues ($n_C \geq 5$), log k was found to be linearly related to alkyl chain length. Extrapolated functional-group retention values τ_w were found to be linearly related to the corresponding π values (with exception of those for ortho substituents).	85
	LiChrosorb RP-18				

Table 1 (continued)

Compound(s)	Stationary phase(s)	Mobile phase(s)	Temperature(s)	Observations	Refs.
Organic pollutants	Zorbax ODS	Methanol-water (ϕ_m = 0.75, 0.85)	Ambient	Log k measured at ϕ_m = 0.75 was found to be excellently correlated with log $K_{d,oct}$, and was used to estimate unknown log $K_{d,oct}$ values. It was observed that from ϕ_m = 0.75 to 0.85, several changes in retention order occurred, implying that highly modified mobile phases should not be used to estimate hydrophobicity-lipophilicity.	86
Phenols	LiChrosorb RP-18	Various combinations of methanol and 0.01 mol dm^{-3} aqueous HCl	30°C	Extrapolated retention parameters, log k_w, were found to be linearly related to log $K_{d,oct}$. Derived τ_w values were found to be linearly related to the corresponding π values.	87
Benzamides	Nucleosil C-18	Acetonitrile - aqueous phosphate - chloride (pH 6.0) (ϕ_m = 0.18)	Not stated	Linear relations were observed between log k and log $K_{d,oct}$ values, that were either measured or calculated using Rekker's f values.	88
Pesticides	LiChrosorb RP-8	Ethanol-water (ϕ_m = 0.35,	45°C	Excellent correlations were found between measured log $K_{d,oct}$,	89

Compound	Column	Mobile phase	Temperature	Comments	Ref.
		0.40, 0.45, 0.50)		and log k values for all mobile phases used. Retention values at different mobile phase compositions were interrelated through k values of standard solute	
Amino acids, peptides	LiChrosorb RP-18	Acetonitrile - (pH 2.1) aqueous phosphate (gradient elution)	Not stated	Rekker's f values were used to predict solute retention in a gradient elution RPLC system.	90
Barbiturates	Partisil ODS Partisil ODS-2	Methanol-water (ϕ_m = 0.30, 0.50)	25°C	Both linear and quadratic functions were applied to relate log k and log K_d,oct. These two parameters were both found useful to describe solute biological activities.	91
Benzenoids, aliphatics	Hypersil ODS	Various methanol-water or aqueous buffer combinations	20°C	A linear relationship was found between log K_d,oct and extrapolated retention parameters, log k_w.	60
Various organic compounds	Permaphase ODS	Various combinations of methanol and water	Not stated	The RPLC system Permaphase ODS/various methanol-water mobile phases was used to estimate log K_d,oct over a wide range of values. The same range could be covered using the other systems—however, with log k and log K_d,oct being linearly related only in a narrow range of parameter values.	92
	Partisil ODS	Methanol-water (ϕ_m = 0.75)			
	GYT C-18	Methanol-water (ϕ_m = 0.75)			

Table 1 (continued)

Compound(s)	Stationary phase(s)	Mobile phase(s)	Temperature(s)	Observations	Refs.
Peptides	LiChrosorb RP-8 LiChrosorb RP-18	Linear gradient: 0.125 mol dm^{-3} pyridine in a pH 3.0 formate buffer to 1-propanol - 1.0 mol dm^{-3} pyridine in a pH 5.5 acetate buffer ($\phi_m = 0.60$)	Ambient	Log $K_{d,oct}$ values, calculated from f values, were used to predict peptide retention in the gradient elution RPLC system. Retention was found not to be strictly correlated to the calculated log K_d values.	93
Benzenoids, aliphatics	Corasil C-18	Various combinations of methanol and water	Not stated	A low correlation was found between log k_w and log $K_{d,oct}$. Using isocratic log k values ($\phi_m = 0.70$), three linear relations with log $K_{d,oct}$ values could be distinguished for: (a) apolar benzenoids and aliphatics, (b) phenols, anilines, and proton-acceptor benzenoids, and (c) benzoic acids.	94
Various organic nonionic compounds	LiChrosorb RP-8 Corasil C-18	Ethanol-water ($\phi_m = 0.7, 0.8$) Acetonitrile-water ($\phi_m = 0.9$) Methanol-water ($\phi_m = 0.85$)	Not stated	Relationships between log t_r and log $K_{d,oct}$ were studied. In most cases, no unique relationship was found.	95

22

Compound	Stationary phase	Mobile phase	Temperature	Comments	Ref.
Isoxazolyl penicillins	LiChrosorb RP-8	Methanol-aqueous acetate (pH 5.5) (75:100 by vol.) Acetonitrile-aqueous acetate (pH 6.6) (34:100 by vol.)	Not stated	Linear relationships were observed between log k and octanol/water log K_d values.	96
Various organic nonionic compounds	Chromosorb LC-7 LiChrosorb RP-8 LiChrosorb RP-18 Hitachi 3053 ODS Unisil Q C-18	Various combinations of acetonitrile and water	25°C	Linear relations were found for all mobile-phase compositions studied between log k and log K_d,oct (calculated from f values). These relations were used to establish a procedure for prediction of RPLC retention when acetonitrile-water mobile phases are used.	97
Benzoylacetic acid - N-phenyl amides	LiChrosorb RP-18	Various combinations of methanol and water, or methanol and 3% (v/v) aqueous acetic acid	20°C	Both isocratic (log k) and extrapolated (log k_w) retention parameters were related to solute biological activities. Excellent results were found with a mobile phase consisting of methanol - 3% (v/v) acetic acid with $\phi_m = 0.70$ (see Fig. 5).	98
Aromatic acids found in urine	Chromosorb LC-7	Various combinations of acetonitrile and 0.04 mol dm^{-3} aqueous phosphoric acid	Not stated	No linear relation was found between log k and log K_d,oct calculated using Rekker's f values. Relations found in a previous study (Ref. 97) between both	99

23

Table 1 (continued)

Compound(s)	Stationary phase(s)	Mobile phase(s)	Temperature(s)	Observations	Refs.
				parameters for nonionic compounds were used to calculate hypothetical log $K_{d,oct}$ values for the acids. Both calculated log $K_{d,oct}$ parameters were subsequently used to predict relative retentions in a gradient elution system.	
Aromatics, aliphatic alcohols and carboxylic acids	Chromosorb LC-7	Methanol-water ($\phi_m = 0.50$) Ethanol-water ($\phi_m = 0.50$) THF-water ($\phi_m = 0.50$) Various combinations of acetonitrile and water	Not stated	Relations were studied between log k and log $K_{d,oct}$ calculated with f values. With acetonitrile as modifier, a deviating relationship (as compared to the other compounds) was found for fatty acids; with methanol or ethanol as modifiers, alcohols showed a deviating behavior. With THF as modifier, all solute classes showed a mutually differing behavior.	100
Benzene derivatives, N-phenylanthranilates, N-phenylsuccimides	Glyceryl-coated controlled pore glass	Various combinations of methanol and water	Not stated	When extrapolated RPLC retention data (log k_W) and log $K_{d,oct}$ values were related, H-bonding and non-H-bonding solutes gave different relationships. This was also observed when isocratic	101

24

Compound	Column	Mobile phase	Temperature	Remarks	Ref.
Peptides	μ-Bondapak C-18	0.1% aqueous trifluoroacetic acid (pH 2) - 0.07% trifluoroacetic acid in methanol: linear gradient to $\phi_m = 0.60$	Not stated	($\phi_m = 0.10$) log k values were used, although the overall correlation was good (see Fig. 6). Hydrophobic fragment values of constituent amino acids could be used to predict peptide retention in a gradient-elution RPLC system.	102
Peptides	Octadecyl-glycerylpropyl silica (bonded phase)	0.02 mol dm^{-3} aqueous acetate (pH 4.5) - acetonitrile: linear gradient to $\phi_m = 0.60$	30°C	For small peptides, retention was found to be linearly related to log $K_{d,oct}$, calculated using Rekker's f values.	103
Aromatic hydrocarbons	Brownlee Labs, Inc. RP-C-18	Acetonitrile-water ($\phi_m = 0.75$)	30°C	Log k was found to be highly correlated with octanol/water log K_d values.	104
Benzenoids, deamino-6-oxytocins, and their sulfoxides	Separon SI-C-18	Various combinations of methanol and aqueous buffers of different pH	Not stated	Linear relations were observed between log k and substituent π values of the solutes. Also log k values for the oxytocins and correspondingly substituted benzenoids were found to be linearly related.	105

Table 1 (continued)

Compound(s)	Stationary phase(s)	Mobile phase(s)	Temperature(s)	Observations	Refs.
N-acetyl-N'-methyl amino acids	LiChrosorb RP-18	0.1 mol dm^{-3} ammonium phosphate buffer (pH 2.1)	20–60°C	RPLC log k values at 25°C, obtained by interpolation of log k values as functions of reciprocal temperature, were found to be highly correlated to log $K_{d,oct}$, calculated using Rekker's f values.	106
Phenyl-indolizines	Octanol coated on silanized Corasil C-18 Silanized Corasil C-18	Octanol-saturated water (system A) Acetonitrile-water ($\phi_m = 0.40$) (system B)	Ambient	Log $K_{d,oct}$ of indolizine was measured using system A calibrated with standard solutes. System B was then used to evaluate functional group values for substituents to phenylindolizine. These were highly correlated with the corresponding π values, and were used therefore to estimate log $K_{d,oct}$ values of the phenylindolizines using the measured log K_d value of the parent solute.	107
Phenothia-zines	μ-Bondapak C-18	Acetonitrile - water - acetic acid (7:2.5:5, by vol.)	Not stated	Good correlations were found between log k and measured octanol/pH 7.5 aqueous buffer log K_d values, despite differences in degrees of solute ionization	108

26

Solute	Sorbent	Mobile phase	Temperature	Comments	Ref.
Benzenoids, polynuclear aromatics	LiChrosorb RP-18	Various combinations of methanol, and water or 0.05 mol dm^{-3} aqueous acetate (pH 4.5)	25°C	between both solute "reference states," and occurrence of ion-pairing effects. An excellent correlation was observed between log $K_{d,oct}$ (measured) and extrapolated retention data (log k_w). Calculated τ_w values were found to be in agreement with the corresponding f values. Log k_w was found only for apolar solutes to correlate with alkane/water log K_d values.	56
Cardiac glycosides	Diphenyl-silica	Various combinations of ethanol and water	Not stated	Relations between log V_r and various solute properties were studied. Log V_r was found to depend nonlinearly on the number of solute hydroxyl groups. Relations were established between log V_r, hydrophobicity (sic), and solute biological activity.	109
Benzenoids	LiChrosorb RP-18	Methanol-water ($\phi_m = 0.50$)	25°C	Log K_d for the liquid/liquid system hexadecane/methanol-water ($\phi_m = 0.50$) was found not to be linearly related to solute log ($V_r - V_0$).	57

27

Table 1 (continued)

Compound(s)	Stationary phase(s)	Mobile phase(s)	Temperature(s)	Observations	Refs.
(Poly)methyl-benzenes	Jasco Finesil C-2 Jasco Finesil C-8 Jasco Finesil C-18	Acetonitrile-water ($\phi_m = 0.7, 0.8$)	Not stated	Log k values were used to estimate log $K_{d,oct}$ after calibration of the LC system with standard solutes of which log $K_{d,oct}$ was calculated using f values.	110
Phenylcar-bamoyl methyl-imino diacetic acids (HIDAs) and their ^{99m}Tc complexes	μ-Bondapak C-18	Methanol - 0.1% aqueous acetic acid ($\phi_m = 0.40$)	Not stated	Log k and measured log $K_{d,oct}$ values of ^{99m}Tc-HIDA complexes were related to the $\Sigma\pi$ values of ligands. For both cases two different linear relations were found: one for m- and p-, and one for di-o-substituted ligands. Mono-o-substituted HIDAs were part of neither relation. Log k was found to be linearly related to the renal clearance of the ^{99m}Tc-HIDA complexes.	111
Nonionic organic compounds	LiChrosorb RP-18	Methanol Acetonitrile Acetonitrile-water ($\phi_m = 0.60, 0.90$)	21°C	Relations were studied between log k and log K_d for the liquid/liquid system hexadecane/mobile phase. No general correlation was found for any of the RPLC systems studied.	58

Solutes	Stationary phase	Mobile phase	Temperature	Remarks	Ref.
Phenols, naphthols	Unisil Q - C-18 (A) Hypersil ODS (B)	Various combinations of acetonitrile and water	30°C	Relations were studied between log k and log $K_{d,oct}$, either measured, or calculated using f values. No satisfactory correlations were observed when stationary phase A was used. Log k for phase B could, however, be predicted from log k for phase A.	112
Organic pollutants	Radial Pak C-18	Methanol-water ($\phi_m = 0.85$)	Not stated	Log t_r was found to be linearly related to the solute logarithmic soil sorption constant, and also to log $K_{d,oct}$.	113
1- and/or 2-substituted 5-nitroimidazoles	μ-Bondapak C-18	Methanol-water, or methanol-aqueous buffer (pH 3.8 or 7.6) ($\phi_m = 0.40$)	Ambient	Fairly high correlations were found between log k and measured log $K_{d,oct}$ values, and between log k and a linear combination of log $K_{d,oct}$ and the sum of molar refractivities of the substituents.	114
Phenylureas, s-triazines, phenoxycarbonic acids	LiChrosorb RP-18	Various combinations of methanol and water or aqueous acetate (pH 2.9), or of acetonitrile and water or aqueous acetate (pH 2.9)	Ambient	Extrapolated retention parameters log k_w, using methanol or acetonitrile as modifiers were calculated. Log k_w data obtained from aqueous methanol log k values were found to be strongly correlated to log $K_{d,oct}$ and could be used to describe solute biological activities.	115

Table 1 (continued)

Compound(s)	Stationary phase(s)	Mobile phase(s)	Temperature(s)	Observations	Refs.
Various non-ionic organic compounds	Hypersil ODS Nucleosil 10 C-18 LiChrosorb RP-18	Various combinations of methanol and water or aqueous buffer	20–25°C	A general relationship for the prediction of RPLC log k (using a system: ODS/aqueous methanol) from log $K_{d,oct}$ values was established.	116
Barbiturates, sulfonamides	Corasil C-18	Methanol - aqueous phosphate buffer ($\phi_m = 0.05$ and pH = 7.45, or $\phi_m = 0.30$ and pH = 7.31)	Not stated	The applicability of log V_r data (Ref. 62), retention index data (Ref. 72), and log k parameters for description of solute biological activities were compared. Excellent correlations were observed, with log k values apparently giving the best results.	117
Liquid chromatographic retention data obtained from various references were used in this paper.				The possibilities for use of RPLC and RPTLC, and of calculatory procedures for estimation of log $K_{d,oct}$ have been reviewed.	118
Alkyl-benzenes	Jasco Finesil C-2 Jasco Finesil C-8 Jasco Finesil C-18	Acetonitrile-water ($\phi_m = 0.70$) Acetonitrile-water ($\phi_m = 0.45$) Acetonitrile-water ($\phi_m = 0.75$)	Not stated	Log k values were found to be linearly related to different physicochemical solute properties, i.e., log $K_{d,oct}$, van der Waals volume, and molecular connectivity index.	119

Solutes	Stationary phase	Temperature	Mobile phase	Comments	Ref.
Polynuclear aromatic hydrocarbons	Jasco Finesil C-2 Jasco Finesil C-8 Jasco Finesil C-18 Develosil-Phenyl	Not stated	Acetonitrile-water (ϕ_m = 0.55) Acetonitrile-water (ϕ_m = 0.55) Acetonitrile-water (ϕ_m = 0.65) Acetonitrile-water (ϕ_m = 0.50)	Log k values were found to be linearly related to different physicochemical solute properties, i.e., log $K_{d,oct}$ calculated using f values, van der Waals volume, molecular connectivity index, length-to-breadth ratio, and the so-called correlation factor (Ref. 121).	120
Various model compounds and drugs	Hypersil ODS	20°C	Methanol-water/ aqueous phosphate (pH = 2.15 or 7.00) (ϕ_m = 0.50, 0.75)	General linear relations were observed between log k values and log $K_{d,oct}$, with better correlation occurring when the mobile phases of ϕ_m = 0.50 were used. Alkane/water log K_d and log k were correlated only for neutral and basic compounds (see Fig. 4).	53

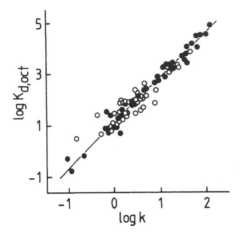

Figure 3 Relationship between log k for various un-ionized compounds, determined using a Hypersil ODS stationary phase and aqueous methanol mobile phases (ϕ_m = 0.50), and octanol/water log K_d. (○) Acids and alcohols; (●) neutrals, bases, and amphiprotics. (Data from Ref. 54.)

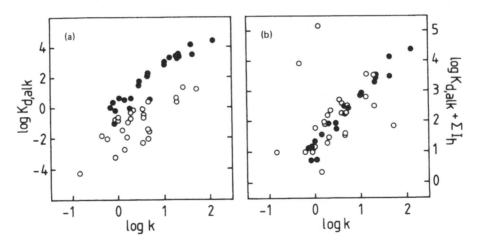

Figure 4 Relations between log k for various un-ionized compounds, determined using a Hypersil ODS stationary phase and aqueous methanol mobile phases (ϕ_m = 0.50), and alkane/water log K_d values (Fig. 4a), and alkane/water log K_d corrected for solute hydrogen-bonding properties using functional group hydrogen-bonding increments I_h (Fig. 4b). Key as for Fig. 3. (Data from Ref. 54.)

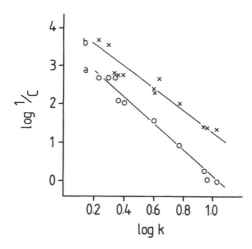

Figure 5 Correlation between biological activity (log C^{-1}) and log k values of N-phenylamides of benzoylacetic acid, determined using a LiChrosorb RP-18 stationary phase and a methanol - 3% acetic acid mobile phase of $\phi_m = 0.70$. (a) Inhibition of prostaglandin-synthetase (○); (b) binding to albumin (×). (Modified from Ref. 98.)

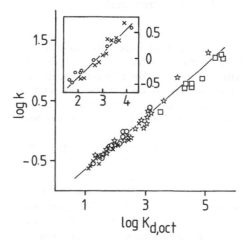

Figure 6 Relationships between log k, determined using a glyceryl-coated controlled-pore glass stationary phase and aqueous methanol mobile phases ($\phi_m = 0.10$), and log $K_{d,oct}$. Main figure: benzoic acids (○), N-phenylsuccinimides (×), N-phenylanthranilates (□), and phenols (☆). Insert: benzenoid non-hydrogen bonders (×) and benzenoid hydrogen-bonders (○). (Modified from Ref. 101.)

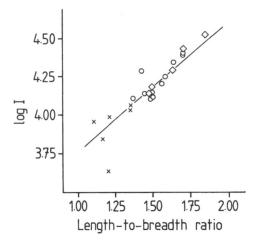

Figure 7 Linear correlation of length-to-breadth ratio and retention for methyl-substituted benzo(*c*)phenanthrenes (×), benz(*a*)anthracenes (○), and chrysenes (◇), determined using the RPLC system Vydac 201 TP / acetonitrile-water (ϕ_m = 0.85). The retention parameter used is the logarithmic solute retention index, based on the retention parameter of unsubstituted, six-membered-ring polycyclic aromatic hydrocarbons as reference solutes. (Modified from Ref. 121.)

index was originally constructed as a topological index of molecular branching [147], which was found to contain, within series of related molecules, quantitative structural information. For related solutes, molecular connectivity indexes were observed to be correlated with, for example, surface areas, liquid solute aqueous solubilities, and 1-octanol/water distribution coefficients (see Ref. 148 and references cited therein).

An interesting study on the relations between RPLC retention data and geometric solute parameters was performed by Tanaka and Thornton [127], who measured values of k for a large series of aliphatic carboxylic acids, alcohols, and alkanes using a μ-Bondapak C-18 stationary phase and aqueous methanol mobile phases. For each solute group, log k was found to be linearly related to the alkyl chain length (e.g., see Fig. 8). From these linear relationships, the free energies of transfer of a methylene group from the mobile to the stationary phase could be calculated. The resulting values of carboxylic acid and alcohol methylene group free energies of retention for totally aqueous mobile phases were found to correspond well with literature values of free energies for distribution of amphiphilic compounds between water and their own micelles.

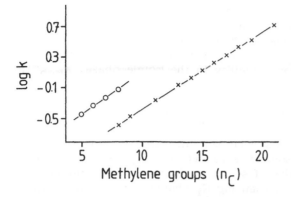

Figure 8 Plots of log k values, obtained using a μ-Bondapak C-18 stationary phase and aqueous methanol mobile phases (ϕ_m = 0.80), versus the number of solute methylene groups, n, for alkanes (○) and aliphatic carboxylic acids (×). (Modified from Ref. 127. Copyright 1977, The American Chemical Society.)

An indication of the coherence between various solvophobic-lipophilic solute parameters was given by the study of Wells et al. [91], who examined the various relations between RPLC retention data of 24 barbiturates (using Partisil ODS and ODS-2 stationary phases and methanol-water mobile phases), and solute octanol/water log K_d, molecular connectivities, and biological activities (IC_{50}, minimum hypnotic dose). The three solvophobic-lipophilic solute parameters were found to show high mutual correlations, and could be successfully used to describe solute biological activities.

In general it has been found that relations between solute liquid chromatographic retention and geometric properties are used to describe and predict solute retention, and only in an isolated instance have liquid chromatographic retention data been used to predict solute geometric properties [123].

3. Aqueous Solubility

A quantitative relationship between solute retention in bonded-phase RPLC and its solubility in the mobile phase was first proposed by Locke [149]. Considering the stationary phase to be a strongly modified adsorbent, Locke applied Everett's thermodynamic adsorption model [150] to describe solute retention. Assuming the magnitude of solute-stationary phase interactions for *similar* solutes to be solute independent, Locke derived an expression for the relative retention:

$$K_{r,i} \cdot K_{r,j}^{-1} = \gamma_{mob,i}^{\infty} \cdot [\gamma_{mob,j}^{\infty}]^{-1} \qquad (4)$$

where γ_{mob}^{∞} is the solute activity coefficient in the mobile phase at infinite dilution, and where the subscripts i and j refer to two similar solutes. For solutes that are poorly soluble in the mobile phase, Locke suggested that γ_{mob}^{∞} may be substituted by the reciprocal of the compound's mole fraction solubility in the mobile phase, X_{mob}^{-1}, to give:

$$K_{r,i} \cdot K_{r,j}^{-1} = X_{mob,i}^{-1} \cdot X_{mob,j} = S_{mob,i}^{-1} \cdot S_{mob,j} \qquad (5)$$

where S_{mob} is the compound's molar solubility in the mobile phase, which, for poorly soluble solutes is related to X_{mob} by a constant factor. Locke concluded that if solute-stationary phase interactions are of a constant magnitude, then $\log K_r$ should vary linearly with $\log S_{mob}^{-1}$. Hiller et al. [151] and Möckel and Masloch [124] studied the experimental validity of Eq. (5) for homologous series of alkylbenzenes, alkylbromides, and alkylsulfides, and for a series of polymethylbenzenes. Using various alkyl-bonded stationary-phase materials and methanol-water ($\phi_m = 0.70$) as mobile phase, they found that within each homologous series, retention could be well described by:

$$\log k = b - a \cdot \log S_{mob} \qquad (6)$$

where b and a are constants, the magnitude of b depending on the homologous series studied, and a being approximately 0.69 for all series. The fact that a differs from unity implies that the product $k \cdot S_{mob}$, and hence the product $K_r \cdot S_{mob}$ (K_r being equal to $k \cdot \phi^{-1}$), will not be constant, as predicted by Eq. (5). However, it must be emphasized that Eq. (5) has been proposed to hold only for similar solutes (not only in chemical nature, but also in size).

Hennion et al. [152] found the product $k \cdot S_{mob}$ for a single solute (phenanthrene) chromatographed using a LiChrosorb RP-8 stationary phase, and various binary aqueous organic mobile phases, to be independent of mobile-phase composition for a given solvent-water combination. Moreover, $k \cdot S_{mob}$ was found to be approximately constant (~ 0.20), when different alcoholic modifiers (methanol, ethanol, 1-propanol) were used, and deviated only for aqueous acetonitrile mobile phases (~ 0.13). If one assumes ϕ, the column-phase volume ratio, to be approximately independent of mobile-phase composition, these findings suggest that for phenanthrene chromatographed on LiChrosorb RP-8, (a) the magnitude of solute - stationary-phase interactions is independent of binary mobile-phase composition for a given organic modifier, and (b) this magnitude is constant for alcoholic modifiers, but different for aqueous acetonitrile mobile phases.

Locke has used Eq. (5) to estimate aqueous solubilities of a number of polycyclic aromatic hydrocarbons, employing literature RPLC retention data obtained on octadecyl- or naphthyl-bonded stationary phases with aqueous methanol mobile phases [149]. The reasonable agreement of his estimated solubilities with literature values suggested to the author that methanol as organic modifier does not interact selectively with these solutes.

Unfortunately, relations such as Eqs. (5) and (6) are applicable only to liquid solutes, because the solubility of solids also depends on the magnitude of intermolecular forces in the crystalline state [153]. By introducing a term that quantifies the magnitude of these forces, and by changing to mole-fraction solubility units, Eq. (6) can be converted into:

$$\log k = b' - a \cdot \log X_{mob} - c \cdot [T_m \cdot T^{-1} - 1] \tag{7a}$$

$$\log k = b' - a \cdot \log X_{mob} - d \cdot \log[T_m \cdot T^{-1}] \tag{7b}$$

where T_m is the solute's melting point (K), T is the absolute temperature, and b', c, and d, are constants. Upon rearrangement, Eqs. (7a) and (7b) become:

$$-\log X_{mob} = \frac{1}{a} \cdot \log k + \frac{c}{a} \cdot [T_m \cdot T^{-1} - 1] - \frac{b'}{a} \tag{8a}$$

$$= \frac{1}{a} \cdot \log k + \frac{d}{a} \cdot \log[T_m \cdot T^{-1}] - \frac{b'}{a} \tag{8b}$$

Equations (8a) and (8b) were applied by Hafkenscheid and Tomlinson [54,60,154,155] to estimate aqueous solubilities by means of RPLC retention parameters. These workers employed extrapolated retention data, i.e., $\log k_w$, obtained with a Hypersil ODS stationary phase and aqueous methanol mobile phases, to estimate aqueous solubilities of benzene derivatives and 2-propyl derivatives having both apolar and polar substituents, and of some polycyclic aromatic hydrocarbons [54, 60]. A typical result is shown in Fig. 9, where estimated values of $-\log X_w$ are plotted against values that were calculated from literature solubility data.

In a similar way, Whitehouse and Cooke [104] applied Eq. (8a) to estimate the aqueous solubilities of aromatic hydrocarbons, using values of $\log k$ obtained with an ODS stationary phase and an aceto-nitrile-water ($\phi_m = 0.75$) mobile phase. For these (strongly related structurally) apolar solutes, good agreement was found between estimated and known (literature) solubility values.

When using isocratic RPLC retention data ($\log k$ for Hypersil ODS as stationary-phase material and aqueous methanol mobile phases of $\phi_m = 0.50$ or 0.75), Hafkenscheid and Tomlinson [54,155] observed that, although for a wide variety of physicochemically different compounds aqueous solubilities could be well estimated, acid and alcohol solutes showed systematic differences in their behavior. These deviations—which were apparently inherent in the use of Eqs. (8a) and (8b)—were found, however, to be simple to correct.

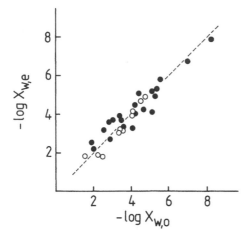

Figure 9 Relationship between observed negative logarithmic mole fraction aqueous solubilities of un-ionized solutes ($-\log X_{w,o}$) and ($-\log$) solubility values estimated using Eq. (8b) ($-\log X_{w,e}$) of 30 aromatic and aliphatic solutes. Log k_w values used in Eq. (8b) were determined using the RPLC system Hypersil ODS/aqueous methanol. (○) Liquid solutes; (●) solids. (Data from Ref. 54.)

Some additional related studies on relationships between aqueous solubility and retention in (reversed-phase) liquid chromatography are presented in Table 2.

B. Electronic Parameters

Unlike solvophobic solute properties, electronic properties of molecules are generally described by a single solute parameter, K_a—the solute acidity constant (often expressed as the logarithm of its reciprocal value, pK_a). The pK_a scale has been used by Hammett [156] to express the electronic effect of a substituent group X on the side-chain (re)activity of an aromatic compound R, i.e., on the (re)activity of a center usually insulated from resonance interactions with the aromatic ring, namely:

$$\log K_{RX} - \log K_{RH} = \rho \cdot \sigma_{m,p,X} \tag{9}$$

where K_{RX} and K_{RH} are the rate or equilibrium constants for the substituted and unsubstituted molecule, respectively, ρ is a solute-/system-dependent reaction parameter, and $\sigma_{m,p,X}$ is defined by:

$$\sigma_{m,p,X} = pK_{a,BX} - pK_{a,BH} \tag{10}$$

where $K_{a,BH}$ and $K_{a,BX}$ are the acidity constants of benzoic acid and X-substituted benzoic acid, respectively, with X located in the

Table 2 Studies on Relationships Between Liquid Chromatographic Retention and Solute Aqueous Solubility

Compound(s)	Stationary phase(s)	Mobile phase(s)	Temperature(s)	Observations	Refs.
Benzenoids, polynuclear aromatics	LiChrosorb RP-18	Various combinations of methanol, and water or aqueous acetate (pH 4.5)	25°C	For fairly insoluble compounds ($X_w \leq 0.01$), both extrapolated retention data (log k_w) and log $K_{d,oct}$ were found to be highly correlated to log γ_w^∞, i.e., the logarithm of (hypothetical) liquid solute aqueous solubility (see Fig. 10).	56
Organic pollutants	Radial Pak C-18	Methanol-water ($\phi_m = 0.85$)	Not stated	Aqueous solubility (in milligrams/liter) was related to a linear combination of log t_r and ($T_m - T$).	113

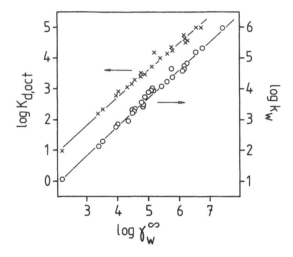

Figure 10 Relationships between log γ_w^∞, the logarithmic activity coefficient in water at infinite dilution, and ($\times$) log $K_{d,oct}$ and ($\circ$) log k_w, determined using a LiChrosorb RP-18/aqueous methanol RPLC system. Solutes studied were benzenoids and polycyclic aromatics. (Modified from Ref. 56.)

meta or para position (by definition, $\rho = 1$ for the benzoic acid reference system). The introduction of the σ parameter has led to numerous studies examining the quantitative relationships between molecular (re)activity and substituent electronic effects.

The pK_a values needed for calculation of σ parameter values are generally obtained via one of the methods described by Albert and Serjeant [157]. In those cases where a low aqueous solubility of the solute prevents direct evaluation of its aqueous pK_a, estimates may be obtained by extrapolation of pK_a values measured in methanol-water or ethanol-water mixtures with different contents of cosolvent. However, analogous to static methods used for the determination of hydrophobic-lipophilic solute properties, such static methods for evaluation of solute pK_a require milligram to gram quantities of pure, stable compound.

Because retention of an ionizable solute in reversed-phase liquid chromatography is directly related to its pK_a (or rather to the difference between mobile phase pH and solute pK_a), dynamic RPLC systems may be used to measure a solute's pK_a. A number of examples exist in the literature in which the relations between RPLC retention and solute pK_a were studied.

Pietrzyk and co-workers [158,159] were the first to model such relations and to establish their practical validity for polymer (XAD) stationary phases, whereas Horváth et al. [142] studies their applicability with ODS phase materials. For monoprotic acids, the relation between retention, solute pK_a, and mobile phase pH may be given as [158]:

$$k = [k^0 + k^-.10^{(pH-pK_a)mob}][1 + 10^{(pH-pK_a)mob}]^{-1} \tag{11}$$

where k^0 and k^- are the capacity factors of the un-ionized and fully ionized solute, respectively. In deriving this equation, it is assumed that (a) all ionic species in the mobile phase show a thermodynamically ideal behavior, and (b) no ion-pair complexation occurs between solute anions and buffer cations present in the mobile phase. The consequences of these assumptions have been discussed in detail and evaluated experimentally by van de Venne [160]. Equations equivalent to Eq. (11) have also been derived for monoprotic bases and for polyprotic weak electrolytes [142].

Equation (11) and its analogues have the form of sigmoidal relationships between k and mobile-phase pH. Such relationships have been observed in several experimental studies [142,158−168]. In some of these studies, the experimentally found relationships have been used to calculate solute pK_a in the mobile phase, usually by setting the second derivative of the k-pH relation to zero.

In this way, Horváth et al. [142] calculated the pK_a values of some phenylalkylcarboxylic and benzoic acids from retention data obtained using a Partisil ODS stationary phase and totally aqueous buffered mobile phases. The resulting values were in good agreement with those measured by potentiometric titration at the same ionic strength as that of the mobile phases used. Similarly, van de Venne [160] found good agreement between pK_a values of a number of benzoic acid derivatives calculated from retention data on LiChrosorb RP-18 with methanol-phosphate buffer mobile phases, and values obtained in corresponding methanol-water mixtures of the proper ionic strength, using potentiometric titration.

Palalikit and Block [164] used retention data for both acids and bases obtained with an Amberlite XAD-2 stationary phase and acetonitrile-phosphate buffer mobile phases ($\phi_m = 0.10$ or 0.20) to evaluate solute pK_a values. Again the dynamic pK_a values were found to be in excellent agreement with values determined for mobile-phase conditions using a spectrophotometric method. Obviously an advantage in the use of XAD-2-like polymeric stationary phase materials is their applicability with mobile phases of relatively high pH (up to $\sim 11-12$), offering the possibility for evaluating pK_a values of strongly basic solutes. However, the application of such phases has been rather limited by their relatively low separation efficiency, and by the effects of different types and concentrations of organic modifiers on their matrix configuration (swelling, shrinking).

Deming and Kong [169] obtained pK_a values for 28 *trans*-cinnamic, phenylalkylcarboxylic, and benzoic acids as a result of model fitting of a large number of retention data in ion-pair RPLC systems. Capacity factors, determined using μ-Bondapak C-18 with nine aqueous methanol mobile phases (ϕ_m = 0.30) differing in pH and in pairing-ion concentration were fitted to the ion-interaction retention model [170,171], resulting in estimates of solute pK_a, k^O, and k^-. Within each group of acids these pK_a values were found to be highly correlated with the corresponding $\sigma_{m,p}$ values, whereas log k^O showed a high overall correlation with Fujita's π values, with solute total carbon number, and with log k^-.

Related studies on the relationships between solute liquid chromatographic retention and electronic properties can be found in Refs. 43, 45, and 71, which have been described in Section III.A, Hydrophobic-Lipophilic Parameters, and in Refs. 57, 134, 167, 168, 172, and 173, which have been abstracted in Table 3.

C. Complex Formation Constants

The evaluation of organic solute pK_a under mobile-phase conditions in fact has been the first application of HPLC for measuring complex formation constants, and adjustment of the mobile-phase pH in reversed-phase liquid chromatography to control the retention of an ionizable solute may be considered as one application of complexation of a solute. Other secondary equilibria [174,175] that are frequently used for controlling retention are (a) ion-pair formation, (b) solute-metal ion complexation, and (c) charge-transfer (mainly π − π*) complexation.

Interestingly, in a study on RPLC retention of a series of pteroyl-oligo-γ-L-glutamic acids, Bush et al. [176] used a general model describing retention as a function of multiple equilibrium complex formation for calculation of solute pK_a values. The equation representing the model is given by:

$$k = [k_f + \sum_i k_i \cdot [A]^i \cdot \prod_j K_{c,j}][1 + \sum_i [A]^i \cdot \prod_j K_{c,j}]^{-1} \tag{12}$$

where k_f is the capacity factor of the uncomplexed solute; k_i, the capacity factor of the complex containing i complexing agent units; [A], the agent concentration in the mobile phase; and $K_{c,j}$, the stability constant of the complex containing j agent units. Using initial parameter estimates, these authors fitted retention data obtained as a function of mobile-phase pH to Eq. (12), and from this were able to calculate the pK_as of the solutes. Equation (12) represents the dependence of the solute capacity factor on complex formation under the assumption that interactions between solute and agent are predominantly localized in the mobile phase. This and other theoretical and practical aspects of measuring stability constants using RPLC

Table 3 Studies on Relationships Between Liquid Chromatographic Retention and Solute Electronic Parameters

Compound(s)	Stationary phase(s)	Mobile phase(s)	Temperature(s)	Observations	Refs.
Halogen-, methyl-, and aminoanilines	Corasil I Corasil II Zorbax Sil Porasil A, Porasil B, or Porasil C	Chloroform Cyclohexane Chloroform Methanol-chloroform ($\phi_m = 0.007$) Chloroform-cyclohexane ($\phi_m = 0.75$)	Not stated	Relations were studied between NPLC retention (log k) and pK_b values of anilines. Linear relations were found between log k and pK_b for each class of anilines (halogen-, methyl-, and amino-) separately.	172
Various anilines	Nucleosil CN	2-propanol - isooctane ($\phi_m = 0.002$)	Ambient	A semilinear relationship was found between log k and pK_b. Different relations were observed for primary, N-methyl-, and N-alkylanilines.	134
Benzoic acids, phenols, anilines, pyridines	LiChrosorb RP-18	Methanol-water or aqueous buffers of various pHs ($\phi_m = 0.50$)	25°C	Linear relationships were found between substituent contributions to retention expressed as log ($V_r - V_o$), and substituent pK_a values.	57
Antipyrines, acetylsalicylic acid	Spherisorb-10 ODS	Methanol-aqueous buffers of various pHs ($\phi_m = 0.50$) Acetonitrile-aqueous buffers of various pHs ($\phi_m = 0.30$)	25°C	RPLC retention was studied with respect to the dependence of mobile-phase pH and solute pK_a, for which methods were proposed for simple determination. Solute pK_a values derived from retention as a function of mobile-phase pH	167

Table 3 (continued)

Compound(s)	Stationary phase(s)	Mobile phase(s)	Temperature(s)	Observations	Refs.
				and potentiometrically determined values were in excellent agreement for aqueous methanol, but corresponded less for aqueous acetonitrile.	
Aromatic carboxylic acids, indandiones	LiChrosorb RP-18	Methanol-aqueous buffers of various pHs (ϕ_m = 0.35, 0.50, or 0.60)	30°C	Apparent solute pK_a values for methanol-water solvents were determined from variations in k values as a function of mobile-phase pH.	168
Acetanilides	Corasil II	Chloroform	Not stated	A linear relationship was found between solute log k and pK_a values for all except the ortho-substituted acetanilides.	173

retention data have been discussed in detail by Horváth et al. [177]. These workers showed that under proper experimental conditions, the complex retention modulus η_c, defined as $k \cdot k_f^{-1}$, determined as a function of the concentration of the complexing agent (Eq. 12), may be used to calculate K_c values that are in excellent agreement with complex formation constants measured using static methods. The complex formation equilibria that were studied were between metal ions (as agents) and nucleotides, 1-nitroso-2-naphthol sulfonate, or dibenzo-crown ethers, in a buffered aqueous solution or in methanol as the solvent. Examples of the relations found between η_c and [A] are presented in Fig. 11, which shows the sigmoidal shape of the η_c-[A] relationship, as is predicted by Eq. (12). The study of metal ion complexes using RPLC is greatly facilitated because (a) complex formation is localized in the mobile phase, and (b) metal ions are easily incorporated in RPLC mobile phases.

Riley et al. [48] circumvented the problem of localizing complex formation by using a general model describing retention in ion-pair RPLC, namely:

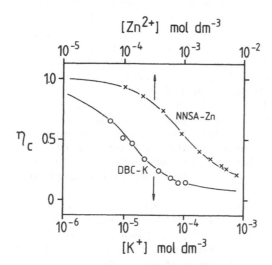

Figure 11 Graph illustrating plots of retention moduli η_c of 2-nitroso-1-naphthol-4-sulfonic acid (NNSA) and dibenzo-18-crown-6 (DBC) as functions of the molar concentrations of Zn^{2+} and K^+, respectively, on logarithmic scales. Stationary and mobile phases were ODS / 0.04 mol dm^{-3} PIPES buffer (pH 6.8) with varying concentrations of Zn^{2+} (NNSA), and perfluoroheptylsilica / methanol with varying concentrations of K^+ (DBC). Solid lines represent relationships calculated using Eq. (12). (Modified from Ref. 177.)

$$k = [k_f + L \cdot [A]][1 + K_c \cdot [A]]^{-1} \cdot [1 + K_{r,A} \cdot [A]]^{-1} \qquad (13)$$

where L is a constant depending on the location of complex formation, and $K_{r,A}$ is the binding constant of the pairing ion to the stationary phase. Using initial parameter estimates based on static ion-pair distribution experiments, k values of various substituted benzoate - tridecylbenzylammonium ion-pairs could be fitted to Eq. (13). As a result, K_c values for the various ion pairs could be calculated. In a subsequent study [166], these authors expanded Eq. (13) to include effects of solute ionization on its retention in ion-pair RPLC. By model fitting of retention data obtained for substituted amino acids and aminobenzoic acids using dodecylsulfate as the pairing ion, solute K_c and pK_a could simultaneously be calculated.

HPLC has been used by other workers in isolated instances to measure various complexation phenomena. These include (a) protein binding of drugs, (b) complex formation between solutes and micelles, and (c) $\pi - \pi^*$ charge transfer complexation. For example, Sebille and co-workers [178,179] studied the equilibrium binding of D- and L-tryptophan, warfarin, phenylbutazone, and furosemide to human serum albumin (HSA) by using HPLC retention data of these ligands. By employing a LiChrosorb Diol stationary phase, for which HSA is unretained, complex formation could be restricted to the mobile phase. In this way, the total affinity of HSA for the ligands studied could be evaluated by using mobile phases consisting of a pH 7.4 phosphate buffer with varying concentrations of HSA, the results being in good agreement with values obtained using static methods.

Furthermore, Yarmchuk et al. [180] studied the effects of micelle formation and micelle concentration of dodecyltrimethylammonium and dodecylsulfate ions on the RPLC retention of some un-ionized aromatic compounds. From variations in solute retention as a function of micelle concentration, solute-micelle complex stability constants and solute distribution constants between water and the stationary phase (Supelcosil LC-1) were calculated.

Armstrong and Carey [181] investigated the influence of the hydrophobic-hydrophilic balance within a series of bile salts on the equilibrium cholesterol-solubilizing capacities of their 0.1 mol dm^{-3} micellar solutions. Using an Ultrasphere ODS stationary phase and a methanol-aqueous phosphate ($\phi_m = 0.75$) mobile phase of pH 5, these workers found the logarithm of the cholesterol-solubilizing capacities (on a mole fraction base) of the micelles to be directly proportional to log k of the corresponding monomeric bile salts (Fig. 12). As can be seen from Fig. 12, different linear relations were obtained for bile salts of different structural types. These linear relations were successfully used to predict the cholesterol-solubilizing capacities for micelles of less common bile salts.

The quantitative effects of $\pi - \pi^*$ charge transfer complexation on the normal-phase liquid chromatographic retention of a series of π-

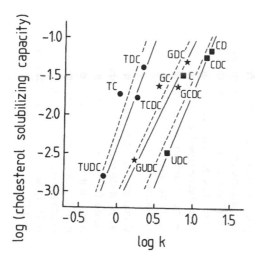

Figure 12 Double logarithmic plots of cholesterol monohydrate (ChM) -solubilizing capacities expressed as the mole fraction of ChM solubilized, versus the RPLC capacity factor k of bile salts, determined using the system Ultrasphere ODS / methanol-aqueous phosphate ($\phi_m = 0.75$, pH = 5.0). Solid lines are regression lines derived for the dihydroxy bile salts ursodeoxycholate (UDC), chenodeoxycholate (CDC), and deoxycholate (DC), and their respective conjugates. Dashed lines represent regressions, which include the trihydroxy bile salts cholate (c) and its conjugates. (Modified from Ref. 181.)

electron donors [i.e., (poly)methylbenzenes and polycyclic aromatic hydrocarbons] were studied by Burger and Tomlinson [182]. Using LiChrosorb SI-60, either plain, or coated with tetracyanoethylene (TCE) as the stationary phase, and *n*-hexane - 1,2-dichloroethane mobile phases, they found solute-TCE charge transfer complexation constants in various solvents to be linearly related to Δk, the difference in solute capacity factors obtained with a coated and a plain stationary phase, respectively.

Three interesting, related studies can be found in Refs. 183—185, and these have been abstracted in Table 4. The two former studies [183,184] were considered to be of sufficient interest to justify a brief description of their contents, notwithstanding the fact, that "low-performance" stationary-phase materials were used.

D. Miscellaneous Studies

The basis for some of the early applications of column-liquid chromatographic retention data for evaluation of physicochemical solute

Table 4   Studies on Relations Between Liquid Chromatographic Retention and Solute Complex Formation Constants

Compound(s)	Stationary phase(s)	Mobile phase(s)	Temperature(s)	Observations	Refs.
1:1 inclusion complexes of α- and β-cyclodextrin with barbiturates, phenothiazines, or sulfonamides	Jasco AV-02-500 Jasco CV-10-500	Aqueous phosphate (various pHs), containing 1–7 mmol dm^{-3} of cyclodextrin	Ambient	Stability constants for 1:1 cyclodextrin inclusion complexes were determined from variations in solute retention as a function of mobile-phase cyclodextrin concentration. The determined K_C values were found to be in agreement with values obtained by other methods.	183
Complexes of tri-m-phosphate with Mg^{2+} or Ca^{2+}	SP-Sephadex C-25	Aqueous solutions of metal ion chloride and tetramethylammonium chloride		Complex stability constants were determined from variations in retention of tri-m-phosphate as a function of mobile-phase metal ion concentration, applying the principles of Donnan exclusion to LC.	184

| Hydroxy-benzene derivatives | μ-Bondapak C-18 | Aqueous micellar solutions of sodium dodecyl sulfate ($0.02-0.20$ mol dm^{-3}) | Ambient | Distribution coefficients of solutes studied between (a) micelle and aqueous pseudo-phases, and (b) stationary and aqueous (mobile) phases were determined from variations in solute retention as a function of mobile-phase micelle concentration. The resulting micelle/aqueous distribution coefficients were used to calculate solute-micelle association constants, which were found to agree closely with known (literature) values, determined using static methods. |

properties was formed by Locke and Martire's quantitative description of retention in liquid/liquid chromatography [186], namely:

$$V_r - V_o = [\gamma_{mob}^{\infty} \cdot (\gamma_{st}^{\infty})^{-1}][n_{st} \cdot (n_{mob})^{-1}] \cdot V_o \qquad (14)$$

where $n_{st} \cdot n_{mob}^{-1}$ is the molar phase ratio of the column (with subscript "st" referring to the stationary phase). This would imply that, when one works isothermally, knowledge of $n_{st} \cdot n_{mob}^{-1}$, V_o, and the solute activity coefficient in one phase (the reference phase) might be combined with experimentally obtained retention data to calculate the solute activity coefficient in the second phase. In this way, Locke [187,188] calculated γ^{∞} both for the first four n-alcohol homologues in glycerol, using n-heptane as the reference (mobile) phase, and for various apolar compounds in acetonitrile, using squalane as the reference (stationary) phase. Static (literature) and dynamic values for γ^{∞} were found to be in good agreement, except for compounds that were poorly soluble in either one, or in both phases of the liquid/liquid LC systems used. Locke attributed the observed deviations mainly to solute adsorption on the liquid/liquid interface. Alessi and Kikic [189], however, ascribed these deviations to small mutual solubilities of the solvent pairs used, and were indeed able to improve Locke's results by correcting for these solubilities. The corrections used were based on the application of the additive properties of the solubility parameters of the pure phase-components.

In a later study [190], Alessi and Kikic observed that problems due to mutual phase solubilities could be efficiently reduced by using stationary phases of high molecular weight. Indeed, γ^{∞} values for a variety of hydrocarbons in aniline or acetonitrile, calculated using Apiezon-L as the reference phase, were found to be in excellent agreement with values reported in the literature.

Oka and Hara [191] studied the relations between equilibrium constants of tautomerization K_{PQ} for several tautomer pairs P-Q of steroidal ketoximes and O-methylketoximes using ethanol as solvent, and solute NPLC retention, expressed as α_{PQ} (= $k_P \cdot k_Q^{-1}$). Capacity factors of the individual tautomers were determined on a Wakogel LC-10 stationary phase, with n-hexane - ethylacetate mobile phases of varying compositions. Capacity factors for any desired composition were obtained by linear inter- or extrapolation of log k as a function of the mole fraction ethylacetate in the mobile phase. For all mobile-phase compositions studied, log α_{PQ} was found to be linearly related to log K_{PQ}, values of which were obtained from static experiments with ethanol as solvent, with an observed optimum ethylacetate mole fraction of 0.40. The relationship between log α_{PQ} and log K_{PQ} for this optimum mobile-phase composition is depicted in Fig. 13.

These findings suggest that tautomerization equilibrium constants can be obtained from solute retention data when appropriate liquid chromatography systems are used.

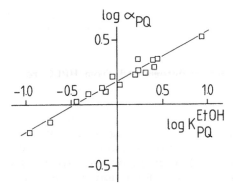

Figure 13 Linear relationship between logarithmic NPLC selectivities of ketoxime tautomers P-Q, log α_{PQ}, and logarithmic equilibrium constants of tautomerization K_{PQ}, measured in ethanol at 20°C. NPLC selectivity values were determined using a silicagel stationary phase and ethylacetate - *n*-hexane (ϕ_m = 0.40) as mobile phase. (Modified from Ref. 191.)

IV. CONCLUDING REMARKS

This overview has shown that under varying experimental conditions liquid chromatographic retention data can be used for estimating various physicochemical solute properties. In general, two main groups of solute parameters can be distinguished, namely (a) hydrophobic-lipophilic parameters, and (b) complex formation constants, the latter group including acidity constants and related electronic parameters. As a consequence of the nature of those parameters of particular interest (e.g., 1-octanol/water distribution coefficients, aqueous solubilities, acidity constants in aqueous solvents), reversed-phase liquid chromatography is found to be by far the most favored technique used.

It may be concluded that both isocratic and extrapolated retention data can serve as excellent measures of solute hydrophobic-lipophilic balance. In addition, the practical advantages in using high-performance liquid chromatography to measure this balance suggest that solute retention data may be used as parameters in structure-biological activity relation studies per se. Stationary phases employed for determining such retention data may either be physically coated, or chemically bonded in nature, with alkyl-bonded silicas being used predominantly. Mobile phases may be either aqueous, or aqueous-organic, with methanol and acetonitrile being primarily used as organic modifiers. Because it has been found that methanol shows little specific interaction effects with various types of solutes as compared to acetonitrile, the use of aqueous methanol mobile phases is to be preferred. However, the use of highly modified mobile phases should be avoided, because reversals in solute retention may occur with increasing mobile-phase modifier contents [18,53,86,98].

Furthermore, other parameters such as electronic and complex formation constants have been accurately determined from HPLC retention data. However, proper evaluation of such parameters, in general, is somewhat laborious in comparison to the determination of hydrophobic-lipophilic parameters, since (a) solute retention has to be measured as a function of mobile-phase complexing agent concentration, (b) the liquid chromatographic conditions used have to meet rather strict requirements [177], and (c) often intricacies are needed to describe retention as a function of complex formation because of the unknown "location" of complexation.

Conversely, knowledge of the appropriate physicochemical solute properties may be combined to predict solute retention in various HPLC systems, indicating, therefore, that it should also be possible to use suitable known retention data for such predictive purposes.

V. APPENDIX: LIST OF SYMBOLS AND ABBREVIATIONS

A. Abbreviations

LC	liquid chromatography
HPLC	high-performance liquid chromatography
TLC	thin-layer chromatography
GC	gas chromatography
SEC	size-exclusion chromatography
NPLC	normal-phase liquid chromatography
RPLC	reversed-phase liquid chromatography
RPTLC	reversed-phase thin-layer chromatography
IP-RPLC	ion-pair reversed-phase liquid chromatography
ODS	octadecylsilica
HSA	human serum albumin
TCE	tetracyanoethylene
THF	tetrahydrofuran
HIDA	phenylcarbamoyl-methylimino-diacetic acid

B. Symbols

C_n	total carbon number
f	hydrophobic fragmental constant

ΔG_r	free energy of retention
I	liquid chromatographic retention index
I_h	functional group hydrogen-bonding increment
k	liquid chromatographic capacity factor
k_w	hypothetical k value for totally aqueous mobile phases obtained by extrapolation
K	an equilibrium or rate constant
K_a	acidity constant
K_b	basicity constant
K_c	complex formation constant
K_d	liquid/liquid distribution coefficient
K_r	equilibrium constant of retention
L	complex formation location constant
n_C	substituent alkyl-chain length
$n_{st} \cdot n_{mob}^{-1}$	liquid chromatographic molar-phase ratio
pH	negative logarithm of hydrogen ion concentration
pK_a	negative logarithm of acidity constant
pK_b	negative logarithm of basicity constant
R	gas constant
R_L	functional group LC retention parameter
R_M	logarithmic thin-layer (paper) chromatographic capacity factor
$R_{M,w}$	hypothetical R_M value for totally aqueous mobile phases obtained by extrapolation
S	molar solubility
T	absolute temperature
T_m	melting point
t_r	chromatographic retention time
t_o	mobile-phase holdup time
V_r	chromatographic retention volume
V_o	mobile-phase holdup volume
$V_{st} \cdot V_{mob}^{-1}$	liquid chromatographic phase-volume ratio

X	mole fraction solubility
α	chromatographic selectivity
γ^{∞}	activity coefficient at infinite dilution
Δ	indicates excess of final over initial value
Φ	liquid chromatographic phase-volume ratio
ϕ_m	mobile-phase organic modifier volume fraction
η_c	complex retention modulus
π	liquid/liquid distribution group contribution constant
Π	continued product operator sign
ρ	reaction parameter
$\sigma_{m,p}$ (σ)	electronic group contribution constant
Σ	continued summation operator sign
τ	liquid chromatographic group contribution constant
τ_w	hypothetical τ value for totally aqueous mobile phases obtained by extrapolation

C. Subscripts

Referring to:

alk	an alkane as organic phase in a solvent/water distribution system
e	an estimated parameter value
f	an uncomplexed solute
i,j	similar solutes i and j
mob	the mobile phase
o	an observed parameter value
oct	1-octanol as organic phase in a solvent/water distribution system
st	the stationary phase
w	water as solvent (mobile phase)

D. Superscripts

Referring to:

o	an un-ionized solute
—	a monovalent solute anion

REFERENCES

1. E. Tomlinson, J. Chromatogr. *113*, 1 (1975).
2. R. Kaliszán, J. Chromatogr. *220*, 71 (1981).
3. I. E. Bush, Methods Biochem. Anal. *13*, 357 (1965).
4. L. R. Snyder, *Principles of Adsorption Chromatography*, Marcel Dekker, New York, 1968.
5. E. Soczewinski, Anal. Chem. *41*, 179 (1969).
6. L. R. Snyder, Anal. Chem. *46*, 1384 (1974).
7. R. P. W. Scott and P. Kucera, Anal. Chem. *45*, 749 (1973).
8. Cs. Horváth, W. Melander, and I. Molnár, J. Chromatogr. *125*, 129 (1976).
9. R. Tijssen, H. A. H. Billiet, and P. J. Schoenmakers, J. Chromatogr. *122*, 185 (1976).
10. B. L. Karger, J. R. Gant, A. Hartkopf, and P. H. Weiner, J. Chromatogr. *128*, 65 (1976).
11. L. R. Snyder and H. Poppe, J. Chromatogr. *184*, 363 (1980).
12. J. Novák, in *Liquid Column Chromatography, a Survey of Modern Techniques and Applications* (Z. Deyl, K. Macek, and J. Janák, eds.), Elsevier, Amsterdam, 1975, p. 45.
13. Cs. Horváth and W. Melander, in *High Performance Liquid Chromatography, Advances and Perspectives*, Vol. 2 (Cs. Horváth, ed.), Academic Press, New York, 1980, p. 266.
14. J. R. Conder and C. L. Young, *Physicochemical Measurement by Gas Chromatography*, Wiley, New York, 1979.
15. W. W. Yau, J. J. Kirkland, and D. D. Bly, *Modern Size-Exclusion Chromatography*, Wiley, New York, 1979.
16. B. G. Belenkii and L. Z. Vilenchik, *Modern Liquid Chromatography of Macromolecules*, Elsevier, Amsterdam, 1983.
17. J. H. Knox, R. Kaliszán, and G. J. Kennedy, Faraday Symp. Chem. Soc. *15*, 113 (1980).
18. G. E. Berendsen, Ph.D. Thesis, Delft University of Technology, 1980.
19. T. L. Hafkenscheid, unpublished observations.
20. H. Meyer, Arch. Exp. Pathol. Pharmacol. *42*, 110 (1899).
21. E. Overton, Studien über die Narkose, zugleich ein Beitrag zur allgemeine Pharmakologie, Fisher, Jena, 1901.
22. J. Ferguson, Proc. Royal Soc. [B]*127*, 387 (19390.
23. R. Collander, Phys. Plant. 7, 420 (1954).
24. C. Hansch and W. J. Dunn, III, J. Pharm. Sci. *61*, 1 (1972).
25. C. Hansch and J. M. Clayton, J. Pharm. Sci. *62*, 1 (1973).
26. C. Hansch, in *Drug Design*, Vol. 1, (E. Ariens, ed.), Academic Press, New York, 1971, p. 271.
27. C. Hansch and A. Leo, *Substituent Constants for Correlation Analysis in Chemistry and Biology*, Wiley, New York, 1979.
28. J. F. M. Kinkel, E. Tomlinson, and P. Smit, Int. J. Pharmaceutics *9*, 121 (1981).

29. C. Hansch, J. E. Quinlan, and G. L. Lawrence, J. Org. Chem. *33*, 347 (1968).
30. S. C. Valvani, S. H. Yalkowsky, and T. J. Roseman, J. Pharm. Sci. *70*, 502 (1981).
31. S. H. Yalkowsky and S. C. Valvani, J. Chem. Eng. Data *24*, 4300 (1979).
32. S. H. Yalkowsky and S. C. Valvani, J. Pharm. Sci. *69*, 912 (1980).
33. G. L. Flynn and S. H. Yalkowsky, J. Pharm. Sci. *61*, 838 (1972).
34. C. T. Chiou, in *Hazard Assessment of Chemicals, Current Developments*, Vol. 1, (J. Saxena and F. Fisher, eds.), Academic Press, New York, 1981, p. 117.
35. T. Fujita, J. Isawa, and C. Hansch, J. Am. Chem. Soc. *86*, 5175 (1964).
36. R. F. Rekker, *The Hydrophobic Fragmental Constant*, Elsevier, Amsterdam, 1977.
37. J. F. K. Huber, C. A. M. Meijers, and J. A. R. J. Hulsman, Anal. Chem. *44*, 111 (1972).
38. J. F. K. Huber, E. T. Alderlieste, H. Harren, and H. Poppe, Anal. Chem. *45*, 1337 (1973).
39. W. J. Haggerty and E. A. Murrill, Res. Devel. *25*, 30 (1974).
40. R. M. Carlson, R. E. Carlson, and H. L. Kopperman, J. Chromatogr. *107*, 219 (1975).
41. J. M. McCall, J. Med. Chem. *18*, 549 (1975).
42. M. S. Mirrlees, S. J. Moulton, C. T. Murphy, and P. J. Taylor, J. Med. Chem. *19*, 615 (1976).
43. S. H. Unger, J. R. Cook, and J. S. Hollenberg, J. Pharm. Sci. *67*, 1364 (1978).
44. I. T. Harrison, W. Kurz, I. J. Massey, and S. H. Unger, J. Med. Chem. *21*, 588 (1978).
45. I. M. Johansson and K. -G. Wahlund, Acta Pharm. Suec. *14*, 459 (1977).
46. A. Hulshoff and J. H. Perrin, J. Chromatogr. *129*, 263 (1976).
47. S. H. Unger and G. H. Chiang, J. Med. Chem. *24*, 262 (1981).
48. C. M. Riley, E. Tomlinson and T. M. Jefferies, J. Chromatogr. *185*, 197 (1979).
49. C. M. Riley, E. Tomlinson, and T. L. Hafkenscheid, J. Chromatogr. *218*, 427 (1981).
50. B. L. Karger, S. C. Su, S. Marchese, and B. -A. Persson, J. Chromatogr. Sci. *12*, 678 (1974).
51. S. C. Su, A. V. Hartkopf, and B. L. Karger, J. Chromatogr. *119*, 523 (1976).
52. B. Fransson, K. -G. Wahlund, I. M. Johansson, and G. Schill, J. Chromatogr. *125*, 327 (1976).
53. T. L. Hafkenscheid and E. Tomlinson, Int. J. Pharmaceutics *16*, 225 (1983).

54. T. L. Hafkenscheid, Ph.D. Thesis, University of Amsterdam, 1984.
55. C. H. Lochmüller and D. R. Wilder, J. Chromatogr. Sci. *17*, 574 (1979).
56. W. E. Hammers, G. J. Meurs, and C. L. de Ligny, J. Chromatogr. *247*, 1 (1982).
57. M. C. Spanjer and C. L. de Ligny, J. Chromatogr. *253*, 23 (1982).
58. H. Colin, A. M. Krstulović, G. Guiochon, and Z. Yun, J. Chromatogr. *255*, 295 (1983).
59. H. Könemann, R. Zelle, F. Busser, and W. E. Hammers, J. Chromatogr. *178*, 559 (1979).
60. T. L. Hafkenscheid and E. Tomlinson, J. Chromatogr. *218*, 409 (1981).
61. P. Seiler, Eur. J. Med. Chem. *9*, 473 (1974).
62. P. J. Twitchett and A. C. Moffat, J. Chromatogr. *111*, 149 (1975).
63. D. Henry, J. H. Block, J. L. Anderson, and G. R. Carlson, J. Med. Chem. *19*, 619 (1976).
64. P. -O. Lagerström, Acta Pharm. Suec. *13*, 213 (1976).
65. J. Crommen, B. Fransson, and G. Schill, J. Chromatogr. *142*, 283 (1977).
66. I. Molnár and Cs. Horváth, J. Chromatogr. *142*, 623 (1977).
67. D. Maysinger, W. Wolf, J. Casanova, and M. Tarle, Croat. Chem. Acta *49*, 549 (1977).
68. K. -G. Wahlund and I. Beijersten, J. Chromatogr. *149*, 313 (1978).
69. K. Miyake and H. Terada, J. Chromatogr. *157*, 386 (1978).
70. N. Tanaka, H. Goodell, and B. L. Karger, J. Chromatogr. *158*, 233 (1978).
71. S. H. Unger and T. F. Feuerman, J. Chromatogr. *176*, 426 (1979).
72. J. K. Baker, D. O. Rauls, and R. F. Borne, J. Med. Chem. *22*, 1301 (1979).
73. J. K. Baker, Anal. Chem. *51*, 1693 (1979).
74. J. K. Baker, R. E. Skelton, and Ch. -Y. Ma, J. Chromatogr. *168*, 417 (1979).
75. G. D. Veith, N. M. Austin, and R. T. Morris, Water Res. *13*, 43 (1979).
76. M. J. O'Hare and E. C. Nice, J. Chromatogr. *171*, 209 (1979).
77. X. Xu, Proc. of the First National Congress of the Pharmacol. Soc. of China, 1979, p. 13.
78. X. Xu, H. Xu, Y. -A. Lu, and J. Chen Yao Hsueh Hsueh Pao *14*, 246 (1979); C. A. *92*:93736g.
79. B. Rittich, M. Polster, and O. Králik, J. Chromatogr. *197*, 43 (1980).

80. C. H. Lochmüller, D. R. Wilder, and W. F. Gutknecht, J. Chem. Educ. *57*, 381 (1980).

81. M. T. W. Hearn and W. S. Hancock, J. Chromatogr. Sci. *18*, 288 (1980).

82. T. Braumann and L. H. Grimme, J. Chromatogr. *206*, 7 (1981).

83. R. M. Smith, J. Chromatogr. *209*, 1 (1981).

84. B. Rittich and H. Dubsky, J. Chromatogr. *209*, 7 (1981).

85. E. Tomlinson, H. Poppe, and J. C. Kraak, Int. J. Pharmaceutics *7*, 225 (1981).

86. B. McDuffie, Chemosphere *10*, 73 (1981).

87. W. Butte, C. Fooken, R. Klussmann, and D. Schuller, J. Chromatogr. *214*, 59 (1981).

88. N. Verbiese-Génard, M. Hanocq, M. van Damme, and L. Molle, Int. J. Pharmaceutics *9*, 295 (1981).

89. H. Ellgehausen, C. d'Hondt, and R. Fuerer, Pestic. Sci. *12*, 219 (1981).

90. I. Molnár and M. Schöneshöfer, in *High-Performance Chromatography in Protein and Peptide Chemistry*, de Gruyter, Berlin-New York, 1981, p. 98.

91. M. J. M. Wells, C. R. Clark, and R. M. Patterson, J. Chromatogr. Sci. *19*, 573 (1981).

92. X. Xu, H. Xu, J. Chen, and Q. Tang, Yao Hsueh Hsueh Pao *16*, 177 (1981); C. A. *95*:104039x.

93. K. J. Wilson, A. Honegger, R. P. Stötzel, and G. J. Hughes, Biochem. J. *199*, 31 (1981).

94. S. Fujiwara and E. Masuhara, Shika Rikogaku Zasshi *22*, 277 (1981); C. A. *96*:91575n.

95. S. Fujiwara and E. Masuhara, J. Biomed. Res. Mater. *15*, 787 (1981).

96. H. H. W. Thijssen, Eur. J. Med. Chem. *16*, 449 (1981).

97. T. Hanai, C. Tran, and J. Hubert, J. High Resol. Chromatogr. Chromatogr. Commun. *4*, 454 (1981).

98. M. L. Bieganowska, J. Liq. Chromatogr. *5*, 39 (1982).

99. T. Hanai and J. Hubert, J. Chromatogr. *239*, 527 (1982).

100. M. d'Amboise and T. Hanai, J. Liq. Chromatogr. *5*, 229 (1982).

101. K. Miyake and H. Terada, J. Chromatogr. *240*, 9 (1982).

102. T. Sasagawa, T. Okuyama, and D. C. Teller, J. Chromatogr. *240*, 329 (1982).

103. C. T. Wehr, L. Correia, and S. R. Abbott, J. Chromatogr. Sci. *20*, 114 (1982).

104. B. G. Whitehouse and R. C. Cooke, Chemosphere *11*, 689 (1982).

105. M. Lebl, J. Chromatogr. *242*, 342 (1982).

106. H. J. A. Wijnne, K. H. van Buuren, and W. Wakelkamp, Experientia *38*, 665 (1982).

107. C. L. K. Lins, J. H. Block, R. F. Doerge, and G. J. Barnes, J. Pharm. Sci. *71*, 614 (1982).

108. F. Barbato, M. Recanatini, C. Silippo, and A. Vittoria, Eur. J. Med. Chem. *17*, 229 (1982).
109. V. Ya. Davydov, M. Elizalde Gonzalez, and A. V. Kiselev, J. Chromatogr. *248*, 49 (1982).
110. K. Jinno, Chromatographia *15*, 723 (1982).
111. A. D. Nunn, J. Chromatogr. *255*, 91 (1983).
112. T. Hanai and J. Hubert, J. High Resol. Chromatogr. Chromatogr. Commun. *6*, 20 (1983).
113. R. L. Swann, D. A. Laskowski, P. J. McCall, K. Vanderkuy, and H. J. Dishburger, Residue Rev. *85*, 17 (1983).
114. M. C. Guerra, A. M. Barbato, G. Cantelli Forti, G. L. Biagi, and P. A. Borea, J. Chromatogr. *259*, 329 (1983).
115. T. Braumann, G. Weber, and L. H. Grimme, J. Chromatogr. *261*, 329 (1983).
116. T. L. Hafkenscheid and E. Tomlinson, J. Chromatogr. *264*, 47 (1983).
117. D. A. Brent, J. J. Sabatka, D. J. Minick, and D. W. Henry, J. Med. Chem. *26*, 1014 (1983).
118. C. V. Eadsforth and P. Moser, Chemosphere *12*, 1459 (1983).
119. K. Jinno and K. Kawasaki, Chromatographia *17*, 337 (1983).
120. K. Jinno and K. Kawasaki, Chromatographia *17*, 445 (1983).
121. S. A. Wise, W. J. Bonnett, F. R. Guenther, and W. E. May, J. Chromatogr. Sci. *19*, 457 (1981).
122. R. B. Sleight, J. Chromatogr. *83*, 31 (1973).
123. J. H. Kindsvater, P. H. Weiner, and T. J. Klingen, Anal. Chem. *46*, 982 (1974).
124. H. J. Möckel and B. Masloch, Fres. Z. Anal. Chem. *290*, 305 (1978).
125. V. Réhak and E. Smolková, J. Chromatogr. *191*, 71 (1980).
126. J. Kriz, L. Vodicka, J. Puncochárová, and M. Kurás, J. Chromatogr. *219*, 53 (1981).
127. N. Tanaka and E. R. Thornton, J. Am. Chem. Soc. *99*, 7300 (1977).
128. K. Callmer, L. -E. Edholm, and B. E. F. Smith, J. Chromatogr. *136*, 45 (1977).
129. W. Markowski, T. Dzido, and T. Wawrzynowicz, Pol. J. Chem. *52*, 2063 (1978).
130. J. F. Schabron, R. J. Hurtubise, and H. F. Silver, Anal. Chem. *50*, 1911 (1978).
131. Gy. Vigh and Z. Varga-Puchóny, J. Chromatogr. *196*, 1 (1980).
132. A. Nakae, K. Tsuji, and M. Yamanaka, Anal. Chem. *53*, 1818 (1981).
133. H. J. Möckel and T. Freyholdt, Fres. Z. Anal. Chem. *368*, 401 (1981).
134. L. -A. Truedsson and B. E. F. Smith, J. Chromatogr. *214*, 291 (1981).

135. O. Podlaha and B. Töregard, J. High Resol. Chromatogr. Chromatogr. Commun. 5, 668 (1982).
136. J. M. diBussolo and W. R. Nes, J. Chromatogr. Sci. 20, 193 (1982).
137. M. J. M. Wells and C. R. Clark, J. Chromatogr. 244, 231 (1982).
138. E. Grushka, H. Colin, and G. Guiochon, J. Chromatogr. 248, 325 (1982).
139. H. Colin, A. M. Krstulović, M. -F. Gonnord, G. Guichon, Z. Yun, and P. Jandera, Chromatographia 17, 1 (1983).
140. P. Dufek and E. Smolková, J. Chromatogr. 257, 247 (1983).
141. H. J. Möckel and T. Freyholdt, Chromatographia 17, 215 (1983).
142. Cs. Horváth, W. Melander, and I. Molnár, Anal. Chem. 49, 142 (1977).
143. M. C. Hennion, C. Picard, and M. Caude, J. Chromatogr. 166, 21 (1978).
144. K. Jinno and A. Ishigaki, J. High Resol. Chromatogr. Chromatogr. Commun. 5, 668 (1982).
145. S. J. Su, B. Grego, B. Niven, and M. T. W. Hearn, J. Liq. Chromatogr. 4, 1745 (1981).
146. R. J. Hurtubise, T. W. Allen, and H. F. Silver, J. Chromatogr. 235, 517 (1982).
147. M. Randic, J. Am. Chem. Soc. 97, 6609 (1975).
148. L. B. Kier and L. H. Hall, *Molecular Connectivity in Chemistry and Drug Research*, Academic Press, New York, 1976.
149. D. C. Locke, J. Chromatogr. Sci. 12, 433 (1974).
150. D. H. Everett, Trans. Faraday Soc. 61, 2478 (1965).
151. K. O. Hiller, B. Masloch, and H. J. Möckel, Fres. Z. Anal. Chem. 283, 109 (1977).
152. M. C. Hennion, C. Picard, C. Combellas, M. Caude, and R. Rosset, J. Chromatogr. 210, 211 (1981).
153. J. H. Hildebrand and R. L. Scott, *Regular Solutions*, Prentice Hall, Englewood Cliffs, N.J., 1962, p. 20.
154. T. L. Hafkenscheid and E. Tomlinson, Int. J. Pharmaceutics 8, 331 (1981).
155. T. L. Hafkenscheid and E. Tomlinson, Int. J. Pharmaceutics 17, 1 (1983).
156. L. P. Hammett, J. Am. Chem. Soc. 59, 96 (1937).
157. A. Albert and E. P. Serjeant, *The Determination of Ionization Constants. A Laboratory Manual*, Chapman & Hall, London, 1971.
158. M. D. Grieser and D. J. Pietrzyk, Anal. Chem. 45, 1348 (1973).
159. C. -H. Chu and D. J. Pietrzyk, Anal. Chem. 46, 330 (1974).
160. J. L. M. van de Venne, Ph.D. Thesis, Eindhoven University of Technology, 1979.
161. H. Takahagi and S. Seno, J. Chromatogr. 108, 354 (1975).

162. D. J. Pietrzyk, E. P. Kroeff, and T. D. Rotsch, Anal. Chem. *50*, 497 (1978).
163. E. P. Kroeff and D. J. Pietrzyk, Anal. Chem. *50*, 502 (1978).
164. D. Palalikit and J. H. Block, Anal. Chem. *52*, 624 (1980).
165. H. Takahagi, K. Asada, A. Nakanishi, K. Unno, Y. Hattori, and S. Seno, Ann. Rep. Sankyo Res. Lab. *33*, 88 (1981).
166. C. M. Riley, E. Tomlinson, and T. M. Jefferies, in *Current Developments in the Clinical Applications of HPLC, GC and MS* (A. M. Lawson, C. K. Lim, and W. Richmond, eds.), Academic Press, London, 1980, p. 35.
167. P. P. Pashankov, P. S. Zikolov, and O. B. Budevsky, J. Chromatogr. *209*, 149 (1981).
168. Gy. Vigh, Z. Varga-Puchóny, A. Bartha, and S. Balogh, J. Chromatogr. *241*, 169 (1982).
169. S. N. Deming and R. C. Kong, J. Chromatogr. *217*, 421 (1981).
170. B. A. Bidlingmeyer, S. N. Deming, W. P. Price, Jr., B. Sachok, and M. Petrusek, J. Chromatogr. *186*, 419 (1979).
171. R. C. Kong, B. Sachok, and S. N. Deming, J. Chromatogr. *199*, 307 (1980).
172. P. R. Young and H. M. McNair, Anal. Chem. *47*, 756 (1975).
173. N. R. Ayyangar and K. V. Srinivasan, J. Chromatogr. *267*, 399 (1983).
174. B. L. Karger, J. N. lePage, and N. Tanaka, in *High-Performance Liquid Chromatography, Advances and Perspectives*, Vol. 1 (Cs. Horváth, ed.), Academic Press, New York, 1980, p. 113.
175. E. Tomlinson, Chem. Ind. *19*, 687 (1981).
176. B. T. Bush, J. H. Frenz, W. R. Melander, Cs. Horváth, A. R. Cashmore, R. N. Dryer, J. O. Knipe, J. K. Coward, and J. R. Bertino, J. Chromatogr. *168*, 343 (1979).
177. Cs. Horváth, W. R. Melander, and A. Nahum, J. Chromatogr. *186*, 371 (1979).
178. B. Sebille and N. Thuaud, J. Liq. Chromatogr. *3*, 299 (1980).
179. B. Sebille, N. Thuaud, and J. P. Tillement, J. Chromatogr. *204*, 285 (1981).
180. P. Yarmchuk, R. Weinberger, R. F. Hirsch, and L. J. Cline Love, Anal. Chem. *54*, 2233 (1982).
181. M. J. Armstrong and M. C. Carey, J. Lipid Res. *23*, 70 (1982).
182. J. J. Burger and E. Tomlinson, Anal. Proc. Chem. Soc. *19*, 126 (1982).
183. K. Uekama, F. Hirayama, S. Nasu, N. Matsuo, and T. Irie, Chem. Pharm. Bull. *26*, 3477 (1978).
184. Y. Tokunaga and H. Waki, J. Liq. Chromatogr. *5*, 2169 (1982).
185. E. Pramauro and E. Pelizzetti, Anal. Chim. Acta *154*, 153 (1983).
186. D. C. Locke and D. E. Martire, Anal. Chem. *39*, 921 (1967).

187. D. C. Locke, J. Gas Chromatogr. *5*, 202 (1967).
188. D. C. Locke, J. Chromatogr. *35*, 24 (1968).
189. P. Alessi and I. Kikic, Chromatographia *7*, 299 (1974).
190. P. Alessi and I. Kikic, J. Chromatogr. *97*, 15 (1974).
191. K. Oka and S. Hara, J. Chromatogr. *202*, 187 (1980).

2

Mobile Phase Optimization in RPLC by an Iterative Regression Design

Leo de Galan and Hugo A. H. Billiet   *Delft University of Technology, Delft, The Netherlands*

Computer-aided procedures for the systematic optimization of the mobile phase composition in reversed-phase liquid chromatography can be distinguished in lattice designs, self-searching routines, and regression analysis. The procedures differ in the possibility of automation, in the number of test runs required, and in the necessary knowledge of the sample composition. Despite a wide variety of optimization criteria, all schemes aim at baseline separation of as many solutes as possible. The iterative regression design developed by the authors is discussed in detail. The indirect approach through a linear model for the retention behavior makes the procedure very efficient, but also requires the recognition of solutes in successive chromatograms. The versatility of the procedure is illustrated by examples of optimizing up to three organic modifiers, the pH, and the concentration of an ion-pairing reagent.

I. INTRODUCTION

A. The Problem

The selection of suitable analysis conditions for a novel sample poses more problems in liquid chromatography (LC) than in most other methods of analysis. Whereas with almost any method (be it voltammetry, infrared spectrometry, or even gas chromatography), the average sample can usually be analyzed under standard conditions, well-known to the operator or described in manuals, this is not true for LC. Indeed, when confronted with a novel sample, the liquid chromatographer must make a series of decisions.

First, he or she must select from a variety of related and often competitive techniques a suitable mode of operation: size-exclusion, ion-exchange, straight-phase, or reversed-phase. Next, he or she must choose the proper instrumentation: pump, column, detector, gradient, or isocratic operation. And finally, the liquid chromatographer must fine-tune the appropriate phase system from the many possible combinations of stationary and mobile-phase compositions.

In all these decisions, the nature of the sample (ionic, macromolecules, solute polarity, saturated or unsaturated compounds, etc.) is the guiding factor, and the more that is known about its composition, the easier the decisions become. Even then, the selection of proper analyzing conditions relies heavily on the chromatographer's experience. There are, at present, no written guidelines to assist the inexperienced analyst with a truly unknown sample. The even more formidable task of designing an expert system for the computerized optimization of all relevant steps in a liquid chromatographic separation is being contemplated [1], but its completion will certainly take many more years.

Fortunately, a completely unknown sample is a rare phenomenon, and better left to the specialist. In the more common situation that a minimum knowledge about the sample composition is available, the appropriate mode of liquid chromatography and the proper instrumentation are relatively easily decided, e.g., by consulting the literature. However, for a novel sample, even the experienced chromatographer may spend a long time in tailoring operating conditions, in particular the mobile-phase composition. It is no exception that a reasonably complex sample requires several days to achieve the first acceptable separation, that eventually is recorded within half an hour.

It is, therefore, not surprising that efforts have been made to delegate this final step in the decision process to computer algorithms and automated procedures, especially because microcomputers are built into virtually every modern liquid chromatograph. During the past few years, several studies on this subject have been published. Although the proposed schemes have a potentially wider applicability, most reports apply to reversed-phase liquid chromatography (RPLC), because this is the most versatile and popular mode of liquid chromatography.

Earlier reviews discussed a particular scheme [2–4] or dealt with mathematical details [5,6]. In this article, we shall attempt a comparison of the main strategies and discuss in some detail the procedure developed in our laboratory. Before we do so, however, it is necessary to define the terminology and the aspects common to all optimization procedures.

B. Terminology

At the start of the optimization, the chromatographer must determine the number, n, and the ranges of the *parameters* to be optimized. Together, they determine the n-dimensional *parameter space*, wherein the optimum is to be found. Preferably, if not principally, the parameters should be continuously and easily variable. The latter consideration excludes an otherwise advantageous parameter like the stationary phase. Parameters considered in the literature are column temperature [7], eluent flow rate [8], pH [2,9,10], and concentrations of various mobile-phase constituents [11,12]. However, the more parameters to be included, the more inefficient and lengthy the optimization procedure becomes. It is, therefore, prudent to restrict the optimization to those key parameters that exert the greatest influence and that mutually interact. It is also wise to distinguish between parameters that influence the elution order or the relative retention and those that influence only the analysis time. In RPLC, elution order is controlled by the concentration of various organic modifiers and, for some samples, by the pH or minor constituents such as an ion-pairing reagent. When the elution order has been optimized, the separation can be adapted by selecting a suitable plate count either through the eluent flow rate or through the column length.

The next step is the formulation of an *optimization criterion* that is a quantitative measure for the quality of a chromatographic separation. It must be single-valued and unambiguously determined from a given chromatogram. The formulation of a suitable criterion is important, because it guides the optimization procedure to its desired end result. For example, if the operator is interested only in the isolation of a single solute, the mutual separation of other solutes should not enter into the criterion. Usually, however, the operator has multiple, often conflicting interests. He or she may put emphasis on the separation of some solutes, without losing sight of all others or without sacrificing too much in analysis time. Consequently, the formulation of a single-valued criterion generally implies a compromise. As a result, it is perfectly possible that different chromatograms exhibit the same criterion value. It is not permissible, however, for one chromatogram to yield different values for the optimization criterion selected.

The variation of the optimization criterion over the parameter space is called the *response surface*. When applied to a particular problem, the optimization procedure could stop when it has located a point on the response surface that exceeds a predetermined threshold value. Because, however, optimization procedures are developed to solve a variety of problems, most schemes aim to locate the *highest* value of the criterion, which is called the *global optimum*. If the response surface has only one maximum, this is also the global optimum. In the more usual situation, the response surface resembles a mountain area with several peaks and valleys. In that case, a common danger for an optimization procedure is to focus onto a lower maximum.

It is obviously possible that even the global optimum fails to yield a satisfactory chromatogram. This is by no means a failure of the optimization procedure, but indicates rather that the aspired separation is not feasible within the constraints (i.e., the parameter space) set by the operator. It is, in fact, a decisive advantage when the optimization procedure rapidly assesses the limitations of a chosen chromatographic approach and activates the operator to seek other solutions. If the global optimum is close to the desired result, a simple increase in plate count may bring sufficient improvement. In more difficult situations, major alterations should be contemplated, such as specific detection, coupled-column techniques, or a change in the mode of operation.

C. Optimization Strategies

A detailed explanation of various optimization procedures is beyond the scope of this paper. A good introduction can be found in the book by Massart, Kaufman, and Dijkstra [13]. However, a brief overview of the procedures used in RPLC seems appropriate. In this presentation we shall also indicate the strong and the weak points of different approaches.

Liquid chromatography is a relatively slow technique, and the cycle time between successive chromatograms including equilibration of a different mobile-phase composition is of the order of 30 min. Consequently, the speed of an optimization procedure, i.e., the number of test runs needed to locate the global optimum, is an important aspect. To some extent it may be balanced by the possibility of automated operation, so that the procedure can run overnight. A final consideration is the required knowledge of the sample, i.e., the number and nature of the solutes present.

A first category of RPLC optimization procedures includes the *lattice designs*. The parameter space is covered by a predetermined lattice of measurement points where chromatograms are recorded (Fig. 1). The best chromatogram as judged either by visual observation or via an optimization criterion, is retained. The precision by which the global optimum is reached and the number of test runs depend upon the grit size of the lattice. Because the number of runs increases with the power of the number of parameters, lattice designs are usually restricted to one or two parameters. Because a standard protocol is followed, the procedure is easily automated and requires

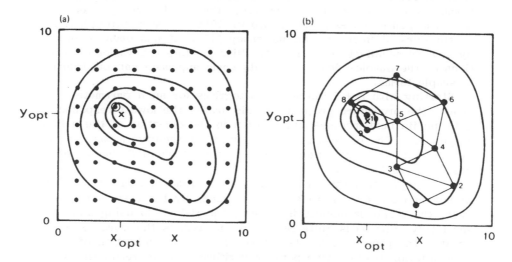

Figure 1   Optimization by lattice design and Simplex. The curves represent contour lines of a two-parameter (x,y) response surface. (Left) The optimum (indicated by ×) is found from a lattice of 81 data.   (Right) The Simplex design is started with three data (1,2,3), and the optimum is found by successive triangles as data point 10.

no knowledge of the sample composition, provided that an adequate optimization criterion has been formulated. A final advantage is that a picture of the complete response surface is obtained.

The simple approach of a full lattice design is utilized in at least one commercial instrument [14]. The major disadvantage is the very large number of test runs required. For example, the full range of water/methanol binaries is covered in steps of 10% with 11 runs. Extension to another modifier adds 10 more runs, and a two-parameter optimization of all ternary mixtures requires a total of 66 chromatograms. When it is realized that most data are collected in relatively uninteresting areas of the parameter space, the drive for more efficient approaches of the optimum is readily understood.

An obvious simplification is to test the parameters sequentially. For example, the approach taken in another commercial instrument [15] is to optimize first one organic modifier in steps that are successively narrowed down. Thereupon the process is repeated for a second modifier, and, if so desired, the resulting optimum binaries can be mixed in equal proportions to give a hopefully better ternary composition. In this way, the number of test runs is significantly reduced, but inevitably some overview of the total response surface is lost.

The idea of using preceding responses to guide the procedure is the principle that underlies a second category of *self-searching* optimization routines. After its successful introduction in analytical chemistry by Deming and Morgan [16], the *Simplex* design has become the prime representative, also in RPLC, notably through the recent work of Berridge and Morrissey [17]. It has been adopted by several manufacturers [18,19]. Disregarding the various modifications of the Simplex design [20], its basic approach can be described in simple terms.

Starting with n + 1 initial chromatograms for an n-parameter space, the procedure proceeds by discarding the lowest response and defining its next measurement location in the direction of the n higher responses. The process is repeated until consecutive measurements give no further improvement, and it is then concluded that an optimum has been reached (Fig. 1).

It is hoped that this is also the *global* optimum, and for a smooth, well-behaved response surface this is generally the case. If, however, the response surface is erratic, the risk of focusing onto a secondary optimum is quite real, and this risk is the more serious because the procedure by its nature provides no indication of a possible inferior result. One popular verification is to rerun the procedure with different starting conditions, but clearly this doubles the effort.

Against this uncertainty, the Simplex design has the advantages that again it is easily automated and requires no knowledge of the sample. It is superior to the lattice design in speed and in the larger number of parameters that can be accommodated. The average number of test runs is a matter of debate and varies significantly with the complexity of the separation problems. Some 15 runs seem to be the minimum for a one-parameter optimization, and this number increases approximately linearly with the number of parameters.

The two categories discussed thus far, share two major advantages: They are easily automated and require no knowledge of the sample. The latter is valid, however, only when the goal of the optimization is the separation of as many solutes as possible. This goal, expressed in a suitable optimization criterion, is indeed generally sought in the literature. It may be argued that the separation of only a few key components (know to be present) represents a more practical situation. Lattice design and Simplex design can then still be used, provided that they keep track of these solutes during the optimization procedure. This is relatively easy when the elution order is maintained throughout the procedure, but, in that case, optimization is usually pointless. In the more profitable situation that the elution order changes, provisions must be included to recognize solutes in successive chromatograms.

This requirement is always essential in the third category of optimization procedures known as *regression analysis*. As the name implies, these procedures attempt to describe the response surface by a mathematical expression (usually a polynomial), the coefficients of which are derived from a corresponding number of data points distributed over the parameter space. Because even a simple response surface can be described only by a polynomial of very high order, a direct approach would require an excessive number of data points. This demand can be significantly relaxed, however, by calculating the response surface *indirectly* from the retention surfaces of individual solutes. Because solute retention varies quite regularly with a change in eluent composition, only a few chromatograms are needed to describe retention surfaces and calculate the optimization criterion over the entire parameter space.

This approach was pioneered for gas chromatography by Laub and Purnell [21,22] and has been adapted to RPLC by Kirkland, Snyder, Glajch, and associates [23,24]. It offers several attractive advantages. It is unsurpassed for speed, requiring only seven chromatograms for a full ternary (i.e., two-parameter) optimization. It provides a complete picture of the response surface, and it allows a rapid verification of the predicted optimum.

There are, of course, also disadvantages. It is obviously mandatory that those solutes whose retention enters into the optimization procedure are accurately recognized in all chromatograms. If, as is widely done until now, this is accomplished by separate injection of all solutes, the advantage of speed is rapidly lost. Moreover, the solutes must be known and available beforehand. Also, the accuracy of the predicted optimum relies heavily upon the mathematical model used to describe solute retention. The commonly used quadratic polynomial needs only a few data points, but is fairly reluctant to alter its shape with the addition of more data.

In an attempt to overcome the latter disadvantage, the present authors have developed an *iterative regression* design [25,26]. Starting like the Simplex with n + 1 initial chromatograms, a first impression of the response surface is obtained by fitting a linear expression through the retention data. The optimum predicted from the derived response surface is verified, and when the result deviates from the prediction, the retention surfaces are improved by entering the new data. This process is repeated until the true response surface is approximated with sufficient precision.

This procedure is equally efficient and probably more accurate than straightforward regression. It shares the disadvantage that all relevant solutes must be recognized in successive chromatograms. Fortunately, there exist possibilities for accomplishing this without resorting to separate injections or even knowledge of the solutes [27]. Finally, the calculation of the response surface over the full-parameter space from solute retentions puts relatively high demands on the computer. As a result, regression analysis has so far been restricted to the optimization of one or two parameters only.

II. OPTIMIZATION CRITERIA

A. General Considerations

If the optimization procedure is to lead from an initially poor chromatogram to an acceptable result, a criterion is needed to distinguish between two successive chromatograms—a better and a poorer one. What constitutes the better chromatogram depends upon the goal of the separation set by the operator. *Better* can mean more symmetrical peaks or a shorter analysis time or an improved separation. When the distinction is based on human judgment, the criterion is frequently an implicit collection of several factors.

When the judgment is referred to a computer algorithm, these factors must be formulated explicitly and blended into a single quantitative value. Strictly speaking, the factors need not enter into the criterion simultaneously. A consecutive consideration of different factors is feasible. For instance, two chromatograms could first be compared for the number of discernible peaks, and when these are equal, a next objective could be considered.

A practical example of such a hierarchical criterion is the simultaneous optimization of pH and organic modifier described by Haddad et al. [9]. The separation of six acids is first optimized for resolution and thereafter, for analysis time. More generally, however, authors wishing to optimize for various objectives attempt to formulate a criterion that contains weighted contributions of all. The proposal of Berridge to be discussed below goes farthest in this direction by including number of peaks, resolution, and analysis time. Inevitably, such a composite criterion involves a tradeoff that is not easily defensible.

Unfortunately, a bewildering number of optimization criteria have been proposed in the literature. Critical comparisons have been published by Debets et al. [28] and by Drouen [29]. Actually, many criteria are closely related and express the same objective in slightly different ways.

An accurate description of the goal of the analysis is certainly the prime demand of the optimization criterion, but it is not the only one. An important practical demand is that the criterion should be easily calculated from an experimental chromatogram. For example, valley-to-peak ratios that are used in many criteria can be computed only when a valley is discernible between two successive peaks in the chromatogram. This puts demands on the minimum resolution, the signal-to-noise ratio, and the peak height ratio of the peaks (cf. Fig. 4 below). When these demands are not met, the corresponding contribution to the optimization criterion cannot be computed. As a result, initial improvements in the separation of strongly overlapping peaks remain unnoticed.

A final demand of the optimization criterion is indicated in Fig. 2, which presents two hypothetical one-dimensional response surfaces. In curve (a), the optimization criterion rises very sharply close to the optimum. Consequently, the optimization procedure spends a long

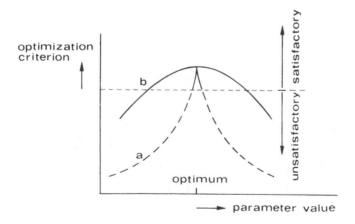

Figure 2 Variation of two hypothetical optimization criteria (a,b) with parameter value. The horizontal dashed line indicates the level of acceptable analytical result. Please see discussion in the text.

time in unsatisfactory regions of the response surface and tends to go into unstable oscillation around the optimum. In the preferred situation of curve b, the procedure moves away rapidly from low regions in the response surface and comes to a smooth stop at or close to the optimum.

B. Two-Peak Separation Criteria

Evidently, the separation between at least two successive peaks enters into all criteria for optimizing full chromatograms. Figure 3 illustrates some descriptions proposed for this separation.

The most simple one is the relative retention $\alpha = k_2/k_1$. Provided that the retention times of the solutes can be measured down to minimal separation, it can be calculated for any peak separation or

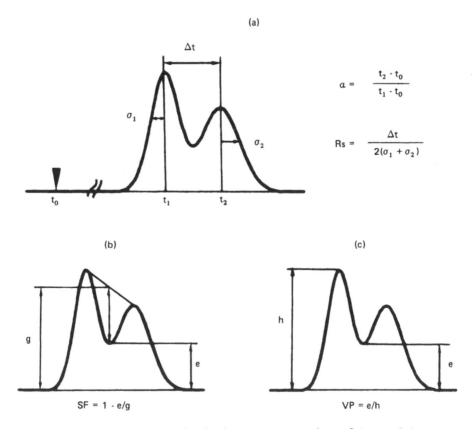

$$\alpha = \frac{t_2 \cdot t_0}{t_1 \cdot t_0}$$

$$R_s = \frac{\Delta t}{2(\sigma_1 + \sigma_2)}$$

SF = 1 - e/g

VP = e/h

Figure 3   Optimization criteria for the separation of two solutes: relative retention, α; resolution, R_s; separation factor, SF, and valley-to-peak ratio, VP.

any peak shape. It is rarely used, because it does not discriminate between narrow and broad peaks. For this reason the chromatographic resolution, Rs, is generally preferred, but the necessary peak widths are difficult to measure experimentally, especially for overlapping or asymmetrical peaks.

If the assumptions are made that the peaks are Gaussian and that the plate count is the same for the two peaks, then the resolution can be derived from retention times or capacity factors only:

$$\text{Rs} = \frac{t_2 - t_1}{2(\sigma_1 + \sigma_2)} = \frac{t_2 - t_1}{2(t_2 + t_1)} \sqrt{N} = \frac{k_2 - k_1}{2(k_2 + k_1 + 2)} \sqrt{N} \tag{1}$$

The condition of constant plate count is invalid when flow rate or column temperature is included in the optimization procedure, but it is reasonable for the majority of situations where the mobile-phase composition is the optimization parameter. The criterion expressed by Eq. (1) is widely used in optimization procedures based on regression analysis where the response surface is derived indirectly from retention data. Mapping designs and Simplex designs allow more freedom in the choice of criterion.

As the relative retention, the resolution has a nonzero value even for minimal separation, but it also keeps increasing with increasing difference, $t_2 - t_1$. To prevent the optimization of a two-solute separation from prolonging beyond the point where an acceptable chromatogram has been produced, an upper limit could be imposed on Rs. However, the appropriate upper limit depends on the shape and the relative heights of the two peaks.

This problem is avoided with two other, closely related criteria for a two-peak separation: the separation factor, SF, introduced by Kaiser [30] and the valley-to-peak ratio, VP, proposed by Christophe [31]. Indeed, both criteria level off to a constant value once baseline separation has been achieved. For two Gaussian peaks of equal height, it can be shown that

$$\text{SF} = 1 - \text{VP} = 0 \qquad\qquad (t_2 - t_1 < 2,3 \ \sigma) \tag{2a}$$

$$\text{SF} = 1 - \text{VP} = 1 - 2e^{-2\text{Rs}^2} \qquad (t_2 - t_1 > 2,3 \ \sigma) \tag{2b}$$

This function has been plotted in Fig. 4, together with the relative retention and the resolution as a function of the peak separation $(t_2 - t_1)$.

Both SF and VP relate closely to the chromatographer's desire for baseline separation, disregarding peak shape or relative abundance. They are easily derived from the experimental chromatogram, as long as a valley can be detected between the two peaks.

From the viewpoint of optimization, the latter condition is also their weakest point. Invariably, chromatographic optimization procedures

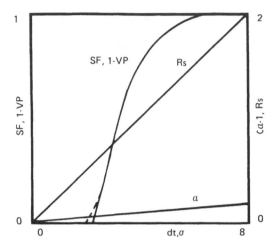

Figure 4 Variation of the two-solute optimization criteria (Fig. 3), with the peak separation, dt, in units of σ for two gaussian peaks of equal height.

are used to improve an initially very poor separation. Modest shifts in the early stages of the procedures can provide an important indication of the correct direction, even when no visible separation has yet been achieved. As is clear from Fig. 4 and Eq. (2a), shifts of up to 2.3 σ remain unnoticed by either SF or VP, even in the favorable case of two equal gaussians.

C. Multipeak Separation Criteria

Real samples contain more than two components. Therefore, even if the separation of only two solutes is sought, their separation from the other components must also enter into the optimization criterion. Actually, all optimization criteria presently developed for RPLC aim at the separation of as many solutes as possible.

The simplest criterion to describe this goal is to consider only the worst-separated peak pair using either of the criteria described in the preceding section. A good example is the minimum resolution observed in a given chromatogram, Rs_{min}, as used by Billiet et al. [32].

Obviously, during the progression of the optimization procedure, the worst pair may represent different solutes. As long as separation of all solutes is our goal, this constitutes no problem. In fact, the built-in hierarchy of the criterion is rather attractive. Once the separation of the worst solute pair has reached an acceptable value, the separation of the other solutes is automatically satisfactory.

However, the exclusion of the separation of other peak pairs from the criterion entails a potential danger. Generally, the chromatographic resolution increases with retention time. It can easily occur that a modest improvement of Rs_{min} from 0.5 to 0.7 for an early eluting peak pair raises the retention of late-eluting peaks to unacceptably high values.

Such adverse consequences are reduced in criteria that consider the separation of all adjacent peak pairs. Giddings was the first to propose the total overlap function, summed over all peak pairs, i:

$$\phi = \sum_i \exp(-2Rs_i) \tag{3}$$

which contains a contribution of one unit for each coinciding peak pair and approaches zero when all peaks are well resolved [33]. The stepwise nature of the criterion makes it less attractive for optimization procedures, and it has been replaced by a variety of mutually fairly similar criteria. Based on resolution (Eq. 1), separation factor (Eq. 2), or another description of the separation of adjacent peaks, the criteria take either the sum or the product of these values over all peak pairs. Typical examples are the chromatographic response function, CRF [34], and the resolution product, R [25].

$$CRF = \sum_i SF_i \tag{4}$$

$$CRF = \sum_i \ln SF_i \tag{5}$$

$$R = \prod_i Rs_i \tag{6}$$

where it is clear that the latter two expressions are similar in the sense that a sum of logarithms is equal to the logarithm of a product.

Like the minimum resolution, all three criteria tend to elongate the chromatogram. This is less serious when the parameter space of the optimization procedure has been restricted to an area of approximately constant analysis time. For example, in the optimization of ternary and quaternary solvent mixtures, the limiting binary mixtures of water with methanol, acetonitrile, or tetrahydrofuran, respectively, can be selected to have the same solvent strength (isoeluotropic) [35].

Although the inclusion of all peak pairs in the optimization criterion might be an advantage, it entails the risk that a poor separation of one pair is compensated by an exaggerated separation of another pair. This is most obvious in the sum function (Eq. 4), where the highly unfavorable situation of two coinciding peaks makes no contribution. Thus, a chromatogram of three solutes with two equal separation factors of 0.4 is considered to be inferior over one with separation factors of 1 (excellent separation) and 0 (unresolved). The product functions (Eqs. 5 and 6) perform better, because both become minimal—infinity and zero, respectively—for any coincidence, provided, of course, that such a poor resolution is detected. Still a compensating effect occurs. For example, the product of two perfectly acceptable resolutions of 1.1 and 1.2 yields a lower value for criterion (6) than the product of two resolutions 3 and 0.5, which is contrary to the chromatographer's preference. The reason is that in the latter chromatogram the resolution between the *first* and the *last* peak has increased from about 2.3 to 3.5. Such errors can be avoided by normalizing the resolution product [26]:

$$r = \frac{\prod\limits_{i} Rs_i}{[\Sigma Rs_i/(n-1)]^{n-1}} \tag{7}$$

In the above example, resolutions of 1.1 and 1.2 would yield a value for r very close to its maximum of r = 1, whereas resolutions of 0.5 and 3 yield r = 0.49. Indeed, in the latter situation a value of r = 1 would correspond to the superior separation where both resolutions are equal to 1.75. The normalized resolution product thus aims to make resolutions between adjacent peak pairs equal, within the time frame dictated by the solvent strength.

It does not tell, however, whether the mutually equal resolutions are also sufficient. Two equal resolutions of 0.2 yield r = 1, just as two equal resolutions of 1.0. Obviously, the improvement is obtained at the expense of an increased analysis time, but in this example, that would be an acceptable price to pay. To stimulate such improvement, but also safeguard against excessively long analysis times, criteria have been proposed that include the analysis time. One example is the modified chromatographic response function:

$$CRF = \frac{1}{t_n} \Sigma w_i \ln SF_i \tag{8}$$

where t_n is the retention time of the last peak. This criterion also includes weighting factors w_i to enhance the influence of the separation of certain peak pairs over others. Expressed as in Eq. (8), the weighting is not satisfactory, because the ith peak pair in one chromatogram need not be the same as in the next chromatogram. Especially in the initial stages of an optimization procedure, peak crossings frequently occur. In fact, it may be argued that without such large shifts in peak positions, optimization is probably fruitless. Weighting factors, therefore, are useful only when the solutes can be identified in successive chromatograms.

One weak point of all criteria discussed so far is that they consider *peak* pairs rather than *solute* pairs. Clearly, if our aim is to separate as many *solutes* as possible, then the appearance of an additional peak, however poorly separated, must be considered an improvement. It is true that such an appearance automatically minimizes the product criteria (Eqs. 5 and 6) of preceding chromatograms, where obviously one solute pair was unresolved. Conversely, however, such a situation often remains unnoticed until the additional peak appears.

Explicit attention to both the analysis time and the number of peaks is expressed by the criterion proposed by Berridge [17].

$$CRF = \sum_i RS_i + n^x - a \, |t_m - t_n| - b \, (t_1 - t_0) \qquad (9)$$

where the exponent x and the coefficients a and b can be chosen to give a desired emphasis on the total peak number n, the absolute difference between the acceptable and actual analysis time $|t_m - t_n|$, and the difference between the retention time of the first peak and the column dead time $(t_1 - t_0)$. Usually, the exponent x is chosen so large (e.g., x = 2) that the appearance of a new peak raises the criterion significantly. Because Berridge's criterion contains the summed rather than the multiplied resolutions, an initially poorly separated novel peak does not negate this improvement.

Elegant as this may seem, the summation of totally different concepts such as resolution, analysis time, and peak number makes the criterion (Eq. 9) difficult to interpret. A more serious objection remains the fact that a novel peak must first appear in the chromatogram, before its influence can be included. It would obviously be highly desirable if the number of *solutes* in the sample were known beforehand.

The next best situation occurs when the number of solutes can be derived from the chromatogram in an initial stage of the optimization procedure. Because the number of peaks affords only a lower estimate, the challenge is to detect multiple solutes in unresolved peaks. Not surprisingly, research in this direction is stimulated by optimization procedures utilizing regression designs. As was discussed earlier, regression designs need to keep track of *solutes* rather than peaks throughout the optimization procedure. If we cannot rely on separate injection of known solutes, then ways must be found to match corresponding solutes in successive chromatograms. Such an identification can then also be used to determine the total number of solutes; in fact, it is usually the first step. We shall return to this problem at a later point.

So far, no criterion has been formulated that has a clear advantage over the other proposals. Fortunately, optimization procedures can accommodate various criteria, although obviously the result achieved depends upon the one chosen. It would, therefore, be desirable when the operator had a choice of criteria. For example, in the optimization of a few similar samples, analysis time is less important and the minimum resolution might be quite acceptable. Analysis time is much more important when a large number of routine samples are to be run. In that case, equal resolutions (Eq. 7) or explicit attention to analysis time (Eqs. 8 and 9) might be more valuable.

III. THE ITERATIVE REGRESSION DESIGN

A. Principle of the Procedure

Over the past five years, the authors have developed an optimization procedure that aims at an accurate description of the response surface with a minimum number of test runs, similar to the regression design discussed above. The response surface is derived indirectly from the retention surfaces of individual solutes. However, the number of test runs and the retention model are not decided a priori, but continuously updated by verifying the predicted optimum. For this reason, the procedure is called an *iterative regression design*. In operation, the procedure proceeds by the following steps:

1. Like a Simplex optimization of n parameters, it starts with n + 1 initial runs, located at the extreme ends of the parameter space.
2. For the example of a one-parameter optimization, the logarithmic retention data of the solutes at the extreme ends are connected by straight lines.
3. The response surface based on a suitable optimization criterion is calculated over the full (one-dimensional) parameter space, and the optimum value and the corresponding parameter value are indicated.
4. In the next run, the optimum is checked, and if the result is not satisfactory, the new retention data are entered into the procedure.
5. In this way, the retention curves are described by a succession of linear segments until the true curves have been found with sufficient accuracy.
6. At that point, the procedure stops and reports the final optimum.

The three major advantages of the procedure are:

1. An impression of the complete parameter space is maintained throughout the procedure.
2. The final verification allows a judgment to be made on the accuracy of the description and, hence, of the result.
3. The number of test runs is adapted to the degree of nonlinearity of the retention surfaces.

Indeed, if the true retention curves are perfectly linear, the optimum predicted from two test runs (one parameter optimization) will be verified in the next run, and the procedure is completed with a total of three chromatograms. When the retention behavior is nonlinear, more runs will be required, but the final verification warrants an accurate result (though not always an acceptable separation!).

Questions of interest are the choice of the initial conditions, the iteration routine, and the stop criterion. These will be discussed in relation to the simple example of a one-dimensional optimization of a ternary solvent mixture composed of two binary eluents. The one parameter to be optimized is thus the mixing ratio.

B. The Gradient Scan

For several reasons it is advantageous, when the parameter space can be restricted to such an area, that at each point the sample solutes elute with capacity factors between 1 and 10; indeed, for $k < 1$, resolution is often inadequate; and for $k > 10$, both the analysis time and the peak shape deteriorate. For reasonable plate numbers, the peak capacity for $1 < k < 10$ is of the order of 30, which should be sufficient for most samples.

It is well known that between 1 and 10 the capacity factor decreases exponentially with increasing volume fraction of the organic modifier [35]:

$$\ln k_i = \ln k_i^{\circ} - S_i \phi \qquad (1 < k < 10) \qquad (10)$$

From a study of over 30 solutes, we have found that the slope factor, S, is approximately constant when acetonitrile is used as organic modifier, whereas for tetrahydrofuran and methanol the slope is linearly correlated with the intercept, that is:

$$S_i = p + q \ln k_i^{\circ} \qquad (11)$$

where p and q are solute independent. Consequently, one experimental result suffices to predict the capacity factor over a wide range of binary compositions from Eqs. (10) and (11).

In principle, any isocratic run could be used to provide the initial experimental data required. For an unknown sample, however, we want to be sure that all solutes are eluted. It is, therefore, preferred to derive the single data from a gradient run, whereby the methanol content is varied linearly from 0 to 100%. Using the relationship between the gradient elution time and the isocratic capacity factor derived by Schoenmakers et al. [35], it is then possible to relate the gradient elution times directly to volume fractions, ϕ, that allow isocratic elution with appropriate capacity factors.

The procedure is exemplified in Fig. 5. The unknown sample is injected onto the column at the start of the linear water-methanol gradient, and the net retention times are recorded. If the first solute elutes at 8.5 min, then a maximum volume fraction of 63% methanol will ensure isocratic elution with $k > 1$. If also the last solute in the gradient run has a retention time of 12.5 min, then volume fractions exceeding 52% methanol will ensure isocratic elution

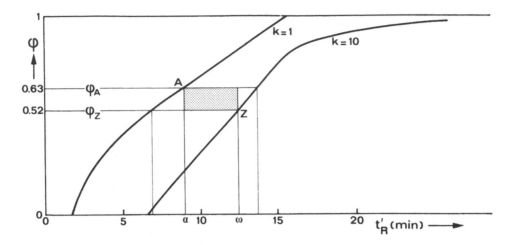

Figure 5   Deduction of binary solvent composition that warrants iso-cratic elution with 1 < k < 10 from retention data obtained with a 0–100% water-methanol gradient.   Discussion in the text.   (From Ref. 35.)

with k_{max} < 10.   Consequently, binary mixtures of water containing between 52 and 63% methanol will provide the desired elution of all sample solutes between k = 1 and k = 10.

If, on the other hand, the gradient run spreads the sample over a range of retention times between 7 and 14 min, then we would require at least 63% methanol to obtain k < 10 for the last solute, whereas no more than 52% methanol is allowed for the first solute to elute with k > 1.   In this case, the gradient run predicts that the variation in solute polarity is too large to allow isocratic elution between 1 < k < 10.   Obviously, the demands could be relaxed (e.g., 0.5 < k < 20), or else the decision should be reached to resort to gradient elution for the analysis.

Once a suitable methanol-water binary mixture has been found, it is fairly simple to derive binary mixtures of other organic modifiers that yield the same feature 1 < k < 10.   The transfer rules can be calculated from solubility parameter theory as:

$$\phi_x = \phi_{MeOH}(\delta_{H_2O} - \delta_{MeOH})/(\delta_{H_2O} - \delta_x) \tag{12}$$

where x is any modifier and δ is the solubility parameter [36].   For ACN and THF, more precise experimental transfer rules have been formulated [35].   For example, 52% MeOH corresponds to 38% ACN and to 34% THF.

As a result of the simple water-methanol gradient we can decide:

1. Whether or not the sample can be eluted isocratically with $1 < k < 10$.
2. Upper and lower boundaries for the methanol content and, hence, an appropriate intermediate value.
3. Binary compositions for other modifiers that promise isocratic elution with similar total analysis time.

These binary compositions are called *isoeluotropic* because they have similar solvent strength and predict equal analysis times. In practice, of course, the analysis times (i.e., k_{max}) are not precisely equal, and in fact, variations up to a factor of two have been observed [35]. However, the optimization procedure does not require a more precise adjustment.

Finally, three practical requirements should be indicated for the above procedure to be applicable. The column dead time must be known to derive the *net* gradient retention times (Fig. 5). The equipment must permit sample injection exactly at the start of the gradient at the top of the column. And, most importantly, the solutes must be detectable under gradient elution.

C. Iteration Routine

We now proceed with the optimization of a ternary mixture composed of two binaries. As an example, we consider the separation of five diphenylamines using MeOH, ACN, and THF as organic modifiers. The gradient run has provided us with the appropriate binary compositions, and the sample is eluted isocratically with these binaries (Fig. 6). The capacity factors of the solutes are entered on a logarithmic scale, and the data for each solute are connected by straight lines (Fig. 7). The initial assumption is thus that the retention of a solute, i, in a ternary mixture can be described by:

$$\ln k_i(x) = (1 - x) \ln k_i(I) + x \ln k_i(II) \tag{13}$$

where $k_i(I)$ and $k_i(II)$ are the capacity factors of solute i in the two limiting binary compositions I and II, respectively, and x is their mixing ratio.

Because the capacity factors $k_i(I)$ and $k_i(II)$ belong to the same solute, it is important that corresponding solutes are identified in the three chromatograms, notwithstanding peak overlap and changes in elution order. It is true that an occasional misinterpretation may be revealed in the progress of the procedure by gross errors in the predicted optimum, but correction is not always easy. For the moment, we assume that solute identification has been accomplished, and we shall return to this point later.

Because the capacity factors can be calculated from Eq. (13) for any ternary composition, the response surface can be easily constructed when an optimization criterion is used that can be derived

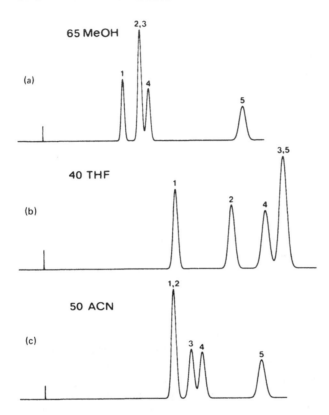

Figure 6 Chromatograms of a sample of five diphenylamines (DPA); a, b, and c represent the initial binary solvents derived from the gradient run and used to start the optimization procedure that subsequently leads to a first (d) and final (e) optimum ternary composition (see Figs. 7 and 8). Solutes: (1) *N*-nitroso-DPA, (2) 4-nitro-DPA, (3) 2,4'-dinitro-DPA, (4) DPA, (5) 2-nitro-DPA. Column length; 15 cm, 4.6 mm i.d.; flow, 1.5 ml/min; holdup time, 62.5 sec; stationary phase, Hypersil ODS 5 μm; number of plates, 5000. (Adapted from Ref. 26.)

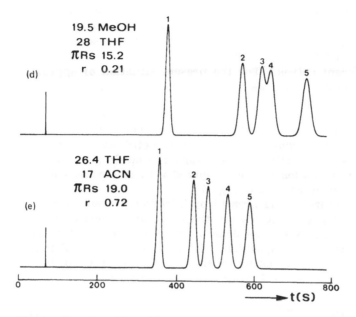

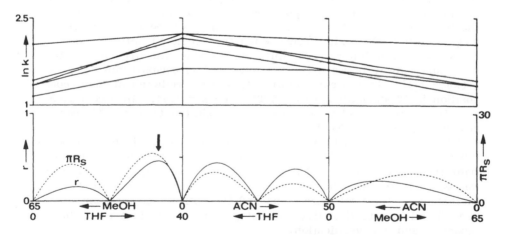

Figure 6 (continued)

Figure 7 Ternary optimization of five diphenylamines. The top diagram represents the straight retention lines connecting the data derived from three binary chromatograms (Fig. 6a, b, c). The bottom diagram represents the derived response surface for the relative resolution product (Eq. 7). The arrow indicates the predicted optimum ternary MeOH-THF-H_2O mixture. (From Ref. 26.)

from the capacity factors. The relative retention could be used,
but as discussed earlier the chromatographic resolution (Eq. 1) is
to be preferred. Either the minimum resolution or the resolution
product (Eq. 6) can be used, as long as it is realized that they lead
to potentially different optima. For the present situation of approximately constant analysis time, we prefer the normalized resolution
product (Eq. 7).

It should be emphasized that the retention data in the three binary
solvents permit the calculation of the response surface over the complete range of all three ternary mixtures: H_2O-MeOH-ACN, H_2O-
MeOH-THF, and H_2O-ACN-THF. As a result, the optimum predicted
in any of the three ternaries can be compared with the maxima found
in the other two ternary mixtures (Fig. 7).

The important feature of the present optimization procedure is
that it does not stop here, but advises the operator to verify the
predicted optimum. This could be done by running a chromatogram
with the proposed ternary composition. If, however, the predicted
optimum lies very close to an initial binary, no new insight will be
gained. Therefore, the advised composition is shifted to:

$$x = x_{opt} + 0.4 \left(0.5 - \frac{x_{opt} - x_1}{x_2 - x_1} \right) \tag{14}$$

where x_{opt} is the predicted optimum composition, and x_1 and x_2 are
the previously measured compositions on either side. At the start of
the procedure, $x_1 = 0$ and $x_2 = 1$. It can easily be seen that the
largest shift of 0.2 units occurs when the predicted optimum coin-
cides with either binary. No shift results when the predicted optimum falls midway between the two surrounding compositions [$x_{opt} =
\frac{1}{2} (x_2 - x_1)$]. In the example of Fig. 6, the predicted optimum com-
position is 10.4% MeOH and 33.6% THF, or $x = 0.84$. In accordance
with Eq. (14), the advised composition is shifted to $x = 0.70$, corresponding to 19.5% MeOH and 28% THF.

This chromatogram is now recorded and the result is used to verify
the prediction in Fig. 6e. The capacity factors observed in the chromatogram are entered into the procedure. If the original assumption
was correct, the new data should fall on the straight lines in Fig. 6.
In that case, the procedure stops and the optimum composition has
been found in a total of five chromatograms: a gradient run, three
binaries, and one verification.

In the common situation that the retention curves are not linear,
the retention data found in the chromatogram do not fall on the
straight lines in Fig. 6. At this point we could resort to parabolic
retention curves, but experience shows that such parabolas are sensitive to experimental errors in the retention data. Therefore, the new
data are again connected by straight lines to the surrounding points,

and the response surface is recalculated. The result is a new prediction of the optimum ternary composition, x_{opt}, different from the previous one.

If the new criterion value has become higher, we stay in the same ternary system. In the example of Fig. 6d, the chromatogram is actually much poorer than had been expected. In fact, the maximum criterion value now calculated for the system H_2O-MeOH-THF has dropped from 0.46 to only 0.21, which is lower than the maximum value of $r = 0.44$ predicted in Fig. 6 for the ternary system H_2O-THF-ACN. Consequently, the procedure advises to investigate this system next.

Whether or not we change to another ternary eluent, the above procedure is repeated: A new (shifted) composition is run, new retention data are entered, the prediction is verified, and so on. Each successive chromatogram provides new information about the actual retention behavior, and the true curves are approximated by a sequence of linear segments connecting measured retention data. Clearly, the number of data, and hence the number of test runs, depends upon the shape of the retention curves. It is important to note that this shape need not be known beforehand, but emerges automatically during the procedure. If the two initial data sets provide an accurate optimum, the retention curves are obviously close to linear. If they are not, additional data points will be called for.

As a typical example, Fig. 8 shows the true retention curves of the diphenylamines and the data points needed to assess the final optimum. Evidently, the iteration procedure discards the H_2O-MeOH-THF ternary after one attempt and correctly ignores the unprofitable H_2O-MeOH-ACN ternary. The smooth, but decidedly nonlinear curves in H_2O-ACN-THF are adequately described by three linear segments, and the final optimum provides an excellent separation (Fig. 6e). It has been found in a total of eight chromatograms, including the initial gradient and the final verification.

D. Stop Criterion

It remains to be discussed when the procedure is satisfied that the response surface has been exhaustively explored. This question is especially important when the search does not yield a satisfactory separation. Before we blame the chromatographic system, we must be sure that the procedure has found the best possible result within the parameter space imposed.

One possible *stop criterion* to terminate the procedure is to compare the solute retention data predicted for the shifted composition x (Eq. 14) with those actually measured in the chromatogram recorded for this composition. When the results agree within a stated error margin (e.g., 0.01 units for ln k or about 1% in the retention time), the procedure might be halted.

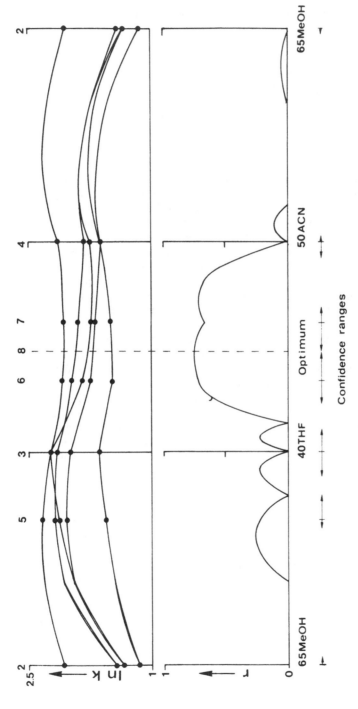

Figure 8 Ternary optimization of five diphenylamines. The top diagram presents the true (but un-known) retention behavior of the solutes as a function of the mixing ratio of two isoeluotropic binaries. The numbered dots represent the successive data that are connected by straight lines (1 is the initial gradient). The bottom diagram is the final response surface of the relative resolution product (Eq. 7). The thin lines below the figure show the confidence ranges in the final situation; the procedure stops when the final optimum (data point 8) falls within these ranges. (Adapted from Ref. 26.)

A more useful stop criterion is obtained when we realize that each data point determines a *confidence range* wherein the capacity factors can be predicted within a certain accuracy. It has been shown that this range can be conservatively defined as [26]:

$$\text{Confidence range} = \tfrac{1}{2}\left[\Delta x - (\Delta x^2 - 4\,\delta\ln k)^{\tfrac{1}{2}}\right] \tag{15}$$

where $\delta\ln k$ is the accepted uncertainty in the predicted capacity factors (e.g., $\delta\ln k = 0.01$) and Δx is the solvent range on either side of the composition x not yet searched. Initially, $\Delta x = 1$ and the first two confidence ranges associated with the starting binary compositions are very small ($0 < x < 0.01$, and $0.99 < x < 1.00$). However, with each new data point, Δx decreases and the confidence range rapidly stretches to cover the entire solvent range ($0 < x < 1$). Note that two confidence ranges touch when $\Delta x = 2\sqrt{\delta\ln k} = 0.2$ in Eq. (15). Usually, the optimization procedure stops before this, when a new predicted composition falls within one of the confidence intervals.

For the example of the diphenylamines, the successive confidence ranges have been entered into Fig. 8. After changing over to the H_2O-THF-ACN ternary mixture, the procedure starts with a parameter value of $x = 0.34$ shifted according to Eq. (14) from the optimum of $x = 0.24$ derived from Fig. 7. It expects an even higher criterion value at $x = 0.55$, but the next measurement at $x = 0.62$ proves this to be wrong. The next optimum is then predicted at $x = 0.46$, which falls within the confidence range of the previous measurement. Consequently, the procedure stops by advising to measure this final composition, which is indeed the global optimum.

E. Extension to Other Parameters

The iterative regression design is not restricted to the optimization of the organic modifier content, but can be applied to any solvent component. For example, in the separation of ionic solutes, the pH and the concentration of an added ion-pairing reagent are important parameters to control retention.

Figure 9 shows the variation of the retention of ten catecholamines and metabolites as a function of the mobile-phase concentration of sodium octanesulfonate added as an ion-pairing reagent. As expected, the retention of anions decreases with increasing sulfonate concentration, whereas the retention of cations initially increases sharply and then levels off. As a result, the elution order changes continuously, and for many sulfonate concentrations two solutes coincide in the chromatogram. In turn, the response surface reveals sharp variations of the optimization criterion, and the determination of the global optimum poses a challenge to any optimization procedure. Indeed, lattice designs would require a very large number of test runs to describe the response surface adequately. A Simplex design subjected

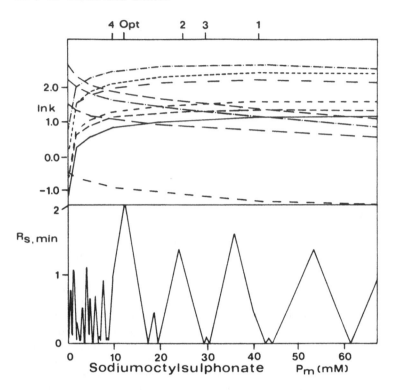

Figure 9 Ion-pair RPLC optimization. The top diagram presents the true retention behavior of 10 catecholamines as a function of the mobile-phase concentration of sodium octylsulfonate. The bottom diagram is the complex response surface of the resolution of the worst-separated peak pair. The numbers refer to the location of successive data points indicated during the course of the optimization procedure. (From Ref. 32.)

to this example frequently focused onto a secondary optimum, depending upon the initial compositions chosen.

When the sample is subjected to the iterative regression design, we must discard knowledge of Fig. 9. Because we expect strong retention of the cations at high sulfonate concentrations, the water-methanol gradient is run at the maximum concentration of 65 mM. It shows that isocratic elution with 10% methanol yields capacity factors between 1 and 10, except for 3,4-dihydroxymandelic acid, which elutes at k = 0.3. As the next solute elutes at k = 2, this is acceptable.

The two initial chromatograms are now run at the two limiting sulfonate concentrations of 0 and 65 mM (Fig. 10) and the corresponding

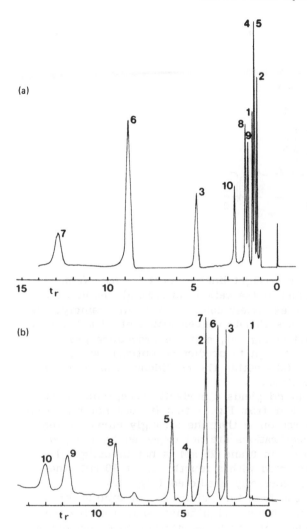

Figure 10 Chromatograms of a mixture of 10 catecholamines; (a) and (b) represent the initial records without and with the maximum concentration of ion-pairing reagent (sodium octylsulfonate). (c) Is the final result of the optimization procedure (Fig. 9). Solutes: (1) 3,4-dihydroxymandelic acid, (2) noradrenaline, (3) 3,4-dihydroxyphenylacetic acid, (4) adrenaline, (5) octopamine, (6) 5-hydroxyindol-3-acetic acid, (7) homovanillic acid, (8) L-DOPA, (9) dopamine, (10) tyrosine. The column (16 cm, 4.6 i.d.) is packed with Hypersil ODS. The mobile phase contains 10% methanol. (From Ref. 32.)

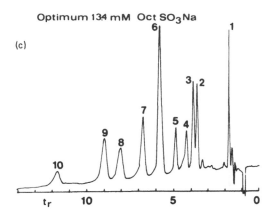

Figure 10   (continued)

lnk values are connected by straight lines. As it is clear from Fig. 9 that the retention variation of the cations is strongly nonlinear, this initial approximation is extremely coarse. Not surprisingly, the first optimum predicted at a sulfonate concentration of 42 mM is completely wrong (Rs_{min} = 0.1 in Fig. 9), and the procedure proceeds with a second and a third attempt to predict an optimum at 37 mM. Because this concentration falls within the confidence range of a previous run, the procedure stops.

Although the optimum found yields a perfectly acceptable separation (Rs_{min} = 1.6), it is clear from Fig. 9 that it does not represent the global optimum. The reason is that the strongly curved retentions at low sulfonate concentrations are not expected by the procedure, and the region between 0 and 25 mM is not investigated. When the procedure is restarted with a fourth run at 10 mM, the global optimum at 13 mM is correctly predicted (Fig. 10).

The warning that emerges from this failure is not unique for the iterative regression design. When a significant portion of the parameter space is left unsearched, it may be worthwhile to invest one more test run to explore this range. For instance, in the example of the diphenylamines discussed above, the H_2O-MeOH-ACN ternary mixture is ignored by the procedure. If we had not been satisfied with the separation achieved in H_2O-ACN-THF, a single attempt would have revealed that the low response surface predicted in Fig. 7 for this ternary agrees with reality (cf. Fig. 8).

In the present procedure, exploration of unsearched areas should be considered more readily when serious nonlinearity can be expected. This is the situation in Fig. 9, where the retention of the cations varies strongly between 0 and 25 mM and much less between 25 and 65 mM. If, however, the sulfonate concentration is plotted on a logarithmic scale, all retention curves become almost linear, as is shown in Fig. 11. The global optimum is now predicted directly from the two initial chromatograms, and the linear behavior is confirmed in the next run.

Another example of nonlinear retention variation is offered by the RPLC separation of five aromatic acids as a function of pH (Fig. 12). In this example, water is used as the solvent, and by adjusting the two compounds of an acetate buffer, a pH range of two units is covered around the average pK_a value of 4.5. Because the pK_a value of 1,4-benzenedicarboxylic acid is only 3.8, its retention decreases strongly with increasing pH. Nevertheless, the optimum pH of 4.31 is found in only two additional chromatograms, and the linearized retention behavior calculated by the procedure (Fig. 13) agrees very well with the true variation shown in Fig. 12.

In Fig. 13, the optimization criterion is again the normalized resolution product, r (Eq. 7), and indeed the solute peaks are distributed evenly over the chromatogram. As we discussed earlier, such an even distribution corresponding to a criteria value close to its maximum of r = 1 does not necessarily imply a good separation. The

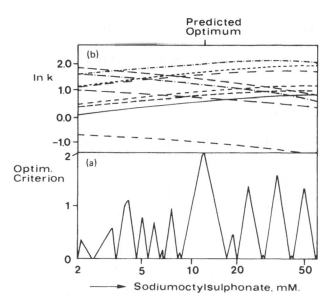

Figure 11   Optimization of ion-pair reagent concentration. When plotted on a double logarithmic scale (ln k versus ln Pm) the retention behavior is so linear that the true optimum is found directly from the two initial data sets taken at 0 and 52 mM sodium octyl sulfonate (compare Fig. 9). (From Ref. 32.)

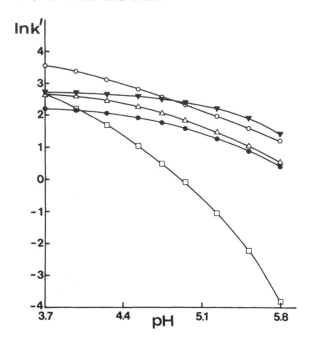

Figure 12 Retention variation of aromatic acids with pH. ○ = benzoic acid, ▼ = 2-aminobenzoic acid, □ = 1.4 benzenedicarboxylic aicd, △ = 4-hydroxybenzoic acid, ● = 4-aminobenzoic acid. (From Ref. 9.)

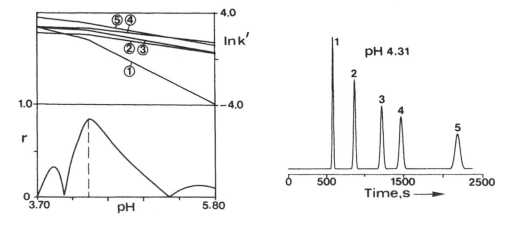

Figure 13 Optimization of aqueous mobile-phase pH for the separation of aromatic acids identified in Fig. 12. The chromatogram at optimal pH is also shown. (From Ref. 9.)

mutually equal resolution may be too low, in which case the plate
number should be increased, for example, by lengthening the column.
In the present example, the final separation is actually too good and
requires a relatively long time (Fig. 13). By increasing the flow
rate or choosing a shorter column, the analysis time could be reduced
at least two fold and still provide an adequate separation.

F. Simultaneous Two-Parameter Optimization

Situations may arise, where two chromatographic parameters exert a
major and mutually dependent influence upon the separation. A case
in point is the influence of pH and ion-pairing reagent concentration
upon the separation of ionic substances. It is, therefore, important
that the iterative regression design can be applied to the simultaneous optimization of two parameters, using the same concepts as discussed above for single-parameter optimization.

Just as *two* points define a straight line in the one-dimensional,
single-parameter space, so *three* points define a triangular plane in
the two-dimensional, two-parameter space. Consequently, three initial chromatograms must be run to provide the retention planes, described in analogy to Eq. (13) by:

$$\ln k_i (x,y) = (1 - x - y) \ln k_i(I) + x \ln k_i(II) + y \ln k_i(III)$$

$$(16)$$

where x and y are the two optimization parameters and $k_i(I)$, $k_i(II)$,
and $k_i(III)$ are the retentions of solute i at the apexes of the triangle
with compositions I, II, and III. On the basis of all solute retentions
initially measured at the extreme limits of the parameter space, a
highest value is again predicted for a suitable optimization criterion
(e.g., Rs_{min} or r). The predicted optimum is verified, and, when
it is proven to be incorrect, the new data are entered into the procedure, thereby dividing the parameter space into four smaller triangles. In this way, the two-dimensional retention surfaces are successively approximated by a net of interconnected triangles. The
two other concepts of the single-parameter procedure are also maintained. In verifying the predicted optimum composition, the measurement is shifted toward a less explored area of the triangle by an
expression analogous to Eq. (14). Each new data point determines
a confidence interval analogous to Eq. (15), and the procedure stops
when a predicted optimum composition falls within an earlier confidence
range.

As an example, we consider the previously considered pH optimization for the RPLC separation of five aromatic acids. It is realized
that the long analysis time apparent in Fig. 13 may also be reduced
by adding an organic modifier. However, because the modifier changes
the effective pH of the solvent [37] and may shift individual solutes
to a different extent, simultaneous optimization is called for.

The result is presented in Fig. 14. Again the acetate buffer components are adjusted so that they cover a pH range from 3.7 to 5.8 in aqueous solution. In addition, the methanol content is allowed to vary from 0 to 40% (v/v). Obviously, the parameter space is now a rectangle, and its subdivision in two triangles is ambiguous. Therefore, a fifth initial data point is located in the center of the rectangle, as indicated in Fig. 14. After three iterations, the procedure stops and indicates an optimum methanol content of 12% and an acetate buffer that yields a pH of 4.33 in aqueous solution.

The resulting chromatogram in Fig. 14 is indeed twice as rapid as the aqueous one in Fig. 13. Because the same optimization criterion (the normalized resolution product) is used, an even distribution of solute peaks is again observed, and a further reduction of the analysis time to about 5 min is possible by reducing the column length from 15 to 5 cm. Similar results are obtained when acetonitrile or tetrahydrofuran is used instead of methanol as organic modifier [9].

Earlier we considered the single-parameter optimization of a ternary solvent obtained by mixing two isoeluotropic binary solvents in a variable ratio (cf. Fig. 7). Actually, the three ternary mixtures represent the edges of a solvent triangle where the apexes refer to the three binary solvents (Fig. 17). Any point inside the triangle constitutes a quaternary solvent containing all three modifiers (MeOH, ACN, and THF) and water in variable proportions. As the sum of the three mixing ratios is constant, there are only two independent parameters. The optimization of a quaternary solvent of constant solvent strength thus constitutes another example of a two-parameter optimization.

Before we discuss this, it might be pointed out that the extension from one to two parameters significantly enhances the computing time. If the parameter space is detailed in steps as small as 1%, the three ternary systems in Fig. 7 cover 300 points, whereas the quaternary triangle in Fig. 17 requires 5000 points to be calculated. Using a current microcomputer and a high-level language, the calculation of the response surface for a single-parameter optimization takes less than 1 min. Depending on the number of solutes and chromatograms entered, it takes some 15 min for a two-parameter optimization. The construction of a pseudo-three-dimensional picture as in Fig. 14 requires an additional 15 min, but this time can be reduced by using equally informative contour plots (Fig. 17). Although these times are not prohibitive, extension to a third parameter is practical only when the step size chosen is much larger than 1%.

Obviously, the next test run in the optimization procedure cannot be started before the result of the previous run has become available. Therefore, in optimizing complex modifier solvents, it is advantageous to complete ternary optimization before proceeding with quaternary optimization. This is the more important, as it is our experience that there are only very few situations where quaternary solvent yields a

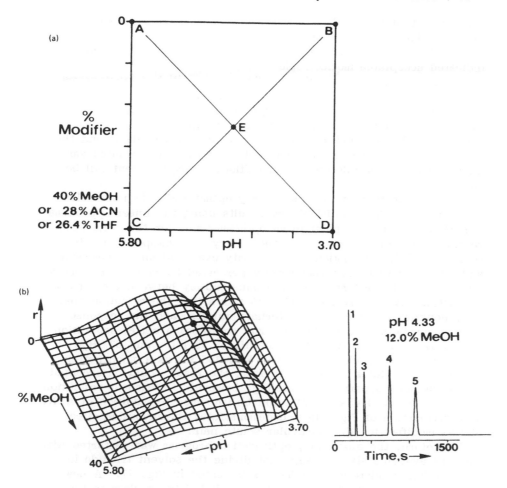

Figure 14   a: Experimental design for the combined optimization of
pH and organic modifier content of the mobile phase.   The letters
A—E represent the mobile-phase compositions used to provide the experimental retention data necessary to initiate the optimization procedure.   b: Two-parameter optimization of pH and methanol content
for the separation of aromatic acids identified in Fig. 12.   The dot in
the response surface represents the optimum combination for which
the chromatogram is shown.   Note the twofold gain in analysis time
in comparison to Fig. 13.   (Adapted from Ref. 9.)

separation that is not already acceptable in a solvent of lesser complexity. This is illustrated in the next example.

For the separation of eight chlorinated phenols, the gradient run indicated acceptable isoeluotropic binary solvents of 50% MeOH, 32% ACN, and 33% THF, respectively. The three initial chromatograms are presented in Fig. 15. Even in acetonitrile, a better separation of the solute pairs 3,4 and 5,6 would be desirable. The prospects for improvement by ternary and quaternary optimization are dim, because the solutes 3 and 4 almost coincide in both the MeOH and the THF binary. Therefore, if the retention of these two solutes varies linearly with the mobile-phase composition, no improvement will be observed.

Indeed, the ternary and quaternary optimization diagrams calculated from the three initial binary results using the linear approximation (Eqs. 13 and 16) point to an optimum value for the normalized resolution product of r = 0.1 at the binary ACN composition. In that case, only a deviation from linearity can yield an improvement, and such deviations are most readily perceived by running chromatograms at the three ternary compositions midway between the binaries.

Fortunately, the result in Fig. 16 shows that the retention does vary nonlinearly over all three ternary ranges, and the response surface predicts (and in the next run confirms) a better separation (r = 0.3) for a mixture of 11.5% MeOH, 25.5% THF, and 63% H_2O. The corresponding chromatogram in Fig. 15d shows a much better resolution of solutes 5 and 6. In fact, the result could be accepted as an adequate separation. Possibly, however, the modest resolution of solutes 3 and 4 (Rs = 0.8) could be improved by addition of ACN, as a comparison of Figs. 15 a–c demonstrates the favorable effect of this modifier for this particular solute pair.

Therefore, the quaternary optimization program was restarted with the now available six data sets that divide the solvent triangle in four smaller equilateral triangles, as indicated in Fig. 17. A new and still higher optimum (r = 0.4) is predicted to be close to the center of gravity of the top triangle and is confirmed in the next run. The final chromatogram in Fig. 15e shows a marginally better resolution of solutes 3 and 4 (Rs = 1.0).

The contour plot of the quaternary optimization in Fig. 17b is easier to plot and to interpret than the pseudo-three-dimensional representation in Fig. 14. The entered curves connect points of equal optimization criterion in steps of 0.05. Note that the values observed in the ternary diagram (Fig. 16) are faithfully reproduced. The best binary is ACN/H_2O, with r slightly over 0.1. Values as high as r = 0.3 are observed along two ternary edges.

G. Solute Recognition

A requirement common to all optimization procedures that derive the response surface of the criterion indirectly from retention surfaces is

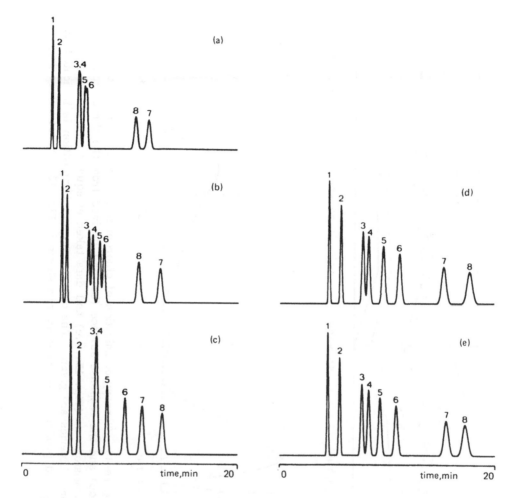

Figure 15 Chromatograms of a mixture of eight chlorophenols recorded in 55% MeOH. (a), 32% ACN, (b), 33% THF, (c), the final ternary optimum (d, 11.5% MeOH and 25.4% THF), and the quaternary optimum (e, 8.4% MeOH, 22% THF, and 5.3% ACN). Solutes: (1) *o*-chlorophenol, (2) *p*-chlorophenol, (3) 4-chloro-3-methylphenol, (4) 2,3-dichlorophenol, (5) *p*-chloro-*o*-cresol, (6) 2,5-dichlorophenol, (7) 2,4-dichloro-5-methylphenol, (8) 3,5-dichlorophenol. (From Ref. 29.)

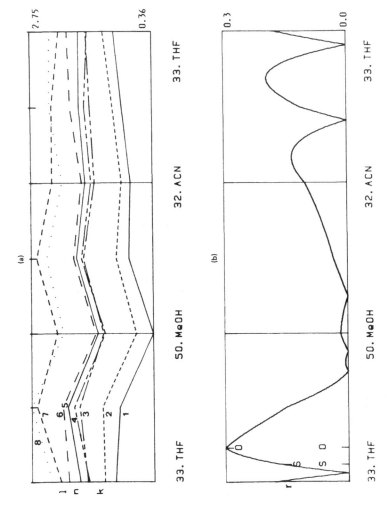

Figure 16 Ternary optimization of chlorophenols. The retention diagram (top) and the response surface (bottom) derived from the three binary chromatograms (Fig. 15a,b,c) and data taken at equal mixtures of the three binaries are given. The optimum ternary composition is indicated by O, and produces the chromatogram shown in Fig. 15d. (From Ref. 29.)

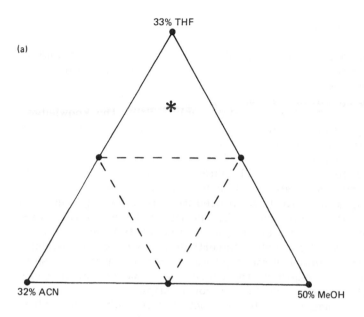

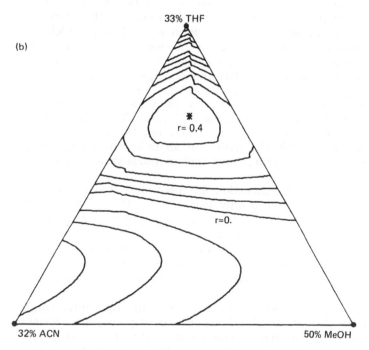

Figure 17 Quaternary optimization of chlorophenols. Location of the data points in the isoeluotropic solvent triangle (a) from which the two-parameter response surface is calculated illustrated by the contour plot. The optimum quaternary composition is indicated by the asterisk and produces the chromatogram in Fig. 15e. (From Ref. 29.)

the need to match corresponding solutes in successive chromatograms. It should be realized that the apparent advantage of more direct optimization procedures, such as the Simplex design, vanishes when the goal of the optimization is no longer the separation of all solutes but the separation of a few key solutes only. Still, the knowledge is vital for regression designs such as the procedure of Glajch et al. [23] and the presently discussed iterative procedure.

If the composition of the sample is known and the solutes are available, separate injection of all solutes provides an easy solution. In almost all cases, separate injection need be done only for the initial chromatograms to start the procedure. In following steps, the elution order of the solutes can safely be matched against their expected retention times, and barring major unexpected shifts, the sample can be injected as such. Nevertheless, even when restricted to the two or three initial chromatograms, separate injection of some ten solutes seriously lengthens the procedures. Moreover, this resource is simply not available for unknown samples.

The problem is aggravated by the fact that the initial chromatograms will be poorly resolved; otherwise optimization need not be contemplated. Matching solutes by their peak areas is, therefore, unprofitable and, hence, any single one-dimensional detector yields insufficient information. What is clearly needed is a detector that provides a recognizable pattern for each solute. As such, spectrometric detectors are paralleled. For the most general sample, a liquid chromatograph - mass spectrometer (LC-MS) combination is probably indispensable. Fortunately, for many samples the currently available multiwavelength UV detectors [27] may offer an equally useful and much less expensive solution. Although the UV spectrum is rarely sufficient to identify the nature of an unknown solute, it can be satisfactory for the present, more modest purpose of matching a limited number of solutes.

Some initial work in this direction has been reported by Drouen et al. The absorbance ratio of only two wavelengths, though useful to recognize peak overlap, is inadequate for solute matching [38]. Full spectra recorded with a photodiode-array spectrometer proved to be more powerful [27]. Unless two solutes coincide exactly, peak overlap is readily detected from varying spectra taken across the unresolved peak, and recognizable spectra can be extracted to allow solute matching.

A simple example is presented in Fig. 18. An apparently single peak is observed during the RPLC separation of polycyclic aromatic hydrocarbons in 65% ACN/H_2O. The dissimilarity of spectra taken at three indicated time slots across the peak clearly demonstrates its inpurity. Elution profiles taken at three representative wavelengths reveal the presence of two solutes. Characteristic spectra extracted at the front end and the tail end of the unresolved peak can be readily matched against standards or against similar spectra extracted from another chromatogram.

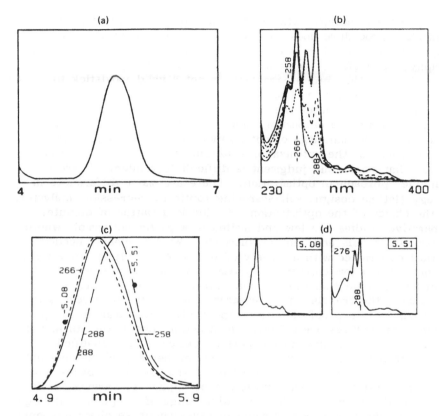

Figure 18 Analysis of two overlapping solutes: (a) the chromato-gram; (b) spectra taken during elution; (c) elution profiles at wave-lengths taken from frame b; (d) representative spectra at elution times taken from frame c.

This brief discussion is not intended to suggest that the problem of solute matching in regression designs has been solved. Many questions need still to be answered. What minimal separation is required? How different should two spectra be? Can the matching be automated? It seems, however, that a solution of the problem is feasible, especially when more advanced mathematical techniques are employed [39].

IV. CONCLUSIONS

Understandably, in the literature on mobile-phase optimization in RPLC, successful separations are more readily reported than failures.

This can be misleading. The mere application of a sophisticated optimization procedure does not guarantee a good separation, because no optimization procedure can be better than the underlying chromatography. Therefore, the achievement of baseline separation, desirable as it is to the chromatographist, is not a valid yardstick to measure the performance of an optimization procedure. Rather, it should be judged whether the procedure finds the best possible separation within the chromatographic constraints imposed by the operator. In the language of optimization, this is defined as the global optimum of the optimization criterion in the parameter space.

Even that more valid judgment is difficult to render. There is no unique approach to optimization. There are major differences in strategy (lattice design, self-searching routines, regression analysis), and the choice of the optimization criterion is a matter of dispute. Comparative studies are few and limited to a particular sample where one procedure fails and another succeeds. What is clearly needed is practical experience with a variety of real samples. It is, therefore, fortunate that a few commercial systems have become available, and it can be expected that other manufacturers will follow suit.

In itself, the translation of research studies into commercial software perhaps is a proof of trust. Apparently, computer-aided optimization procedures are effective in providing useful results in many situations. Even if they do not perform better than a qualified chromatographist, they certainly make his or her experience available to the novice. The examples discussed in the preceding pages have shown that optimization procedures can be applied to various mobile-phase conditions and can address a wide range of samples. Although the present discussion is restricted to reversed-phase chromatography, there is no reason why the procedures could not be equally useful in other modes of liquid chromatography. Indeed, some examples for straight-phase chromatography have already been published [24,40], and ion-exchange or size-exclusion chromatography should follow soon.

The development of optimization procedures is still young. Further progress can be expected. Some improvements likely to emerge in the near future are:

1. An evaluation of existing and new optimization criteria and the freedom for the operator to choose the criterion most suitable to the problem at hand.
2. Strategies focused on the separation of a few key solutes rather than all components in the sample.
3. Better tools to determine the number of solutes in an unknown sample, which is an essential figure in all procedures. Techniques such as factor analysis [39] and target testing [41] can be explored for this purpose.

ACKNOWLEDGMENT

The authors wish to thank Millipore, Waters Chromatography Division, Milford, Mass. U.S.A., for financial support.

REFERENCES

1. J. Karnicky, T. Schlabach, Sj. van der Wal, and S. Abbott Construction of an expert system for liquid chromatography, Lecture 4a-B4, Eighth International Symposium on Column Liquid Chromatography, May 20–25, 1984, New York.
2. J. L. Glajch and J. J. Kirkland, Anal. Chem. 55, 319A (1983).
3. S. N. Deming, J. G. Bower, and K. D. Bower, *Advances in Chromatography*, Vol. 24, Marcel Dekker Inc., New York, 1984, p. 35.
4. H. J. Issaq, *Advances in Chromatography*, Vol. 24, Marcel Dekker Inc., New York, 1984, p. 55.
5. J. C. Berridge, Chromatogr. *244*, 1 (1982).
6. J. W. Weyland, C. H. P. Bruins and D. A. Doornbos, J. Chromatogr. Sci. *22*, 31 (1984).
7. J. R. Gant, J. W. Dolan, and L. R. Snyder, J. Chromatogr. *185*, 153 (1979).
8. J. Rafel, J. Chromatogr. *282*, 287 (1983).
9. P. R. Haddad, A. C. J. H. Drouen, H. A. H. Billiet, and L. de Galan, J. Chromatogr. *282*, 71 (1983).
10. S. N. Deming and M. L. H. Turoff, Anal. Chem. *50*, 546 (1978).
11. W. Lindberg, E. Johansson, and K. Johansson, J. Chromatogr. *211*, 201 (1981).
12. B. Sachok, J. J. Stranahan, and S. N. Deming, Anal. Chem. *53*, 80 (1981).
13. D. L. Massart, L. Kaufman, and A. Dijkstra, *Evaluation and Optimization of Laboratory Methods and Analytical Procedures*, Elsevier Science Publishers, Amsterdam, 1978.
14. PESOS, Perkin-Elmer, solvent optimization system, Norwalk, Conn., U.S.A.
15. OPTIM II, System for automatic HPLC-optimization, Spectra Physics.
16. S. N. Deming and S. L. Morgan, Anal. Chem. *45*, 278A (1973).
17. J. C. Berridge and E. G. Morrissey, J. Chromatogr. *316*, 69 (1984).
18. Bruker, Automated Optimization of HPLC Separations Using a Simplex Procedure.
19. LDC/Milton Roy, Computer aided solvent optimization for reversed-phase chromatography.

20. P. F. A. van der Wiel, Anal. Chim. Acta, *122*, 421 (1980).
21. R. J. Laub and J. H. Purnell, J. Chromatogr. *112*, 71 (1975).
22. R. J. Laub, *Physical Methods in Modern Chemical Analysis*, Vol. 3 (Th. Kuwana, ed.), Academic Press, New York, London, 1983, p. 249.
23. J. L. Glajch, J. J. Kirkland, K. M. Squire, and J. M. Minor, J. Chromatogr. *199*, 57 (1980).
24. J. L. Glajch, J. J. Kirkland, and L. R. Snyder, J. Chromatogr. *238*, 269 (1982).
25. P. J. Schoenmakers, A. C. J. H. Drouen, H. A. H. Billiet, and L. de Galan, Chromatographia *15*, 688 (1982).
26. A. C. J. H. Drouen, H. A. H. Billiet, P. J. Schoenmakers, and L. de Galan, Chromatographia *16*, 48 (1982).
27. A. C. J. H. Drouen, H. A. H. Billiet, and L. de Galan, Anal. Chem. *57*, 962 (1985).
28. H. J. G. Debets, B. L. Bajema, and D. A. Doornbos, Anal. Chim. Acta *151*, 131 (1983).
29. A. C. J. H. Drouen, Computerized Optimization and Solute Recognition in Liquid Chromatographic Separations, Thesis TH, Delft, 1985.
30. R. Kaiser, Gaschromatographie, Portig, Leipzig, 1960, p. 33.
31. A. B. Christophe, Chromatographia *4*, 455 (1971).
32. H. A. H. Billiet, A. C. J. H. Drouen, and L. de Galan, J. Chromatogr. *316*, 231 (1984).
33. J. C. Giddings, Anal. Chem. *32*, 1707 (1960).
34. M. W. Watson and P. W. Carr, Anal. Chem. *51*, 1835 (1979).
35. P. J. Schoenmakers, H. A. H. Billiet, and L. de Galan, J. Chromatogr. *205*, 13 (1981).
36. P. J. Schoenmakers, H. A. H. Billiet, and L. de Galan, J. Chromatogr. *218*, 261 (1981).
37. A. Leitold and Gy. Vigh, J. Chromatogr. *257*, 384 (1983).
38. A. C. J. H. Drouen, H. A. H. Billiet, and L. de Galan, Anal. Chem. *56*, 971 (1984).
39. B. G. M. Vandeginste, R. Essers, Th. Bosman, J. Reijnen, and G. Kateman, Anal. Chem. *57*, 971 (1985).
40. L. R. Snyder, J. L. Glajch, and J. J. Kirkland, J. Chromatogr. *218*, 299 (1981).
41. B. G. M. Vandeginste, W. Derks, and G. Kateman, Anal. Chim. Acta *173*, 253 (1985).

3

Solvent Elimination Techniques for HPLC/FT-IR

Peter R. Griffiths and Christine M. Conroy   *University of California, Riverside, California*

I. INTRODUCTION

Infrared spectrometry has several important characteristics which
make it a potentially ideal detector for high-performance liquid
chromatography (HPLC). Most organic compounds and many inorganic compounds have strong, relatively narrow absorption bands in
the mid infrared. These absorptions are highly specific and give
detailed structural information about the compound. Even when an
unequivocal identification is impossible, important functional group
and isomeric substitution pattern information can be obtained. The
real-time measurement of the infrared spectra of compounds eluting
from a chromatograph is made possible through the use of a Fourier
transform (FT-IR) spectrometer. FT-IR spectrometry allows spectra
to be measured at good signal-to-noise ratio (SNR) in reasonably
short periods of time. In addition a great deal of data manipulation,
such as spectral subtractions, can be done with ease. For these and
other reasons, FT-IR spectrometers have been successfully used in
a variety of chromatographic interfaces.

The most successful interface between chromatography and IR
spectrometry has been the linking of a gas chromatograph to a FT-IR
spectrometer (GC/FT-IR) [1,2]. GC/FT-IR is gradually beginning
to compete with gas chromatography/mass spectrometry (GC/MS) as
one of the more widely used hyphenated techniques. The HPLC/
FT-IR interface, however, is not at such an advanced state of development as GC/FT-IR, and some of the major reasons for this can
be seen from a comparison of GC and HPLC with respect to their
mobile phases and the widths of the eluting peaks.

The most commonly used mobile phases in GC are helium and nitrogen, neither of which absorb in the mid infrared. Therefore, the
nature of the mobile phase has no effect on the design or dimensions
of the light-pipe (a flow-cell with a highly reflective gold coating on
the inner walls of the tube) used in GC/FT-IR interfaces. The light-
pipe dimensions can be optimized to give the maximum absorption due
to each peak and a minimum noise in the baseline of the spectrum
when the volume of the light-pipe is approximately equal to the full
width at half height (FWHH) of the narrowest peak in the chromatogram [3].

In the analogous HPLC/FT-IR interface, the effluent from the
column is passed through a flow-cell. However, the volume of the
flow-cell must generally be much smaller than the FWHH of the eluting

Table 1   Cell Dimensions

	H$_2$O/MeOH	NP-HPLC solvents	CHCl$_3$
Path length (μm)	25	100	500
Diameter (mm)	3	3	3
Volume (μl)	0.18	0.71	3.5

peak because of the large absorption by the mobile phase at long pathlengths. The cell pathlength must be short enough to ensure that sufficient energy reaches the detector from all wavelengths so that reasonable baseline noise levels are achieved throughout the spectrum. Obviously, a single pathlength flow-cell cannot be used for all mobile phases. For good infrared solvents, such as CCl$_4$, cell pathlengths can be 10–50 times longer than for highly absorbing solvents like water. Table 1 summarizes the maximum cell pathlengths and volumes for typical mobile phases used in HPLC. Assuming that the volume of a peak between its half-height points is approximately 250 μl, the percentage of the peak in the beam at any one time can be calculated to range from about 1.5% (for CHCl$_3$) to 0.05% (for water). Therefore, the sensitivity of the measurements will be fairly poor, because only a fraction of the solute is in the beam at any one point in time.

Another problem with flow-cell measurements is that the absorption bands due to the solvent must be removed from the solution spectrum in order to obtain the full amount of spectral information due to the solute. In FT-IR, this is achieved using spectral subtraction routines. However these subtractions must be sufficiently accurate so that the residual features remaining after the subtraction have an amplitude of no more than a few parts per million; otherwise they will be larger than the absorption bands due to the solute. These stringent requirements virtually eliminate the use of gradient elution techniques because accurate reference spectra would be extremely difficult to obtain.

II.   MEASUREMENTS WITH FLOW-CELLS

A. Size-Exclusion Chromatography

Even with the above limitations and restrictions, flow-cell measurements have been made in a variety of applications. Size-exclusion chromatography (SEC) is one of the applications best suited to flow-

cell measurements for two distinct reasons: First, the mobile phase
is not crucial in the separation process, and therefore good IR solvents can often be used without degrading the chromatography.
Also, column capacities are higher for SEC columns than for most
other types of HPLC columns, so that overloading of the column and
the subsequent loss of chromatographic resolution is usually not a
problem. For example, Vidrine and Mattson [4] used tetrahydrofuran
to elute a series of polymers from a SEC column. The components of
a mixture containing 1 mg of poly(butyl acrylate) and 4 mg of polystyrene could be identified as they eluted from the chromatograph.
However, several regions of the infrared spectra were "blacked out"
by the mobile phase, and it would have been difficult to identify
these compounds solely on the basis of the reported spectra.

B. Normal-Phase Chromatography

In normal-phase (NP) adsorption chromatography, the choice of mobile phase is critical to the separation process. Therefore, solvents
with good IR transmission are not often used, and certain areas of
the spectrum are often "blacked out" by solvent absorption bands.
In addition, the capacities of silica columns are lower than SEC columns, which further limits their use for HPLC/FT-IR. In spite of
these limitations, a few applications have been addressed by HPLC/
FT-IR. The first HPLC/FT-IR measurements were reported by Kizer,
Mantz, and Bonar [5], who used a flow-cell with AgCl windows to
measure spectra of diphenyl phthalate, testosterone, and p-nitrophenol
eluting from a silica column, with chloroform as the mobile phase.
They were able to measure identifiable spectra when 500 μg of each
component was injected into the chromatograph. Another example of
NP-HPLC/FT-IR measurements using a flow-cell was reported by Scha-
fer et al. [6], in which priority pollutants were identified as they eluted
from a NP column with n-hexane as the mobile phase. They were
able to identify several isomers that could not be differentiated via
MS. Detection limits were high, however, and several important areas
of the spectra were obscured by solvent bands.

Another potentially useful application of flow-cell HPLC/FT-IR
measurements is in the separation and characterization of complex
mixtures of energy-related materials such as solvent-refined coals
(SRCs). In principle, the presence and concentration of certain
functional groups in different SRC fractions can be monitored after
separation by liquid chromatography. Brown and Taylor [7] were
able to determine information about the aliphatic and aromatic content
of the intermediate polarity fraction of SRCs, as well as classes of
oxygen- and nitrogen-containing compounds in this fraction. The
analyte was separated using a bonded-phase polar amino-cyano (PAC)
column with a $CDCl_3$ mobile phase. Although detection limits were

not very good, a great deal more structural information was procured than could be obtained from most conventional HPLC detectors.

C. Microbore HPLC/FT-IR Measurements Using a Flow-Cell

Recently, the use of microbore HPLC (MHPLC) columns has alleviated several of the problems associated with flow-cell HPLC/FT-IR measurements which were discussed above. The lower flow rates result in the increased concentration of minor components, and the use of costly deuterated solvents is much less prohibitive. Teramae and Tanaka [8] used a micro-SEC column to separate a simple, two-component mixture using CCl_4 as the mobile phase, and were able to measure spectra at about the 10-μg injected level.

More recently Taylor and co-workers [9–11] have used 1-mm i.d. PAC columns to study a variety of phenols, amines, and azaarenes. The major component of the mobile phase was $CDCl_3$, and small amounts of modifiers such as CCl_4 and triethylamine were added. Injected minimum detectable quantities (IMDQs) were about 1 μg for the stronger bands in the spectra of several of the solutes. This work was subsequently extended to the identification of similar compounds found in SRC fractions separated under similar conditions [11]. By using MHPLC columns and good infrared solvents, Johnson and Taylor [12] have shown that the flow-cell dimensions can approach the optimum values and that higher SNR spectra can be obtained.

The first reversed-phase (RP) MHPLC/FT-IR data using flow-cells were obtained by Jinno et al. [13], who employed a 0.5-mm i.d. C_{18} column with a polytetrafluoroethylene (PTFE) flow-cell with CD_3CN and D_2O as the mobile phase. Absorption bands due to a few specific functional groups such as C–H and C=O were monitored, but the bulk of the fingerprint region was obscured by solvent and PTFE bands. Detection limits were about 10 μg of sample injected into the chromatograph, which well exceeded the capacity of the column.

The major limitation to the sensitivity of all HPLC/FT-IR and MHPLC/FT-IR measurements using flow-cell measurements is the absorption of radiation by the solvent. A logical conclusion would be to eliminate the solvent prior to the measurement of the infrared spectrum. This approach for HPLC/FT-IR has been pursued by several groups, and their work is discussed in the rest of this review.

III. SOLVENT ELIMINATION TECHNIQUES: EARLY WORK

A. Normal-Phase Chromatography Measurements

Some of the earliest solvent elimination approaches were "off-line" techniques, whereby the solutes were collected, the solvent was evaporated, and the spectra were measured by a conventional sampling technique. The first successful "on-line" technique was developed by Kuehl and Griffiths [14,15], using diffuse-reflectance (DR) spectrometry as the sampling technique. They passed the effluent from a NP-HPLC column through an ultraviolet (UV) detector; they then sprayed it with a gentle stream of nitrogen into a heated tube, which concentrated the effluent about ten fold by differential evaporation. The deposition of the solute relied on a trigger from the UV detector which, along with a time delay, was stored in the microprocessor controlling the interface. If no "peaks" were sensed by the UV detector, the effluent was aspirated through a solenoid valve to waste. When a peak was eluting, the solenoid valve was shut (after a preset time delay) and the solute was deposited on a 4.5-mm-diameter, 4-mm-deep cup, which was filled with powdered KCl. Thirty-two cups were held around the periphery of a carousel which was rotated to a new position after the eluate had been deposited. The remaining solvent was removed by aspiration, and the DR spectrum was measured automatically. Detection limits for non-volatile compounds eluted with volatile solvents were generally 1 μg or less, between one and two orders of magnitude better than previous NP-HPLC/FT-IR measurements using flow-cells. Kuehl and Griffiths also showed that chromatographic resolution was not degraded in the interface, and gradient elution was no problem in the measurement of the eluting samples. Two additional advantages accrued because the sample was deposited onto KCl powder. After the chromatographic run was finished, the spectrum of a given sample could be signal-averaged for a longer period of time than was permitted for an on-line measurement, in order to improve the SNR. In addition, the sample could be extracted from the KCl powder and analyzed by other analytical techniques (MS, NMR, etc.), if desired.

Obviously, there are several limitations to this technique. First, because the deposition of the solute relied on a signal from the UV detector, compounds without a UV chromophore were not collected. Second, because there were only 32 sample cups, data acquisition had to be stopped and the cups refilled quite often. Third, if high percentages of polar organic modifiers (such as 2-propanol or ethyl acetate) had been added to the mobile phase, they were often difficult to eliminate and could therefore obscure the spectrum of the solute. Finally, this interface could not be used with RP-HPLC because the latent heat of vaporization and the surface tension of water is too high to permit rapid evaporation. This is a rather severe limitation because nearly 75% of all HPLC separations are performed in the RP mode. Nevertheless, many of these limitations can be overcome. The principles developed in the preliminary work of Kuehl and Griffiths [14,15] have been expanded to include most types of HPLC, and some recent results are described in subsequent sections of this article.

B. Reversed-Phase Chromatography Measurements

In an attempt to extend the HPLC/FT-IR concept developed by Kuehl and Griffiths to RP-HPLC, Kalasinsky and co-workers [16] have suggested a reaction scheme which eliminates water from the mobile phase as it elutes from the column. By means of a low dead-volume tee, 2,2'-dimethoxypropane (DMP) is added to the aqueous effluent from the column. In the presence of acid, DMP reacts with water to give acetone and methanol. This effluent is then concentrated by differential evaporation, and the highly concentrated solute is deposited on KCl powder. After an additional drying step, the DR spectrum is measured. Although good results have been obtained for several compounds, this technique has several limitations: First, DMP is expensive, and fairly high flow-rates (>1 ml/min) are required when 4.6-mm i.d. columns are used. Second, if the mobile phase contains a buffer or an ion-pairing reagent, these will be concentrated and deposited on the KCl powder. The bands from these species dominate the spectrum of each separated "peak." On the other hand, the technique is fairly simple, and data obtained using a minor modification of this technique to permit it to be applied to RP-MHPLC columns are reported in this review. An alternative technique involving a continuous extraction of each eluate into an immiscible organic phase is described later (see Section V.A).

C. Microbore Chromatography Measurements

When solvent elimination techniques are used, the large volumes of solvent to be removed may become a major limitation; MHPLC columns have been used to alleviate this problem. Jinno et al. [17,18] were the first to publish a solvent elimination technique using MHPLC columns in an interface they termed the *buffer memory* technique. The effluent from a NP-MHPLC column flows down a stainless-steel capillary and is deposited on a KBr salt plate which is continuously translated during the chromatographic run. The solvent is evaporated by a stream of warm nitrogen, leaving the solutes deposited on the crystal. After the chromatography is complete, absorption spectra are measured as a function of distance moved along the KBr crystal. Detection limits for these measurements range from hundreds of nanograms to several micrograms (comparable to or slightly greater than previous DR work using 4.6-mm i.d. columns, which had a greater capacity than the microbore columns used in these separations). Problems with evaporating less volatile solvents (alcohols, water) still represent a major limitation to the technique.

If HPLC/FT-IR is to be accepted as an important analytical tool, it must be able to be used with all types of chromatographic separations and provide significant amounts of information about major and minor components of mixtures and, where possible, provide unequivocal identification of each peak. It has been a goal of our research group to extend the earlier solvent elimination techniques described previously so that they can be used with both conventional NP and RP columns, SEC columns, and NP and RP MHPLC columns. Progress toward the development of a completely automated HPLC/FT-IR interface to provide a continuous, sensitive, universal detector which can give detailed structural information and identification of components eluting from a liquid chromatograph will be described in the rest of this chapter.

IV. SOLVENT ELIMINATION TECHNIQUES: RECENT WORK

A. Introduction

In designing a hardware interface whose function is to reconcile the contradictory output limitations of one instrument to the input limitations of another, each of the techniques utilized may have to be adapted to some extent. The more successful the interface, however, the less compromise will be needed. Therefore, several important aspects of HPLC and FT-IR spectrometry were studied in an attempt to design the optimum interface from both a chromatographic and spectroscopic point of view.

B. Optimization of Chromatographic Parameters

Although HPLC is a fairly powerful separation technique, the efficiencies of the best columns are still more than an order of magnitude less than those found in capillary GC. Therefore, if complex mixtures are to be studied and charcaterized, chromatographic efficiency cannot be compromised in the interface. Factors that affect chromatographic efficiency include the linear velocity of the mobile phase, the particle size of the packing material, the amount of dead volume in the system, and the capacity of the column. Of these four factors, the latter two usually have the greatest impact on the design of an interface between a liquid chromatograph and a FT-IR spectrometer. Studies [19,20] have shown that losses in efficiency of 10—50% can occur when significant amounts of dead volume are present in the system. Therefore, any connecting tubing, flow-cells, etc., must introduce only minimal amounts of dead volume if extracolumn band broadening is to be prevented.

Possibly the most important point to consider, and one that has been largely neglected in the development of many previous HPLC/ FT-IR systems, is the capacity of the columns, because the detection limits of the FT-IR measurement must be at least one, and preferably two, orders of magnitude less than the column capacity. For a conventional (4.6-mm i.d.) column packed with 5 or 10 μm silica, a 10% decrease in efficiency occurs when about 5 μg of a component is injected, as compared with much smaller injected quantities. For microbore columns, the capacity is reduced approximately in proportion to the relative cross-sectional area of the column. Thus the capacity of a 1-mm i.d. column is about 20 times less than that of a 4.6-mm i.d. column.

In a complex mixture, many of the more important and interesting components are present as only a small percentage of the total mixture. Therefore, the ideal HPLC/FT-IR system must be able not only to detect and identify components injected at the column capacity (major components in the mixture) but, more importantly, to yield identifiable spectra of components present at 0.1—1.0% of the column capacity. As a design goal, therefore, identifiable spectra should be measurable from components injected from the low-nanogram level to the column capacity, if the detection technique is to be of general utility for identifying unknown compounds.

C. Microsampling with Diffuse-Reflectance Spectrometry

Kuehl and Griffiths [15] postulated that one reason why their HPLC/FT-IR technique was so sensitive was the fact that the solute in a nonpolar solvent is deposited in only the top fraction of a millimeter of the KCl powder. However, this is not always the case when solvents more polar than n-hexane are used. Work by Duff et al. [21] showed that the polarity of the solvent greatly affects the depth to which the solute is washed by the solvent. Using dye solutions in different solvents, they showed that solvents of intermediate polarity, such as dichloromethane, wash the bulk of the solute to the bottom of the sampling cup, thereby decreasing detection limits by at least an order of magnitude. A very simple solution to this problem was discovered. With simple warming of the KCl powder under an infrared heat lamp just prior to sample deposition, the solvent flash-evaporated at the surface of the KCl, leaving the solute on the top millimeter of the powder. Once again detection limits of several hundred nanograms were easily achieved. However, if this type of interface is to be used with microbore HPLC columns, the detection limits need to be at least an order of magnitude less. Studies were therefore undertaken to determine whether the detection limits of DR spectrometry could be reduced to the low-nanogram range.

Fuller and Griffiths [22] showed that the sample cup dimensions need to be similar to the beam dimensions in order to maximize the sensitivity of the DR measurements. After investigating sample cups of several different sizes, they determined that for very small samples, cups with a 2-mm diameter and a 2.5-mm depth gave spectra with the best SNR measured. Using these cups, Fuller and Griffiths performed a series of off-line experiments in an effort to determine the detection limits of DR spectrometry. A series of solutions containing decreasing amounts of nitrobenzene in CCl_4 were deposited with a microliter syringe on warm KCl powder. After the CCl_4 evaporated, the spectrum of the nitrobenzene remaining was measured. Figure 1 shows the spectra of 100, 40, and 4 ng of nitrobenzene deposited on warm KCl powder. It would appear from the SNR of the 4-ng spectra that subnanogram detection limits should be possible; however, this was not found to be the case. Spectra corresponding to 1 ng or less could not be measured, even though the SNR of certain bands in the spectrum of 2 ng of nitrobenzene was nearly 10:1. The probable explanation for this observation is that at these concentrations (1–2 µg/ml), the nitrobenzene is volatilized (steam-distilled) with the solvent. When the spectra of a series of solutions containing tetraphenylporphyrin, a very nonvolatile compound, were measured in an analogous manner, it was found that one or two bands showed an absorption of greater than 1%, and could therefore be observed with measurement times of only a few seconds. It is possible that a combination of an equally nonvolatile but stronger mid-IR absorber would allow subnanogram detection limits to be achieved. However, a 1-ng detection limit for strong IR absorbers is certainly adequate for many HPLC/FT-IR experiments.

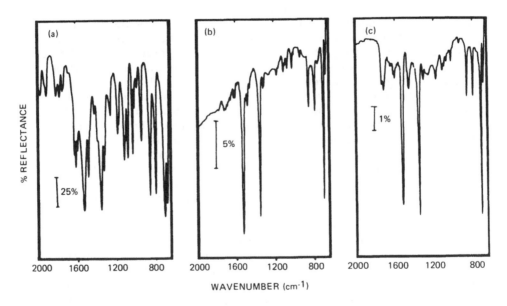

Figure 1   Spectra of (a) 100, (b) 40, and (c) 4 ng of nitrobenzene measured after deposition on warm KCl powder with a microliter syringe (CCl_4 used as solvent).

D. Design of the Concentrator

For a solvent-elimination technique to be universal in nature, the efficiency of the concentration step is very important. The concentrator used in the original Kuehl and Griffiths [14] interface very effectively concentrated the column effluent, provided that volatile, nonpolar mobile phases were used. In some cases, however, the modifier (2-propanol or ethyl acetate) was concentrated along with the solute and became a major interference in the DR spectrum. A great deal of work was done by Duff [23] to design a concentrator that was better suited to a wider variety of mobile phases. One approach was to use a resistively heated nichrome ribbon as the concentrator. The ribbon was 20 cm long and 3 mm wide and was oriented about 40 degrees from the horizontal. The effluent was gently concentrated as it flowed down the heated ribbon until it was deposited on the heated KCl powder, where any remaining solvent was flash-evaporated. Although this concentrator worked fairly well for many solvents, problems with splashing and unevenness of flow at higher flow rates (>0.8 ml/min) limited its usefulness.

A compromise between this design and the original design of Kuehl and Griffiths proved to be effective at a variety of flow rates (>2.0 ml/min) and was suitable for a wide range of solvents. A glass tube, about 10 cm long, was cut in half for nearly two-thirds of its length. The tube was wrapped with nichrome wire and heated to different temperatures by varying the current passing through the wire. A representation of the concentrator is shown in Fig. 2. The tube was oriented at about 50 degrees from the horizontal to prevent sample holdup and consequent loss of chromatographic resolution. When a concentrator of this design was used, most solvents could be eliminated at temperatures of only 30°−60°C.

E. Automation of the Interface

One of the major goals in the development of the interface was to provide a continuous, automated, yet simple system. Obviously, reliance on a signal from an ultraviolet (UV) detector, as in the original interface, precluded its utility as a universal interface. Therefore, a system independent of detectors other than FT-IR spectrometry was devised. Two important changes from the original interface were made. First, the sampling carousel was redesigned. In place of the solid aluminum carousel, a lightweight aluminum annulus, containing 162 wells (4.5-mm diameter, 4.5-mm depth), was mounted on three stands. The ring rotated on ball bearings, and the position was fixed by a solenoid-actuated locking pin. The ring was turned by a direct-drive DC motor, and timing of the locking pin determined how far the ring rotated. The second change was that the ring motion was actuated upon receiving a signal from a drop monitor.

from
column

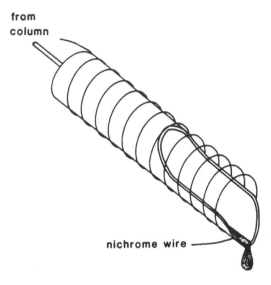

nichrome wire

Figure 2   A schematic representation of the concentrator.

The drop monitor used a photocell to monitor the output of a He-Ne
laser situated in the path of the drops falling from the concentrator.
When a drop fell, the beam was broken and the controlling electronics
signaled the motor and pin to advance the ring to the next sample
cup.   In this manner, spectra could be collected continuously without relying on the signal from an auxiliary detector.

Unfortunately, there were several problems with this interface.
First, the DC motor and locking pin were not always well synchro-
nized, and occasionally a sample position would be missed or the ring
would not advance.   In addition, the locking pin tended to cause vi-
brations in the ring so that spectra could not be measured if the
locking pin had just moved.   These problems were solved by sub-
stituting a stepping motor for the direct-drive DC motor.   The stepping motor, with appropriate controlling electronics, could accurately
step the ring from one position to the next with little or no vibration
and excellent accuracy, so that a locking pin was no longer necessary.
Once again, the wheel movement relied on a signal from a drop monitor so that continuous monitoring of the column effluent was possible.
A schematic representation of the interface is shown in Fig. 3.

F. Adaptation of Interface for MHPLC Measurements

A new sampling ring was designed for the MHPLC experiments; the
ring contained 180 wells, which were 2 mm in diameter and 2.5 mm
deep.   The wells were 2 degrees apart so that the ring movement was

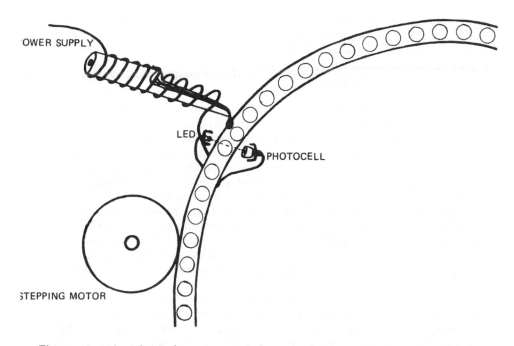

POWER SUPPLY

LED

PHOTOCELL

STEPPING MOTOR

Figure 3 A schematic representation of the interface, consisting of a concentrator, a drop monitor, a stepping motor, and a sampling wheel.

still easily controlled by the stepping motor. The drop monitor was ineffective at very low flow rates (1—10 μl/min), and the ring movement was controlled by a simple timing sequence when this was the case. The other change from the previous interface was the elimination of the concentrator. The effluent was deposited directly on warm KCl powder and the solvent was flash-evaporated, leaving the solute on the top layers of the KCl powder.

In its final form, this interface allowed the continuous monitoring of the effluent eluting from SEC columns, NP conventional and microbore columns, and, with some modifications (described later), RP conventional and microbore columns.

V. NEW DESIGNS FOR THE RP-HPLC/FT-IR INTERFACE

A. Continuous Extraction Interface

1. Design Considerations

The combination of RP-HPLC with FT-IR spectrometry has always been a very difficult problem because water has such strong absorption bands in the mid-IR. Several attempts were made by Duff [23] to develop a system analogous to the Kuehl—Griffiths interface that could be used with aqueous solvents. It was apparent from these experiments that if reasonable detection limits were to be achieved for RP-HPLC measurements, each eluting solute needed first to be separated from the aqueous phase and then isolated prior to the spectroscopic characterization. A technique developed by Karger and colleagues for RP-HPLC/MS [24,25] appeared to be particularly well suited for RP-HPLC/FT-IR. In this interface, each eluate was continuously extracted into an immiscible solvent, usually dichloromethane, after which the aqueous phase was separated and passed to waste, whereas the organic phase was transferred to a moving ribbon where the solvent was eliminated prior to the introduction of the solutes to the mass spectrometer. A similar technique was developed by Duff [23] and Conroy et al. [26], in which the dichloromethane solution was passed into a concentrator, and the solutes were deposited on KCl powder so that their DR spectra could be measured. A description of both the basic extraction interface and a new interface for RP-HPLC/FT-IR measurements are presented in this section.

The continuous extraction interface can be considered to consist of four components: (a) a mixing tee through which a modifier may be added to the mobile phase; (b) a mixing tee that introduces the organic phase to the aqueous column effluent; (c) an extraction coil, which facilitates the transfer of sample into the organic phase; and (d) a separator tee, in which the lighter aqueous phase is aspirated to waste. All of the initial experiments involved dichloromethane as the extraction solvent and methanol:water mixtures as the aqueous mobile phase. Not all aqueous compositions of water and methanol are immiscible with dichloromethane. When the mobile phase contains at least 80% methanol, water must be added through the first mixing tee to ensure that a two-phase system is passing through the extraction coil. This may be achieved by metering a small volume of water (typically 0.5 ml/min) into the first mixing tee. After separation from the aqueous phase in the separator tee, the dichloromethane solution is concentrated and deposited onto a KCl substrate for identification by DR spectrometry. A schematic representation of the extraction system is shown in Fig. 4.

A major concern in the design of this interface was the possibility of postcolumn broadening of the chromatographic peaks. Broadening is minimized by keeping the internal diameter and lengths of the tubing used to connect the extractor components as small as possible, and by avoiding abrupt changes in internal diameters. Probably the most substantial reduction in dispersion is achieved by coiling the tubing where possible. Coiling is particularly important in the extraction step, not only to prevent broadening, but also to facilitate sample transfer from the aqueous phase into the organic phase. This is because secondary flow patterns are established perpendicular to

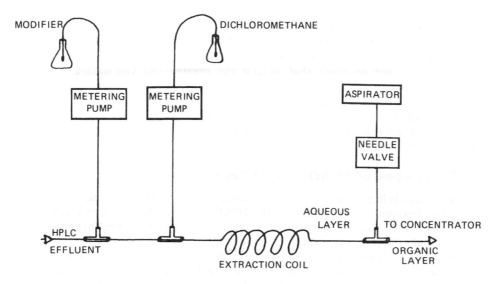

Figure 4 A schematic representation of the continuous extraction interface.

the main direction of flow when a liquid flows through a coiled tube [27]. This type of flow pattern minimizes dispersion along the axis of the tubing and promotes mass transfer between the boundary of the aqueous segments and the organic layer.

The optimum flow rate of dichloromethane metered into the extractor was found to depend on both the flow rate of the chromatographic effluent and on the composition of the aqueous phase. When the composition of the aqueous phase is 50% water or greater, an organic flow rate higher than the aqueous rate is desirable. A lower organic flow rate causes an increase in the concentration of the sample and, therefore, an increase in sensitivity. However, the flow rate of the organic phase can be decreased only to a certain point. Below that point the segments of the organic layer fail to reunite in the separator, and clean and complete phase separation is not achieved. It was found that for mobile phases containing at least 60% methanol, an organic flow rate of approximately 90% that of the aqueous phase is desirable, whereas at lower methanol content, a high ratio of organic to aqueous flow rate is preferred [23,26].

In the separator tee, the segments must reunite in a manner such that the lighter aqueous phase can be aspirated from the top branch of the tee while the heavier dichloromethane can be passed to the concentrator tube with as little of the aqueous phase as possible present. This effect is facilitated by providing a wide surface area along the bottom of the tee which is wettable by dichloromethane.

This goal is achieved by inserting a PTFE tube into the tee, which is spliced about 4 mm before the point at which the aqueous phase is aspirated, while the entire aqueous layer can be removed. It is important to provide enough surface area so that the segments can reunite, but not so much that mixing can occur. At the outlet branch, tubing with a small outside diameter (1 mm) and a fairly large internal diameter is placed along the bottom, which allows the dichloromethane to flow through easily and also helps any aqueous phase from interfering. A schematic representation of the separator is shown in Fig. 5.

2. Examples of RP-HPLC/FT-IR Measurements

The feasibility of using this interface for RP-HPLC as well as for ion-suppression and ion-pairing RP-HPLC was explored. Examples of common RP measurements are illustrated with the separation of isomers of nitrobenzoic acid and naphthol. Isomeric species are of interest both from a chromatographic point of view, because they are often difficult to separate, and from a spectrometric point of view, because one of the major advantages IR has over other detection techniques (especially MS) is its capability to distinguish between isomers.

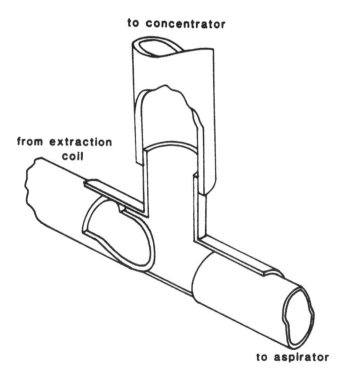

to concentrator

from extraction coil

to aspirator

Figure 5   A schematic representation of the separator tee.

The separation of *o*- and *m*-nitrobenzoic acid proved quite difficult unless ion suppression was used, as seen in the chromatograms shown in Fig. 6. The two isomers were not resolved unless an acidic buffer was added to the mobile phase. Spectra corresponding to 1 µg (injected) of each isomer are shown in Fig. 7; the aromatic C-H out-of-plane deformation modes in the 900- to 700-cm^{-1} region, which are highly sensitive to the substitution pattern, easily distinguish the two isomers.

The separation of 1- and 2-naphthol, using a mobile phase of 40:60 methanol: water with 0.1 M acetic acid added, is shown in Fig. 8. In order to prevent the extraction of the weak organic acid, 0.1 M NaOH was added into the first tee at 0.2 ml/min, so that the pH of the solution was about 12.2. At this pH, the acetic acid is ionized and is therefore not extracted. Surprisingly perhaps, in view of their pK_as, the naphthols are incompletely ionized and a small percentage of each is extracted. The measured spectra are shown in Fig. 9. The weakness of these spectra illustrates the fact that careful pH control is necessary if the acetic acid modifier is to be retained in the aqueous phase while the eluates (which in this case were even weaker acids) are partially extracted. Further work to optimize similar measurements is obviously called for.

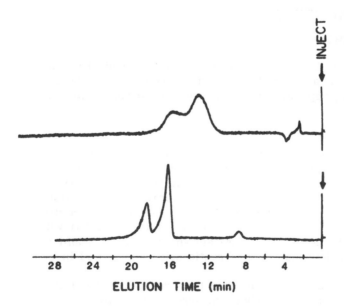

Figure 6 Chromatograms of the separation of *o*- and *m*-nitrobenzoic acid measured with the on-line UV detector. The upper trace shows the separation of 1 µg each of the two isomers on a C_{18} column, using a mobile phase of 40:60 methanol:water at a flow rate of 1.0 ml/min. The lower trace shows the same separation, except that 0.05 M phosphoric acid has been added to the mobile phase.

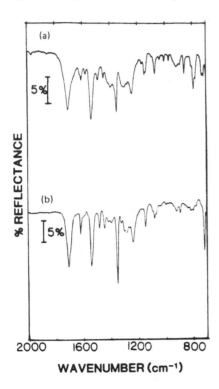

Figure 7 Spectra of 1 μg (injected) each of (a) o- and (b) m-nitro-benzoic acid measured as they eluted from the continuous extraction interface after being separated on a C_{18} column via ion-suppression chromatography.

Finally a mixture of phenols was separated using both a 2:98 methanol:water mobile phase and a 2:98 methanol:water mobile phase with 0.1 M acetic acid added; the two chromatograms are shown in Fig. 10. Once again the ability to use ion suppression chromatography greatly aided the resolution of the components in the mixture. Spectra of 500 ng of p-nitrophenol measured (a) with and (b) without acetic acid in the mobile phase are shown in Figs. 11 a and b respectively. Essentially no trace of acetic acid is observed in Figs. 9 and 11a.

These results are significant because they represent the first HPLC/FT-IR measurements that allow the use of ion-suppression reagents. Therefore, by using this extraction technique, the chromatography is not limited by the spectrometric detection technique. Unfortunately, the extraction technique is fairly complicated, and operation of the interface is not a trivial task. In addition, in some

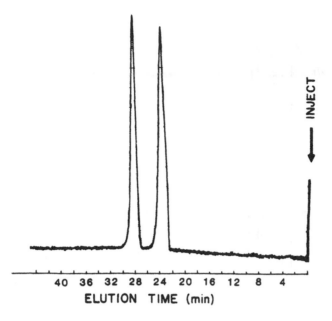

Figure 8 Chromatogram (UV detection, 254 nm) of the separation of 1- and 2-naphthol on a C_{18} column. The mobile phase was 40:60 methanol:water with 0.1 M acetic acid added, and the flow rate was 1.2 ml/min.

cases some prior knowledge of the nature of the sample is necessary to ensure that significant amounts of the solute are extracted. In an attempt to simplify the operation of the interface, a new design, also fundamentally based on an existing LC/MS interface, was investigated for RP-HPLC/FT-IR measurements. This device was based on the thermospray that had been developed for HPLC/MS by Vestal and Blakley [28]

B. Thermospray Interface for RP-HPLC Measurements

The thermospray uses a heated vaporizer to volatilize the effluent from a RP-HPLC column. Under the vacuum conditions of the mass spectrometer, a supersonic jet is created, and large volumes of mobile phase can be successfully input to the ionization chamber. Initially we believed that the thermospray could be used in the RP-HPLC/FT-IR interface by salting the effluent with a concentrated KCl solution prior to the volatilization of the mobile phase. It was believed that a layer of KCl could be formed and used directly as the substrate for DR infrared spectrometry. Although several concentrations (0.5–1.0 M) of KCl were used, it soon became apparent that the layer of KCl that

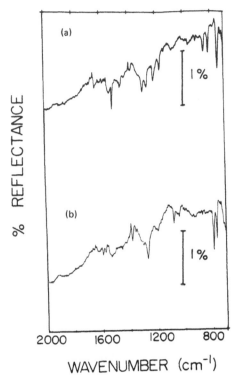

Figure 9 Spectra of 800 ng (injected) each of (a) 1-naphthol and (b) 2-naphthol, measured as they eluted from the continuous extraction interface.

was deposited was only a small fraction of a millimeter (>0.1 mm) and was unsuitable for DR measurements.

The next approach involved spraying the effluent onto heated KCl (analogous to earlier work), with the assumption that the solute would be adsorbed by the KCl and the mobile phase would be evaporated. Even when relatively large amounts (e.g., at least 20 μg) of strong infrared absorbers were injected into the chromatograph, few bands attributable to the analytes could be observed in the spectrum. On the other hand, absorption bands due to methanol and water were very strong. As is often the case, the affinity of water for KCl is so great that although the bulk of the mobile phase is evaporated, trace amounts of water remain, obscuring much of the spectrum of the solute (see Fig. 12).

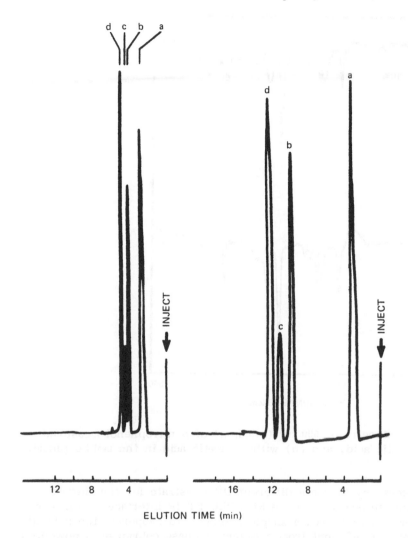

Figure 10   Chromatograms of the separation of four phenols, (a) 2,4-dinitrophenol, (b) *p*-methoxyphenol, (c) phenol, (d) *p*-nitro-phenol, measured by the UV detector. The mobile phases were (a) 2:98 methanol:water and (b) 2:98 methanol:water with 0.1 M acetic acid added; the flow rate was 0.8 ml/min in each case.

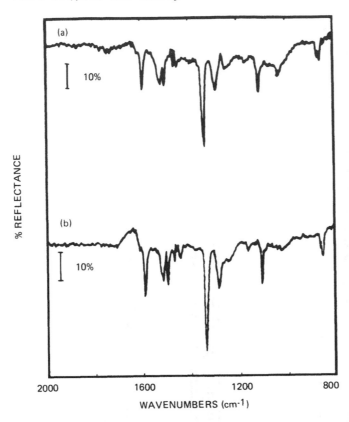

Figure 11 Spectra of 500 ng (injected) of *p*-nitrophenol measured
(a) with acetic acid, and (b) without acetic acid in the mobile phase.

It is necessary to use a different DR substrate for the thermospray device to become a viable RP-HPLC/FT-IR interface. Azarraga
and co-workers [29] used a simple nebulizer to evaporate the solvent
and aspirate the effluent from a reversed-phase column onto powdered
diamond, trapping the solutes and then measuring their spectra.
Preliminary investigations that used powdered diamond as a substrate
for DR measurements employing the thermospray device have verified
the potential of this approach.   This is demonstrated in Fig. 13,
where the upper trace is the spectrum of 5 µg of 2,4-dinitrophenol
deposited from CCl_4 onto KCl powder, and the bottom trace is the
spectrum of 10 µg of 2,4-dinitrophenol injected into the column,
passed through the thermospray, and deposited on powdered diamond.
   A separation of a mixture of phenols injected into the chromatograph at levels of 10 µg was used to test the interface (see Fig. 10).

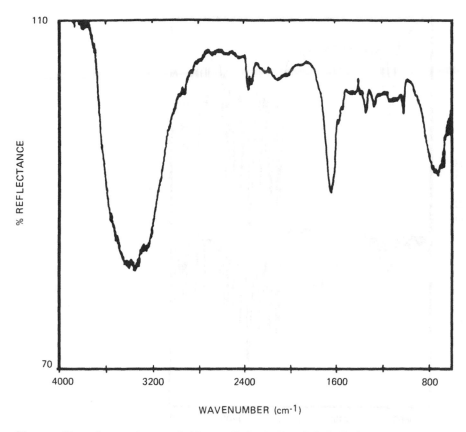

Figure 12　A spectrum of 20 µg (injected) of 2,4-dinitrophenol deposited on KCl with the thermospray, after eluting in a mobile phase of 2:98 methanol:water.

All of the components of the mixture except phenol could be measured at good SNR, and examples of the spectra are shown in Fig. 14. Because phenol is the most volatile compound of the group, the weakest IR absorber, and the most poorly resolved from the other components, it certainly represents a "worst case" for the interface. This result points to one of the problems of the thermospray interface. Because the temperature at the outlet is fairly high (140°–150°C), volatile compounds may evaporate with the mobile phase, and thermally labile compounds may decompose. Another problem occurs at flow rates exceeding 1.5 ml/min. At these high flow-rates, the jet begins to widen so that the solute can be spread over a fairly large area, thereby decreasing the detection limits of the DR measurements by at least a factor of two.

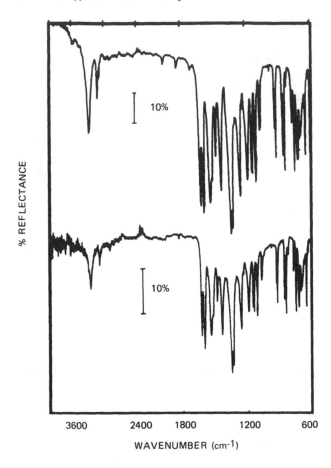

Figure 13 Spectra of 2,4-dinitrophenol. The upper trace corresponds to 5 μg deposited from CCl_4, with a microliter syringe, on KCl powder. The lower trace corresponds to 10 μg (injected) deposited on powdered diamond with the thermospray.

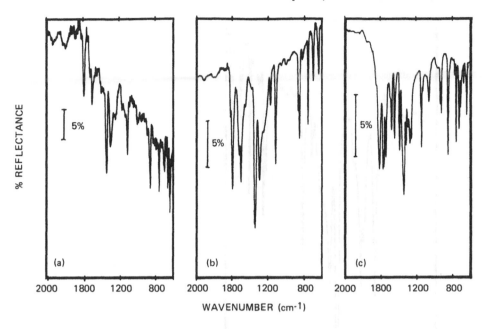

% REFLECTANCE

WAVENUMBER (cm⁻¹)

Figure 14   Spectra of 10 μg (injected) each of (a) *p*-methoxyphenol, (b) *p*-nitrophenol, and (c) 2,4-dinitrophenol deposited with the thermospray on diamond powder after eluting in a mobile phase of 2:98 methanol:water.

A simple experiment was performed to investigate the effect of including ion-suppression reagents in the mobile phase, on the spectra of samples deposited via the thermospray. A mobile phase containing 0.1 M acetic acid was used to elute 2,4-dinitrophenol (see Fig. 10b). The DR spectrum of 10 μg (injected) of *p*-nitrophenol eluted in this manner is shown in Fig. 15. No trace of acetic acid is apparent in the region around 1700 cm⁻¹, and the sample is still easily identified as *p*-nitrophenol. When smaller quantities of analytes are injected, the remaining acetic acid limits the sensitivity, even though a large percentage of it is volatilized with the solvent.

In general, the thermospray interface is not suitable for ion-pairing or ion-suppression reagents that are not fairly volatile. Therefore, even though the extraction technique is more complex, the results are much better for samples requiring these reagents. The thermospray device does appear to be suited to a wider variety of aqueous modifiers (such as acetonitrile) than the extraction technique, and detection limits appear to be comparable to previous results. More research on the nature of diamond as a DR substrate in view of

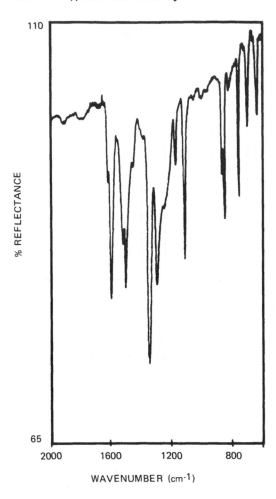

Figure 15 The spectra of 10 μg (injected) of *p*-nitrophenol deposited on diamond powder with the thermospray. The mobile phase was 2:98 methanol:water with 0.1 M acetic acid added.

potential spectral distortions due to its high refractive index, as well as the use of other substrates, appears to be necessary. Nevertheless, the thermospray appears to have some potential for use as a HPLC/FT-IR interface.

VI. NEW MICROBORE HPLC/FT-IR INTERFACES

A. Normal-Phase Chromatography

The use of microbore HPLC columns would appear to alleviate some of the problems associated with the high-volume flow rates encountered in conventional HPLC/FT-IR interfaces. Although superficially the construction of such an interface involving the elimination of solvents eluting from MHPLC columns appears fairly trivial, the effects of a number of important factors, such as column dimensions, mobile-phase flow rates, and choice of solvents, have important consequences on the design of the MHPLC/FT-IR interface.

Preliminary work has indicated that 1-mm i.d. columns appear to be optimally suited for MHPLC/FT-IR for two reasons: First, their capacity ($\sim$250 ng) is well above the detection limits of DR spectrometry. Second, the elimination of solvents of low and medium polarity, such as n-hexane and chloroform, at flow rates of about 50 μl/min, proves to be fairly simple. Therefore, no major modifications to the basic interface designed for 4.6-mm i.d. columns are necessary for NP chromatography.

In the early NP-HPLC/FT-IR studies reported by Kuehl and Griffiths [14], Stahl's dye solution was used to show the feasibility of their solvent elimination approach. Similarly, Stahl's dye was one of the first samples to be studied by MHPLC/FT-IR. The three dyes in this solution were separated on a 50-cm-long, 1-mm i.d. microbore column packed with 10 μm silica, using a mobile phase of 2% 2-propanol in hexane at a flow rate of 40 μl/min. Spectra of two of the components—namely, 250 ng each of Butter Yellow and Indophenol Blue—are shown in Fig. 16. The difference in the intensity of these two spectra illustrates the fact that HPLC/FT-IR detection limits are dependent on both the nature of the analyte and its retention time. In the spectrum of Butter Yellow, not only are the absorptivities of the stronger bands larger than those for Indophenol Blue, but under these chromatographic conditions the full width at half height (FWHH) of the peak due to Indophenol Blue is about 1.5 times that of the Butter Yellow peak. Because the broader peaks can elute over several drops, different quantities may be collected per cup, even though equivalent amounts were injected. Ideally, the FWHH of all the peaks can be made identical through the use of gradient elution techniques,

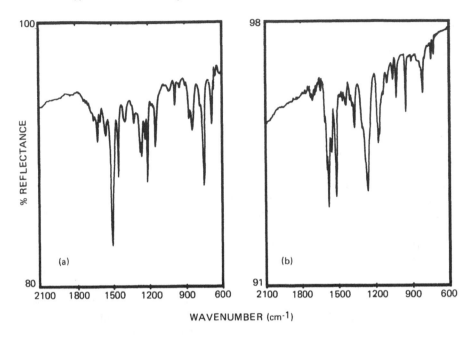

Figure 16 Spectra of 200 ng each of (a) Butter Yellow and (b) Indophenol Blue, measured as they eluted from a silica column in a mobile phase of 2% methanol in hexane.

but this is not readily achieved with the equipment available for MHPLC. Fortunately, DR spectrometry is sensitive enough to detect most solutes, even if they elute over several cups. Nevertheless, overall sensitivity of the method is decreased under these circumstances because this is a mass-sensitive measurement, and care must be taken to optimize the chromatography if very low detection limits are needed.

One of the major reasons for using solvent elimination techniques with DR spectrometry is that no portions of the spectrum are "blacked out" by solvent bands, and the full spectrum can be used in the identification of unknown compounds. However, when one is working with submicrogram amounts of material, impurities and atmospheric absorption bands can lead to severe spectral interferences. In many cases, the detection limits of the MHPLC/DR interface were imposed by interferences (atmospheric, impurities, etc.) rather than absolute limits due to the SNR of the spectrum of the solutes, and care must be taken to obtain the best possible purge of water vapor from the optical path.

Jinno et al. [17,18] found that their buffer memory technique is best used with columns of very narrow internal diameter in view of the low flow-rates required and the concomitant ease of solvent elimination. This result is due to the fact that the tip of the capillary from which the column effluent emerges touches the plate where it is deposited. Conversely, with DR spectrometry it was found that the surface of the powder cannot be disturbed, and great care must be taken to ensure that the substrate is not disturbed. Indeed the actual deposition of the solute on the KCl powder was found to be the most crucial and difficult step in the interface. When flow rates were very low (<15 µl/min), it was found to be extremely difficult to deposit the solute efficiently by dripping the eluent onto the heated substrate. Apparently, the eluent tends to creep up the outside of the stainless-steel capillary which is positioned over the cup containing the warm KCl powder on which the solute is to be deposited. The solvent appears to evaporate at the tip of the capillary, leaving a fraction of the solute deposited on the outer surface of the capillary [30]. It is also possible that solutes having an appreciable vapor pressure can evaporate along with the solvent during the deposition step.

B. RP-MHPLC/FT-IR

1. Introduction

As stated earlier, nearly 75% of the work done in HPLC is performed in the reversed-phase mode; therefore, if the interface is to truly be a viable detector for HPLC, the interface must be adapted to handle RP solvents. Several attempts have been made to deal with aqueous solvents, including the reaction technique of Kalasinsky et al. [16] and the extraction technique initially investigated by Duff et al. [21,23,26]. The extraction technique applied to conventional (4.6-mm i.d.) columns has been discussed in some detail earlier in this chapter, but it was found that the amount of dead volume associated with this system precluded its use with microbore columns. However, both the reaction technique of Kalasinsky and the application of a thermospray can be used for RP-MHPLC/FT-IR measurements. This work is at a very preliminary stage, but a few results will be described.

2. Reaction with DMP

The reaction technique of Kalasinsky and co-workers is based on the reaction of water with acidified 2,2'-dimethoxypropane to form methanol and acetone. These two products are relatively easy to eliminate so that DR spectra of the solute can be measured. Preliminary experiments in our laboratory quickly showed that the flow rate of DMP relative to that of the aqueous effluent was critical if all the water was to be reacted. Provided that the ratio of DMP to that of the mobile phase was between 1.0 and 1.3, good results were usually found. Because the reaction is endothermic, the mixing tee and the stainless-steel capillary from which the reacting mixture emerges must be held at a temperature of ∿45°C. The elevated temperature both speeds up the reaction and increases the ease of evaporation of the effluent as it elutes from the stainless-steel capillary.

The deposition step is again the most critical step in the interface. Simultaneously depositing the solute and eliminating the solvent is particularly difficult because the effluent contains methanol, acetone, and unreacted DMP. Careful heating of the capillary and warming of the KCl substrate eliminated the bulk of the solvent in most cases. In certain cases, the KCl had to be heated after deposition to completely eliminate the solvents. The major limitation to the sensitivity of the measurements was interferences from the solvents and the DMP, rather than absolute limits in the SNR.

The spectrum corresponding to 800 ng of anisole injected into the chromatograph is shown in Fig. 17. Even though anisole is fairly

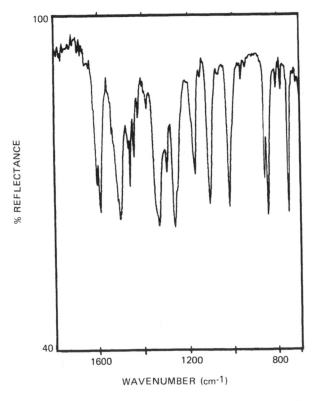

Figure 17   Spectrum of 800 ng (injected) of anisole, measured after reaction with DMP to eliminate the aqueous mobile phase.

volatile, this spectrum is more than adequate for identification purposes. It is apparent that solvents pose no major problems, provided that fairly large amounts of solute are deposited. Detection limits for this technique are presently on the order of 100 ng (injected) for most compounds, and are limited by interferences due to the solvent and unreacted DMP.

3. Thermospray Interface

The thermospray was fairly easily adapted for RP-MHPLC/FT-IR measurements. Indeed the main problem was the fact that the particular thermospray device was rather large and awkward and therefore was not easily attached to the column exit tubing. The microsampling wheel

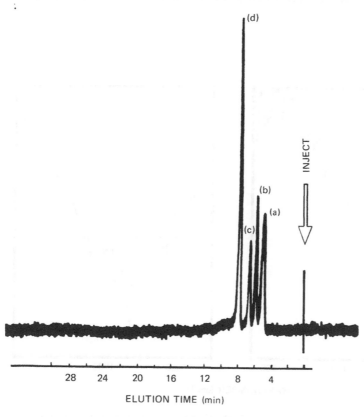

Figure 18 Chromatogram of the separation of four phenols, (a) 2,4-dinitrophenol, (b) *p*-methoxyphenol, (c) phenol, and (d) *p*-nitrophenol, on a 1 mm × 250 mm C_{18} column, using a 15:85 methanol: water mobile phase.

described earlier in this chapter was used for these measurements, and, as with the thermospray experiments using conventional RP columns, powdered diamond (3000 grit) was used as the DR substrate.

Detection limits were nearly an order of magnitude better than the reaction technique described above. The major interferences to the measurements were again the residual solvents left on the substrate and the difficulties of eliminating ion-suppression and ion-pairing reagents. The chromatogram of a mixture of phenols separated on a 10-μm C_{18} column (1 mm × 250 mm), using a mobile phase of 15% methanol in water, is shown in Fig. 18. Spectra, corresponding to 20 ng injected of each component, were collected, and Fig. 19 shows representative spectra of *p*-nitrophenol and 2,4-dinitrophenol. The spectrum of phenol was not measured, again probably because it is the most volatile of the series and is evaporated with the mobile phase.

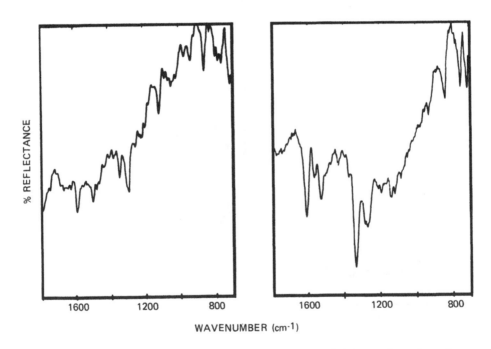

Figure 19 Spectra of 20 ng (injected) each of (a) *p*-nitrophenol and (b) 2,4-dinitrophenol deposited with the thermospray after separation using a mobile phase of 15:85 methanol:water.

VII. SUMMARY

It is apparent that the only way to achieve normal-phase and reversed-phase HPLC/FT-IR measurements is by rapidly and continuously evaporating the solvent prior to measuring the infrared spectra of each solute. We believe the work described in this chapter gives a clear indication of the future directions of this field. Identifiable infrared spectra of components of mixtures present at a concentration of at least 1% of the concentration of the major solute are measurable for size-exclusion, normal-phase, and reversed-phase chromatography, including both ion-suppression and ion-pairing HPLC. Although large reductions in the detection limits reported in this article will probably not be realized within the next year, the degree of automation should be improved. We would forecast that an HPLC/FT-IR interface based on this technology will become available commercially within several years.

ACKNOWLEDGMENT

Although the information described in this document has been funded in part by the United States Environmental Protection Agency under assistance agreement CR-810430 to the University of California, Riverside, it does not necessarily reflect the views of the Agency, and no official endorsement should be inferred.

REFERENCES

1. M. D. Erickson, Appl. Spectrosc. Rev. *15*, 261 (1979).
2. P. R. Griffiths, J. A. de Haseth, and L. V. Azarraga, Anal. Chem. *55*, 1387A (1983).
3. P. R. Griffiths, Appl. Spectrosc. *31*, 284 (1977).
4. D. W. Vidrine and D. R. Mattson, Appl. Spectrosc. *32*, 502 (1978).
5. K. Kizer, A. W. Mantz, and L. C. Bonar, Am. Lab. 7(5), 85 (1975).
6. K. Shafer, S. Lucas, and R. J. Jakobsen, J. Chromatogr. Sci. *17*, 471 (1979).
7. R. S. Brown and L. T. Taylor, Anal. Chem. *55*, 723 (1983).
8. N. Teramae and S. Tanaka, Spectrosc. Lett. *13*, 117 (1980).
9. C. C. Johnson and L. T. Taylor, Anal. Chem. *55*, 1492 (1983).
10. R. S. Brown and L. T. Taylor, Anal. Chem. *55*, 1492 (1983).
11. P. G. Amateis and L. T. Taylor, Anal. Chem. *56*, 966 (1984).
12. C. C. Johnson and L. T. Taylor, Anal. Chem. *56*, 2642 (1985).

13. K. Jinno, C. Fujimoto, and G. Uematsu, Am. Lab. *16*(2), 39 (1983).
14. D. Kuehl and P. R. Griffiths, J. Chromatogr. Sci. *17*, 471 (1979).
15. D. T. Kuehl and P. R. Griffiths, Anal. Chem. *52*, 1394 (1980).
16. K. S. Kalasinsky, J. T. McDonald, and V. F. Kalasinsky, Paper No. 357 Pittsburgh Conf. Anal. Chem. and Appl. Spectrosc., Atlantic City, N.J. (March 1983).
17. K. Jinno and C. Fujimoto, J. High Resol. Chromatogr. Chromatogr. Commun. *4*, 532 (1981).
18. K. Jinno, C. Fujimoto, and D. Ishii, J. Chromatogr. *239*, 652 (1982).
19. K. Jinno and C. Fujimoto, J. High Resol. Chromatogr. Chromatogr. Commun. *3*, 313 (1980).
20. T. Tsuda and M. Novotny, Anal. Chem. *50*, 632 (1978).
21. P. J. Duff, C. M. Conroy, P. R. Griffiths, B. L. Karger, P. Vouros, and D. P. Kirby, Proc. Soc. Photo-Opt. Instrum. Eng. *289*, 53 (1981).
22. M. P. Fuller and P. R. Griffiths, Appl. Spectrosc. *34*, 533 (1980).
23. P. J. Duff, Ph.D. Dissertation, Ohio University, Athens, Ohio (1984).
24. B. L. Karger, D. P. Kirby, R. Vouros, R. Foltz, and B. Hidy, Anal. Chem. *51*, 2324 (1979).
25. D. P. Kirby, P. Vouros, B. L. Karger, B. Hidy, and B. Peterson, J. Chromatogr. *203* , 139 (1981).
26. C. M. Conroy, P. R. Griffiths, P. J. Duff, and L. V. Azarraga, Anal. Chem. *56*, 2636 (1984).
27. R. Tijssen, Anal. Chim. Acta, *114*, 71 (1980).
28. C. R. Blakley and M. L. Vestal, Anal. Chem. *55*, 750 (1983).
29. L. V. Azarraga, Paper presented at Eastern Analytical Symposium, New York, N.Y. (November, 1983).
30. C. M. Conroy, P. R. Griffiths, and K. Jinno, Anal. Chem. *57*, 822 (1985).

4

Investigations of Selectivity in RPLC of Polycyclic Aromatic Hydrocarbons

Lane C. Sander and Stephen A. Wise   *National Bureau of Standards, Gaithersburg, Maryland*

Certain commercial equipment, instruments, or materials are identified in this report to specify adequately the experimental procedure. Such identification does not imply recommendation or endorsement by the National Bureau of Standards, nor does it imply that the materials or equipment identified are necessarily the best available for the purpose.

I. INTRODUCTION

In 1971, Schmit et al. [1] first described the separation of polycyclic aromatic hydrocarbons (PAH) by liquid chromatography (LC) using a chemically bonded octadecylsilane (C_{18}) stationary phase. Since Schmit's report, reversed-phase LC on chemically bonded C_{18} stationary phases has become by far the most popular LC mode for the separation of PAH. The popularity of reversed-phase LC is due, in part, to the excellent selectivity of this technique for the separation of PAH isomers. The complex mixtures of PAH encountered in environmental samples contain numerous isomeric structures. Because certain isomers are often more mutagenic and/or carcinogenic than other isomers, the separation of individual PAH isomers is necessary to assess accurately the potential health effects of exposure to various pollutant sources. Even when high-efficiency open-tubular gas chromatography (GC) is used, a number of isomeric PAH are still difficult to separate on conventional nonpolar stationary phases [2], e.g., chrysene and triphenylene; benzo[b]fluoranthene, benzo[j]-fluoranthene, and benzo[k]fluoranthene; and dibenz[a,c]anthracene and dibenz[a,h]anthracene. As a result of the excellent selectivity of reversed-phase LC on C_{18} phases, this technique has been specified as the method of choice for the analysis of aqueous effluents for determination of the 16 PAH on the U. S. Environmental Protection Agency's priority pollutant list [3] (see Fig. 1).

A. Retention Indexes

In addition to the environmental interest in the separation of PAH by LC, PAH have been used extensively as solutes in the investigation of reversed-phase LC retention mechanisms. In this report we shall summarize the investigations in our laboratory since 1979, which have been directed toward understanding selectivity in reversed-phase LC separations of PAH. In several of these studies [4,5], retention data for a number of PAH on various columns are presented as the logarithms of the retention indexes. These retention indexes, which are similar to Kovats indexes in GC, were determined as described by Popl et al. [6].

In this retention index system, the elution volume of the solute was measured simultaneously with the elution volumes of standards (benzene, naphthalene, phenanthrene, benz[a]anthracene, benzo[b]chrysene), representing one- to five-condensed-ring PAH. The retention index, I, was calculated using the following equation:

$$\log I_x = \log I_n + \frac{\log R_x - \log R_n}{\log R_{(n+1)} - \log R_n} \tag{1}$$

where x represents the solute, n and (n+1) represent the lower- and higher-eluting standards, and the R values are the corresponding corrected retention volumes. The standards were assigned the following values (log I): benzene (1), naphthalene (2), phenanthrene (3), benz[a]anthracene (4), and benzo[b]chrysene (5). Thus, a PAH with a log I value of 3.5 elutes between phenanthrene and benz[a]anthracene. Lee et al. [2,7] have described a similar system for GC, using PAH as standards rather than n-alkanes as in the traditional Kovats retention index system. The retention index system of Lee et al. [2,7] has found widespread use in GC for the comparison of retention data among different laboratories. However, the use of this and other retention index systems in LC has been limited, mainly because of the differences observed in retention indexes on various C_{18} columns. In fact, it was this attempt to develop a retention index system for PAH in 1979 that first led to our observation of differences in selectivity for PAH on different C_{18} columns.

B. Differences in Commercial C_{18} Columns

As a result of the environmental interest in the separation of PAH using reversed-phase LC, several studies in 1980–81 [5,8–11] compared the differences in commercial C_{18} columns from various manufacturers for the separation of PAH. In this laboratory [5], the

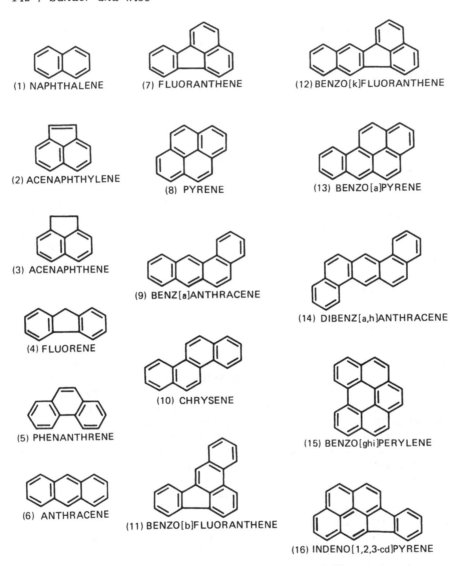

Figure 1 Components of Standard Reference Material (SRM) 1647, consisting of the 16 PAH on the Environmental Protection Agency's priority pollutant list.

selectivity characteristics of 16 PAH on seven different columns were compared, as summarized in Table 1. Ogan and Katz [8] evaluated the selectivity characteristics of seven PAH on eight different columns; Colmsjö and MacDonald [10] studied the retention characteristics of 11 PAH on three columns; Amos [11] investigated the retention of 15 PAH on 10 different C_{18} columns. In these studies, attention was focused on three groups of PAH isomers which were difficult to separate: benz[a]anthracene and chrysene; benzo[e]pyrene, benzo[a]-pyrene, benzo[b]fluoranthene, and benzo[k]fluoranthene; and benzo-[ghi]perylene and indeno[1,2,3-cd]pyrene. In addition to C_{18} columns, Amos [11] studied eleven other reversed-phase columns (e.g., C_2, C_6, C_8, C_{22}, and phenyl), but concluded that the C_{18} phases provided better PAH separations.

All of these studies found that different C_{18} columns provided somewhat different retention and selectivity characteristics for PAH. These differences in chromatographic retention and selectivity are a

Table 1   Retention Indexes for Reversed-Phase LC Separation of PAH on Different Columns

Compound	Column materials[a]						
	E	C	D	H	A	B	G
Naphthalene	2.00	2.00	2.00	2.00	2.00	2.00	2.00
Fluorene	2.76	2.70	2.71	2.74	2.73	2.70	2.73
Phenanthrene	3.00	3.00	3.00	3.00	3.00	3.00	3.00
Anthracene	3.11	3.12	3.11	3.14	3.14	3.19	3.24
Fluoranthene	3.45	3.43	3.42	3.44	3.42	3.44	3.38
Pyrene	3.61	3.60	3.59	3.66	3.62	3.69	3.56
Chrysene	3.96	3.98	3.98	3.99	4.00	4.04	4.10
Benz[a]anthracene	4.00	4.00	4.00	4.00	4.00	4.00	4.00
Benzo[b]fluoranthene	4.47	4.47	4.47	4.45	4.40	4.41	4.30
Benzo[k]fluoranthene	4.53	4.48	4.52	4.52	4.48	4.49	4.45
Benzo[e]pyrene	4.48	4.50	4.50	4.48	4.40	4.43	4.25
Benzo[a]pyrene	4.64	4.65	4.64	4.68	4.63	4.66	4.52
Dibenz[a,h]anthracene	4.92	4.89	4.90	4.85	4.78	4.74	4.69
Benzo[b]chrysene	5.00	5.00	5.00	5.00	5.00	5.00	5.00
Indeno[1,2,3-cd]pyrene	5.13	5.14	5.09	5.13	5.03	5.04	4.83
Benzo[ghi]perylene	5.14	5.18	5.17	5.16	5.05	5.05	4.71

[a]Column designations are from Ref. 5; columns used were LiChrosorb RP-18; MicroPak CH-10; MicroPak MCH-10, Nucleosil 10 C_{18}, Partisil 5-ODS, and Vydac 201TP. Retention data presented as Log I.
Source: Refs. 5 and 12.

result of the utilization of different silica substrates as supports and a variety of reagents and procedures to produce the C_{18} bonded phase. Several studies [12–14] have shown that the capacity ratio, k', generally increases with increasing carbon content. However, as pointed out by Unger et al. [15], carbon content alone is often misleading because differences in the surface area of the silica substrate result in different surface coverage (or concentration) of the bonded alkyl groups (micromoles per square-meter). Most of the data in the literature for surface concentration (density) of bonded alkyl groups [15–21] have been calculated by using the specific surface area of the underivatized silica and the percent carbon of the bonded-phase material (see discussion to follow). Ogan and Katz [8] used surface area and percent carbon data obtained from the column manufacturers' literature to calculate surface coverages for the seven columns evaluated in their study, but observed little correlation between k' for benzo[a]pyrene, and surface coverage. In an attempt to determine which characteristics produce the differences in column selectivity illustrated in Table 1, the phase type (as denoted by the manufacturers) and the physical characteristics of these seven columns were determined (see Table 2). In all of the column comparison studies previously mentioned [5,8–11], one particular material was successful in resolving all of the difficult-to-separate isomers (see Table 1). Based on the data in Table 2, the characteristic that set this material apart from the other phases studied was the polymeric C_{18} phase on a wide-pore silica (330 Å), as opposed to monomeric phases on narrow-pore silicas (<100 Å).

After our initial studies of differences in selectivity for various commercial stationary phases [4,5], it became evident that such studies were somewhat limited because the exact details concerning the silica substrate and the bonded-phase syntheses were unavailable. As a result of this limitation, studies were initiated in our laboratory to understand more fully the influence of factors such as bonded-phase type (monomeric or polymeric), silica substrate characteristics (e.g., surface area and pore size), and C_{18} ligand density on selectivity in reversed-phase LC of PAH. These studies have included both the synthesis and characterization (physical and chromatographic) of C_{18} stationary phases. As part of our studies, reversed-phase LC retention data for approximately 100 PAH on a polymeric and a monomeric C_{18} phase have been published [4,5]. This data set has recently been expanded to include retention data for over 160 PAH on both a polymeric and a monomeric C_{18} column; these data are included in this chapter in the Appendix [22]. In addition to the investigations of the influence of stationary-phase parameters on selectivity, we have also examined the influence of solute characteristics (e.g., shape and planarity) on selectivity and retention. In this paper, the results of these studies concerning the factors that affect selectivity in reversed-phase LC of PAH will be described in detail.

Table 2   Reversed-Phase C_{18} Column Characteristics

Column material[a]	Phase type	Surface area, m²/g		Pore diameter (Å)	Percent carbon (w/w)[d]	Coverage, μmol/m²		Reversed-phase k'[e]	
		Silica[b]	Bonded silica[c]			(Eq. 5) N*	(Eq. 3) N	BaP	BbC
E	monomeric	400	235	55–60	10	1.8	1.2	4.0	7.2
C	monomeric	500	259	60	12.2	2.0	1.2		8.1
D	monomeric	300	220	100	12.4	2.4	2.1	5.7	7.5
H	monomeric	275–300	166	70	13.1	3.3	2.4–2.2	12.5	14
A	monomeric	300	149	100	19.6	5.5	3.6	13.2	14
B	polymeric	500	120	60	19.0	7.3	2.5	13.2	27
G	polymeric	90	57.7	330	8.2	6.6	4.8		32

[a]Column designations from Ref. 5, see Table 1 for columns used in this study.
[b]Nominal values reported by the manufacturers.
[c]Determined by BET.
[d]Average of duplicate analyses.
[e]Mobile phase: 80% acetonitrile in water. BaP = benzo[a]pyrene and BbC = benzo[b]chrysene.
Source: Refs. 12 and 35.

II. BONDED-PHASE SYNTHESES

The chromatographic properties of bonded-phase sorbents are dependent on the conditions of synthesis and the physical and chemical properties of the underlying silica substrate. Because of the importance of column selectivity in liquid chromatographic separations, it is understandable that alkyl bonded phase materials have received considerable attention in the literature [13–18,23–25]. Historically, at least four types of bonding schemes have been used to immobilize organic molecules to silica: ester bonds, direct silicon to carbon bonds, amine bonds, and silane bonds. Phases with ester bond linkages are relatively simple to prepare by reaction of an alcohol with dry silica [24]. Because the silanol groups of silica are slightly acidic, alcohols react with silica to form an ester bond linkage (Si-O-C). This reaction is reversible, and consequently this type of phase is subject to degradation by hydrolysis. A much more stable type of bonded phase results from the reaction of Grignard reagents with chlorinated silica [24]. Chlorinated silica is first prepared by reaction of silica with thionyl chloride. Reaction of this silica with a suitable Grignard reagent results in a highly stable phase with silicon-carbon bond linkages. A by-product of the reaction is $MgCl_2$, which is solid and may be difficult to remove from the silica pore network. Amine bond linkages are formed by the reaction of primary or secondary amines with chlorinated silica [24]. The stability of this bond linkage is intermediate to the ester and carbon bond linkages. Because the by-product of this reaction is a gas (HCl), wash procedures are much simpler than with Grignard syntheses. By far the most popular type of bonded phase produced today is made by reaction of active silane reagents with silanols at the silica surface. The resulting siloxane-bond linkage is stable to most common solvents even under elevated temperatures.

A number of terms are often used interchangeably to describe the concentration of bonded ligands on the silica surface. In this work, the terms *surface coverage*, *ligand density*, and *surface concentration* refer to a measure of the extent of surface modification in units of either micromoles per square-meter or groups per square-nanometer (see Eq. 3). In comparison, *carbon loading* is a measure of the percentage of carbon on a bonded-phase sorbent. Carbon loading does not take into account the surface area of the substrate. *Phase loading* is essentially the same as *carbon loading*, but is usually used in more general terms.

A. Monomeric Syntheses

Depending on reaction conditions and the type of silane reagent used, both monomeric and polymeric stationary phases can be prepared. Monomeric stationary phases result from the reaction of monofunctional silanes (e.g., monochloro- or monoalkoxysilanes) with silanols on the silica surface [18] (see Fig. 2). In a typical synthesis, an excess of the silane reagent is dissolved in a nonpolar solvent (e.g., chloroform, toluene, or carbon tetrachloride). The quantity of silane required is calculated from the weight and surface area of the silica, and from the expected surface coverage of the bonded phase. For example, to prepare a monomeric C_{18} phase using 5 g of a silica substrate with a specific surface area of 350 m^2/g:

$$(5g)(350 \ m^2/g)(3.3 \ \mu mol/m^2)(347.1 \ g/mol)(10^{-6}) = 2.00 \ g \ silane \tag{2}$$

A *stoichiometric quantity* (2.00 g) of dimethyloctadecylchlorosilane would be required, assuming a maximum phase coverage of 3.3 $\mu mol/m^2$. We define the *stoichiometric quantity* of silane as the quantity of silane reagent that would be required to give maximum phase coverage if all of the silane in solution were to react with the silica. In practice, only a portion of the silane reacts with the silica, so to obtain maximum phase coverage a large excess of the reagent must be used. Excesses of up to 10 fold are sometimes used in an attempt to increase phase loadings. Ligand density can also be increased by removing reaction by-products during bonded-phase synthesis. For example, pyridine is commonly added to syntheses with chlorosilanes to remove the HCl by-product from solution. Alcohol by-products from alkoxysilane syntheses can be removed by distillation.

Phases with lower loadings can be prepared simply by using less than stoichiometric quantities of the silane. Such phases are often useful in the study of retention mechanisms. At low silane concentrations (relative to the stoichiometric quantity), carbon loading increases linearly with silane concentration as all of the silane in solution reacts with the silica (see Fig. 3). In the example above, if only 0.20 g of dimethyloctadecylchlorosilane were used (instead of 2.00 g), the resulting phase coverage would be close to 0.33 $\mu mol/m^2$.

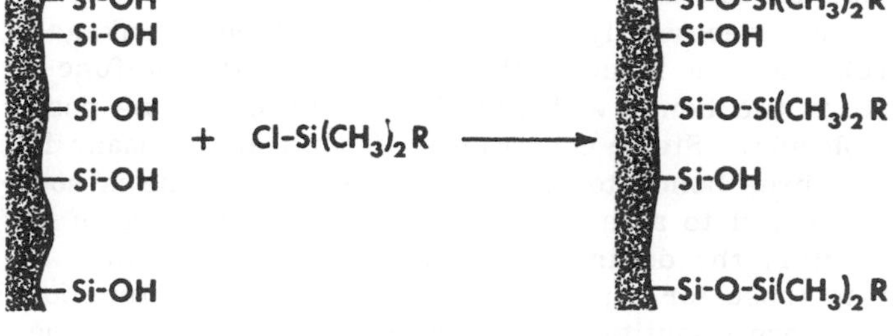

Figure 2   Synthesis scheme for monomeric surface modification of silica, using monofunctional silane reagents. (From Ref. 37.)

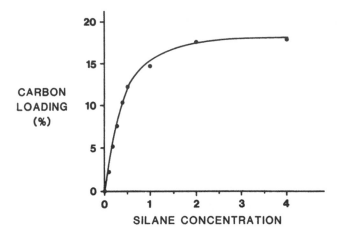

Figure 3 Effect of silane reagent concentration on carbon loading for dimethyloctadecylchlorosilane on Adsorbosil silica (nominal pore diameter, 60 Å; surface area, 361 m^2/g).

At high silane concentrations, increases in surface coverage fall off rapidly so that only fractional increases occur even with large excesses of silane. Phase loading is ultimately limited by steric constraints of the bonded phase. Unbonded silica has about 4.8 silanols/ nm^2 at the surface. Even after modification under rigorous monomeric reaction conditions, over half of these silanol groups remain unbonded [17,26]. Alkyl chains bonded to the surface shield adjacent unbonded silanols from reaction. The actual surface coverage limit is somewhat dependent on alkyl chain length, silane reactivity, and reaction conditions. For dimethyloctadecylchlorosilane, the maximum surface coverage value is about two alkyl groups/nm^2 (3.3 μmol/m^2) [18,26].

Monomeric stationary phases can also be prepared by reaction of silica with di- or trifunctional silanes, if care is taken to exclude water from the reaction mixture (Fig. 4). Reaction of a monofunctional silane with silica necessarily results in a single bond linkage for each silane molecule that reacts with the silica. With polyfunctional silanes, one or two bonds with the silica surface may be possible per silane molecule. Steric constraints of bond angles make it highly unlikely for three bonds to form. This type of phase is sometimes incorrectly referred to as a polymeric phase. When care is taken to exclude water, the dominant properties of the resulting phase are similar to those for a phase prepared from a monofunctional silane reagent. If trace quantities of water are present in the reaction slurry with polyfunctional silanes, polymerization reactions may

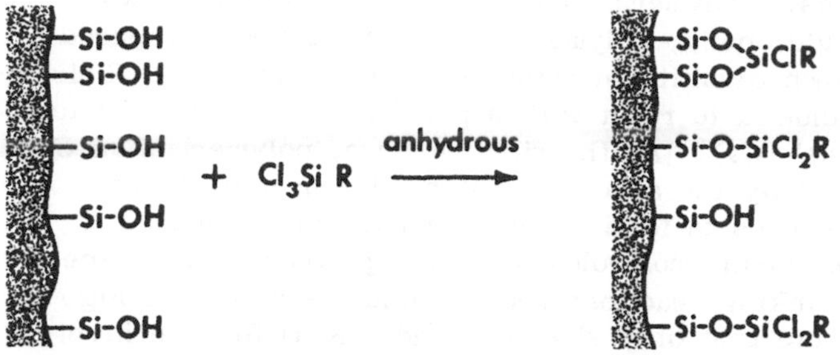

Figure 4 Synthesis scheme for monomeric surface modification of silica, using polyfunctional silane reagents (From Ref. 37.)

occur; however, the total contribution to phase properties is probably small.

B. Polymeric Syntheses

Polymeric phases are synthesized by intentionally introducing a measured quantity of water into a synthesis involving polyfunctional silanes [27–30] (Fig. 5). The term *polymeric phase* should not be confused with *polymer-based substrate phases*. The latter designation refers to phases that are prepared using polymer-based substrates instead of silica substrates. Such phases can be used over a wide pH interval and are particularly useful for separations requiring basic eluents. In contrast, *polymeric phases* refer to phases usually (but not exclusively) synthesized on silica substrates by polymeric surface

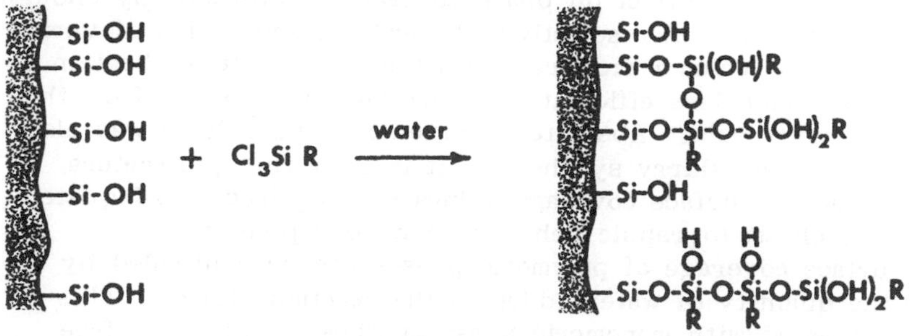

Figure 5 Synthesis scheme for polymeric surface modification of silica. Water is added either to the silica (*water preequilibration synthesis*) or to the reaction mixture (*water-slurry synthesis*). (From Ref. 37.)

modification reactions. Polymeric phases can be prepared in two distinct ways. Traditionally, polymeric phases have been prepared by adsorbing a known quantity of water onto dry silica [29]. This *wet* silica is then allowed to react with a polyfunctional silane (*water preequilibration synthesis*). At the silica surface, polymerization of the silane is initiated by the adsorbed water. Water hydrolyzes chloro- and alkoxysilanes to form silane silanols. These molecules can react with other silane molecules to form a polymer. Both linear addition and cross-linking reactions are possible, as shown in Fig. 5. It is important to note that only siloxane bonds (Si-O-Si) are involved in polymerization. Alkyl side chains, e.g., octadecane, are incorporated as substituents on the polymer backbone.

Alternatively, water can be added directly into the reaction slurry to initiate the polymerization reaction (*water slurry synthesis*) [27,30]. It is believed that the silane polymer forms in solution and then bonds to the silica surface. With this process, silica pore size is an important parameter, and phase loading has been shown to increase with increasing pore diameter (see below) [30].

A comparison of these methods of polymeric phase synthesis has been carried out using a wide-pore (330 Å diameter) silica substrate [27]. Prior to reaction, the silica was dried under reduced pressure at 150°C. In the first case (water-slurry synthesis), an aliquot of water was added directly to a slurry of silica in carbon tetrachloride, also containing a measured quantity of octadecyltrichlorosilane. The slurry was refluxed for 4 hr and then filtered, washed, and dried. In the second case (water preequilibration synthesis), the same quantity of water was added to the silica in a sealed bottle and allowed to equilibrate for 48 hr. The synthesis was carried out as before, but no additional water was added to the reaction slurry. Surface coverage values for the two phases were nearly identical (4.73 versus 4.77 $\mu mol/m^2$). This suggests that the order of addition of water to the reaction has little effect on phase coverage. Interestingly enough, the two phases had similar selectivity toward the separation of various PAH, but the column produced from the water preequilibration synthesis was much less efficient than the column prepared from the water-slurry synthesis. Absolute retention for this column was also less than the water-slurry synthesis. It is clear that parameters other than simple surface coverage values are required to adequately describe the chromatographic behavior of bonded phases.

The surface coverage of polymeric phases can be controlled by varying the quantity of water added to the reaction slurry. This can be contrasted with monomeric phase syntheses, where surface coverage is controlled by varying silane reagent concentration. In a typical water-slurry polymeric phase synthesis, 10 ml octadecyltrichlorosilane is added to 100 ml CCl_4 containing 4 g silica; 0.5 ml H_2O is added, and the slurry is refluxed for 4 hr. At the conclusion of the reaction, the bonded silica is filtered, washed, and dried. Phases

produced in this way have surface coverage values approaching
5 μmol/m^2. By changing the quantity of added water, higher or
lower phase loadings can be achieved. The effect of added water on
surface coverage for polymeric C$_{18}$ syntheses is illustrated in Fig. 6.
Phase coverage values asymptotically approach a limit as a function
of water added to the reaction slurry. This suggests that surface
modification is limited by the number of silanol sites available for
reaction (as with monomeric phase syntheses). In general, maximum
polymeric phase loadings are about twice that obtainable with monomeric syntheses, all other factors being equal. Thus, if silane polymer forms in solution and then bonds only to sites on the silica, a
limit would be reached as the sites are filled.

The reproducibility of water-slurry polymeric syntheses is excellent if reasonable care is taken to control the reaction conditions.
In a study of phase reproducibility, polymeric phases were prepared
on five different wide-pore substrates [27]. Surface areas for the
substrates ranged from 45 to 250 m^2/g. Despite the large differences
in surface area, very similar surface coverages were obtained: 5.13,
5.34, 5.00, 4.73, and 5.31 μmol/m^2 (relative standard deviation,
4.9%). The small differences can probably be attributed to impreci-
sion in the manufacturer's nominal surface area values for the substrates. Even better precision is obtained for the synthesis of polymeric phases on a single lot of one type of silica. Four polymeric

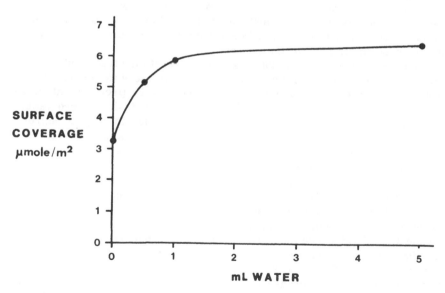

Figure 6   Effect of the amount of water on the resulting surface
coverage for polymeric phase syntheses. (From Ref. 27.)

phases were prepared under similar conditions on Vydac TP silica from the same production lot. Surface coverages were 4.84, 4.79, 4.73, and 4.77 $\mu mol/m^2$ (relative standard deviation, 0.96%). A fifth synthesis prepared on a different lot of Vydac TP silica had a surface coverage value of 5.21 $\mu mol/m^2$, again indicating the importance of accurate surface area measurements.

C. Oligomeric Syntheses

A third type of alkyl bonded phase has been produced with properties intermediate to monomeric and polymeric phases. This phase, termed *oligomeric*, is synthesized stepwise by a sequential reaction scheme (see Fig. 7) [27]. The reaction is essentially a controlled polymerization, with one C_{18} silane monomer being added with each reaction step. Thus, the maximum degree of linear polymerization is equal to a number of reaction steps used.

The reaction consists of four parts: (a) anhydrous reaction with octadecyltrichlorosilane, (b) filtering and washing, (c) hydrolysis of unreacted chloro groups, and (d) drying. Each cycle is essentially a monomeric synthesis that introduces additional silanol groups for further modification. Two types of silanols can be distinguished: silica silanols at the silica surface, and silane silanols, produced by hydrolysis of chloro groups (or other reactive groups) on the silane reagent. Only silica silanols are modified in the first reaction step. The C_{18} chains added in subsequent steps are bonded to silane silanols. Although cross-linking reactions are not shown in Fig. 7, this type of reaction is not necessarily excluded. However, because the reaction occurs at the silica surface rather than in solution, the extent of cross-linking is probably small.

A plot of percent carbon loading versus reaction step number is illustrated in Fig. 8. A nine-step oligomeric phase was prepared on Vydac TP silica as described above. Columns were prepared at steps 1, 3, 5, 7, and 9, and small samples of the bonded silica were taken at each step for carbon analysis. The open circles in Fig. 8 represent carbon loading values of silica aliquots further silanized (endcapped) with hexamethyldisilazane (HMDS). Although measurable increases in carbon loading occur for each reaction step, these increases are minimal after about the third step. The reaction appears to be sterically limited, as with monomeric bonded-phase syntheses. In all but the first step, reaction with HMDS gives rise to larger increases in carbon loading than do subsequent C_{18} silane reactions. Because the HMDS molecule is so much smaller than the bulky octadecylsilane (ODS) molecule, modification by HMDS is sterically more favorable than by ODS. For example, in step 7, the endcapping reaction resulted in an increase in carbon of 0.25%. If the same number of silanols were modified with ODS, the increase in carbon would be 1.7%. The actual increase in carbon from step 7 to step 8 was only

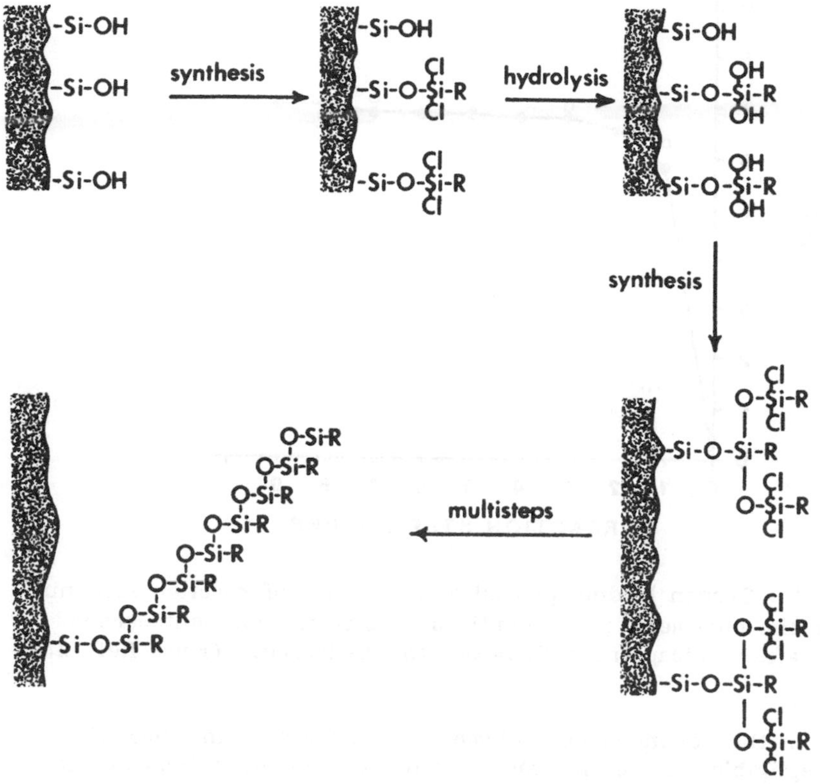

Figure 7   Oligomeric phase synthesis scheme. Anhydrous silaniza-
tion and hydrolysis reactions are sequentially alternated so that the
bonded phase is built up one unit at a time. The process can be
considered a controlled, step-wise polymerization. (From Ref. 27.)

0.15%, so only about one out of every eleven silanols accessible to
HMDS was modified by ODS.

The chromatographic effect of endcapping bonded phases has been
investigated in a number of studies [18,31,32], but reported results
have often been in conflict. Berendsen et al., [18] have suggested
that for densely bonded monomeric phases, further reaction with tri-
methylsilane reagents is negligible. Others have demonstrated significant differences in chromatographic properties of endcapped and nonendcapped phases [32]. The ability of polymeric C_{18} phases, endcapped
and nonendcapped, to separate PAH has been studied [27]. A polymeric C_{18} phase was prepared on Vydac TP silica, and half of the
material was endcapped using HMDS. A small but measurable increase
in carbon loading resulted from the endcapping procedure (i.e., 8.77%
versus 8.93% after endcapping). Retention of various PAH was slightly

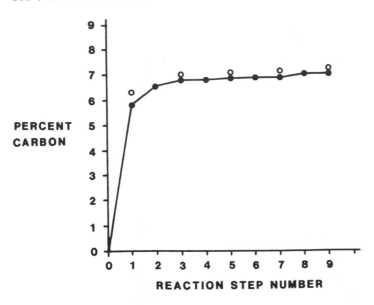

Figure 8 Carbon loading plotted as a function of reaction step number for the oligomeric phase synthesis. Circles represent carbon loading after endcapping with hexamethyldisilazane. (From Ref. 27.)

longer on the nonendcapped column, but differences in selectivity were negligible. No effort was made to examine the effects of endcapping on the separation of polar compounds, but more significant differences between the phases might be expected.

III. BONDED-PHASE CHARACTERIZATION

A. Phase Loading

One of the most useful properties for characterizing bonded phases is the surface coverage value of bonded ligands. This value, also referred to as ligand density, is commonly expressed in units of micromoles per square-meter or in bonded groups per square-nanometer. Surface coverage values are calculated from the percent carbon loading of the bonded phase and the surface area of the unbonded substrate:

$$N(\mu mol/m^2) = 10^6 \frac{P_c}{[1200n_c - P_c(M - 1)]S} \qquad (3)$$

where P_c is the percent carbon loading of the bonded phase, n_c is the number of carbon atoms in the bonded ligand, M is the molecular

weight of the bonded ligand, and S is the surface area of the unbonded substrate [18]. For example, a monomeric C_{18} phase with
16% carbon loading prepared using dimethyloctadecylchlorosilane and
a 350 m^2/g silica substrate has a surface coverage of $(10^6)(16)$ /
$[1200(20) - 16(310)](350) = 2.40$ μmol/m^2. Note the molecular weight
of the dimethyloctadecylsilyl group, $CH_3(CH_2)_{17}Si(CH_3)-$ (MW 311)
is used for M, rather than the molecular weight of the chlorosilane
reagent. Conversion between micromoles per square-meter and groups
per square-nanometer is simple:

$$N(\text{groups}/nm^2) = N(\mu mol/m^2)(0.6) \tag{4}$$

Using the example above, 2.40 μmol/m = 1.44 bonded groups/nm^2.
Equation (3) is easily applied to the calculation of surface coverage
values for monomeric phases. For polymeric or oligomeric phases,
however, an assumption must be made about the molecular weight of
the bonded species. For phases prepared from trichlorosilanes,
$CH_3(CH_2)_{17}Si(OH)_2O-$ (MW 331) has been used as the representative
unit comprising the phase [27]. The oxygen atoms in the molecule
are introduced as a result of silane hydrolysis during phase synthesis, and as such are considered part of the bonded phase rather than
part of the silica. Also note that for octadecyltrichlorosilane, $n_c = 18$
rather than 20.

Because small differences in carbon loading can give rise to large
changes in chromatographic properties, precise measurements of
bonded phase carbon loading are essential for meaningful phase characterization. The precision of carbon loading measurements on alkyl
bonded phases has been studied for an elemental carbon analysis technique [33]. With this technique, bonded silica is burned in an oxygen
environment, and the carbon dioxide that is produced is quantified
by infrared spectroscopy. Replicate carbon determinations were made
on a single lot of bonded silica, and the relative standard deviation
of eight measurements was determined to be 0.8% (average carbon loading, 8.26%). The high precision of the technique permits reliable
evaluation of the extent of endcapping reactions. For bonded substrates with low absolute carbon loadings (e.g., as a result of low
ligand densities or low substrate surface areas), a correction must
be made for the inherent carbon in the unmodified silica. This value,
referred to as *background carbon*, has been observed to vary from
trace levels to nearly 1% (w/w) carbon [33].

B. Chromatographic Characterization

Differences in the chromatographic behavior of monomeric and polymeric C_{18} phases are apparent for the separation of PAH. In general,
better separations of complex mixtures of PAH are possible with polymeric C_{18} phases than with monomeric C_{18} phases [27,30,33—37]. In

order to classify phases as having either monomeric or polymeric properties, a simple empirical LC test has been devised to gauge phase characteristics [27]. As will be discussed later, the retention behavior of planar and nonplanar PAH are markedly different. These differences are greatest for polymeric alkyl phases. A three-component mixture consisting of phenanthro[3,4-c]phenanthrene (PhPh), 1:2,3:4,5:6,7:8-tetrabenzonaphthalene (TBN), and benzo[a]pyrene (BaP) has been developed that makes use of these differences to permit classification of monomeric or polymeric character of a particular phase (see Fig. 9). Depending on the elution order of this mixture, phases can be rapidly screened and column selectivity toward more complex PAH mixtures predicted. Under mobile-phase conditions of 85% acetonitrile in water, BaP elutes before PhPh or TBN on monomeric C_{18} columns (see Fig. 10). This retention order has been observed for a large number of commercial and homemade columns. For polymeric C_{18} columns, BaP is retained longer than either PhPh or TBN. This retention behavior is somewhat unexpected, because the reversed-phase retention of PAH is usually thought to correlate with molecular weight (or number of aromatic carbon atoms). However, BaP has five aromatic rings, whereas PhPh and TBN both consist of six aromatic rings. BaP is retained the longest relative to TBN on heavily loaded polymeric phases. As might be expected, the elution order of the three components on the oligomeric phases is PhPh, BaP, TBN—i.e., intermediate to monomeric and polymeric phases. In general, the selectivity factor, α, (i.e., relative retention $\alpha = k'_1/k'_2$ for BaP/TBN depends on the surface coverage of the bonded phase.

The retention behavior of the three-component mixture on a given column is also indicative of the overall selectivity toward PAH isomers. Separation of a 16-component PAH mixture, Standard Reference Material (SRM) 1647 (see Fig. 1), on the three different phase types is shown in Fig. 10. The incomplete separation shown in the upper chromatogram is characteristic of monomeric phases. Most commercial

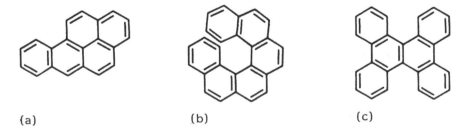

(a)   (b)   (c)

Figure 9 Structures of the three-component, planar/nonplanar test mixture. (a) Benzo[a]pyrene is planar, whereas (b) phenanthro-[3,4,c]phenanthrene and (c) tetrabenzonaphthalene have nonplanar properties.

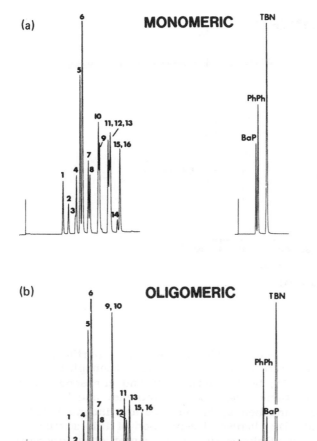

Figure 10 Separation of SRM 1647 and the three-component (planar/nonplanar) test mixture on representative (a) monomeric, (b) oligomeric, and (c) polymeric phases prepared on the same wide-pore silica substrate. Separation of the 16-component mixture was performed by using gradient elution, 40—100% acetonitrile in water over 30 min at 2 ml/min. The three-component mixture was chromatographed isocratically with 85% acetonitrile in water. Peak assignments are as listed in Figs. 1 and 9. (From Ref. 27.)

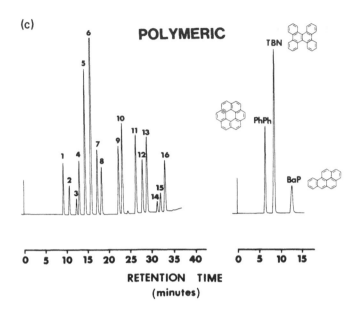

(c)

POLYMERIC

RETENTION TIME
(minutes)

Figure 10   (continued)

C$_{18}$ columns fall into this category. The unresolved or partially resolved components in this mixture are isomers, e.g., acenaphthalene
and fluorene (peaks 3 and 4); benz[a]anthracene and chrysene (peaks
9 and 10); benzo[b]fluoranthene, benzo[k]fluoranthene, and benzo-
[a]pyrene (peaks 11−13); and benzo[ghi]perylene and indeno[1,2,3-
cd]pyrene (peaks 15 and 16). Improved separation of this mixture
is obtained for the seven-step oligomeric phase (center chromatogram).
Using a C$_{18}$ polymeric phase (lower chromatogram), complete separation of all 16 components was achieved under the same conditions.
In general, better separations of isomeric PAH are obtained by using
polymeric stationary phases than the more conventional monomeric
phases. However, because changes in selectivity and elution order
can occur with different types of columns, monomeric phases may, in
some instances, provide a better separation of a critical solute pair
in a complex PAH mixture.

IV. SUBSTRATE AND BONDED-PHASE PROPERTIES

A. Substrate Properties

Silica substrates prepared by various manufacturers differ in a number of important ways [38]. Surface area, particle size and size
distribution, pore diameter, and pore volume are interrelated variables that affect solute retention and chromatographic performance

[39–43]. Column efficiency is directly related to particle size, whereas solute retention is more closely dependent on the specific surface area of the substrate. The effect of pore size on the chromatographic properties of bonded phases is not well understood. For constant pore volume, specific surface area increases for decreases in pore diameter. Similarly, for a given pore size, surface area is dependent on pore volume and thus increases with substrate porosity. For these reasons, high surface area, wide-pore substrates are generally more friable than low-pore-volume substrates, and more care must be exercised in handling the silica and in preparing the columns. The formation of submicrometer "fines" from careless handling invariably results in high column back-pressure. Caution should be exercised when using nominal or average values for surface area or pore diameter as reported by the manufacturer. In past studies, nominal surface-area values have been shown to differ from measured values by as much as 50% [30]. Other substrate parameters such as chemical composition undoubtedly further influence chromatographic behavior [44,45]. Although the importance of these parameters is clear, this information is often difficult to obtain.

A listing of some physical properties for commercial silica substrates is given in Table 3 [30]. The particle and pore diameter values are nominal values specified by the manufacturers. Specific surface areas were measured in our laboratory for each of the silicas via a multipoint nitrogen adsorption technique. The surface areas range from 32 to 433 m^2/g, the highest areas originating from narrow-pore substrates. Pore volume and density measurements were made on certain substrates by packing the silica into tared column blanks. The pore volume was taken as the difference in retention volumes of toluene and a high-molecular-weight polystyrene standard, using tetrahydrofuran as the mobile phase. Packing density was measured by subtracting the tubing tare weight from the dry, packed column weight. Columns were dried by flushing with supercritical carbon dioxide at 70°C. A carbon determination was made on each of the unbonded substrates using a technique previously described [33]. This value, designated as *background carbon*, may be important for accurate assessment of bonded-phase loading, especially with low loaded phases. Finally, the pH of a 10% aqueous slurry of the silica was measured. One gram of silica (as received from the manufacturer) was added to 10 ml deionized water (pH 7 ± 0.5), and the pH was measured immediately upon mixing.

B. Surface Coverage

A number of monomeric and polymeric C_{18} phases have been synthesized on the substrates listed in Table 3. Surface coverage values for the monomeric phases were about half of those for polymeric phases (see Table 4). Carbon loadings ranged from about 2 to 24% carbon (w/w), and surface coverages ranged from 1.3 to 5.3 $\mu mol/m^2$. In general, the highest surface coverages were obtained for polymeric

Table 3 Physical Properties of Silica Substrates

Silica	Particle diameter (μm)	Particle shape[a]	Pore diameter (Å)	Surface area (m²/g)	Pore volume (ml/g)	Density (g/cm³)	Background carbon (%)	pH
Nucleosil	10	s	50	316				
Adsorbosil	10	i	60	382				
LiChrosorb (60)	10	i	60	398	0.66	0.44	0.17	7.2
Polygosil	10	i	60	245	0.71	0.43	0.07	6.5
RSil	10	i	60	433	0.75	0.42	0.85	8.0
Vydac HSB (60)	5	s	60	287	0.44	0.59	0.21	4.8
Zorbax (60)	10	s	60	432	0.50	0.57	0.16	5.6
Adsorbosphere	10	s	80	174				
Econosphere	5	s	80	171	0.52	0.62	0.22	9.0
RoSil	8	s	80	357	0.71	0.51	0.14	8.4
Partisil	10	i	85	429	0.72	0.44	0.27	4.9
Vydac HSB (90)	10	s	90	423			0.82	4.2

LiChrosorb (100)	10	i	100	297	1.11	0.36	0.06	6.7
LiChrospher (100)	10	s	100	266			0.14	4.6
Zorbax (100)	7	s	100	139	0.45	0.66	0.08	3.8
Sperisorb S5W	5	s	120	160				
Hypersil	10	s	125	149				
Zorbax (150)	7	s	150	99	0.39	0.68	0.06	4.8
Hypersil WP-300	5	s	300	57			0.10	5.2
LiChrospher (300)	10	s	300	207			0.18	4.7
Protesil	10	i	300	257			0.11	4.6
Zorbax (300)	8	s	300	39	0.28	0.72	0.06	5.4
Vydac TP	10	s	330	82	0.73	0.50	0.38	4.1
LiChrospher (500)	10	s	500	59			0.06	8.6
LiChrospher (1000)	10	s	1000	32			0.05	8.1

[a] s = spherical and i = irregular.
[b] Nominal values for average pore diameter provided by manufacturer.
Source: Ref. 30.

Table 4 Column Properties

Silica	Phase type	Pore dia (Å)	Area (m²/g)	Carbon loading (%)	Surface coverage (μmol/m²)	α TBN/BaP	k' BaP
Zorbax (300)	monomeric	300	39	2.16	2.31	2.12	3.46
Vydac TP	monomeric	300	96	5.80	2.53	2.00	1.57
Zorbax (100)	monomeric	100	139	8.30	2.76	1.97	4.56
Zorbax (150)	monomeric	150	99	5.00	2.22	1.82	2.34
Zorbax (60)	monomeric	60	432	11.32	1.26	1.78	3.18
Zorbax (60)	polymeric	60	432	17.70	2.57	1.33	8.42
Zorbax (100)	polymeric	100	139	10.05	3.92	1.25	8.32
RSil	polymeric	60	433	22.22	3.39	1.21	10.93
Econosphere	polymeric	80	171	11.50	3.69	1.18	8.55
Partisil	polymeric	85	429	21.91	3.49	0.97	11.11
LiChrosorb (100)	polymeric	100	297	19.82	4.41	0.97	9.41

LiChrosorb (60)	polymeric	60	398	22.00	3.81	0.93	12.68
LiChrospher (1000)	polymeric	1000	32	3.16	4.72	0.92	1.56
RoSil	polymeric	80	357	20.10	3.72	0.90	12.61
LiChrospher (500)	polymeric	500	59	5.38	4.54	0.89	2.77
Polygosil	polymeric	60	245	17.08	4.34	0.86	9.93
Zorbax (300)	polymeric	300	39	4.28	5.35	0.86	4.21
Vydac HS (90)	polymeric	90	423	23.98	3.92	0.84	13.28
Zorbax (150)	polymeric	150	99	8.89	4.77	0.83	9.88
Hypersil	polymeric	120	149	13.53	5.26	0.79	11.73
Nucleosil	polymeric	50	316	18.38	3.74	0.75	10.71
LiChrospher (100)	polymeric	100	266	19.48	4.78	0.71	11.42
LiChrospher (300)	polymeric	300	207	16.43	4.83	0.71	6.57
Vydac HS (60)	polymeric	60	288	16.11	3.38	0.69	13.60
Hypersil	polymeric	300	57	5.71	5.03	0.66	4.93
Protosil	polymeric	300	257	20.06	5.17	0.65	13.07
Vydac TP	polymeric	330	94	9.15	4.99	0.62	6.77

Source: Ref. 30.

phases prepared on wide-pore substrates. A plot of surface coverage as a function of nominal pore size for polymeric phases is shown in Fig. 11. The plotted curve is shown for the 60, 100, 150, and 300 Å pore diameter Zorbax substrates, which were manufactured under similar conditions. Although the relationship of surface coverage to pore size is not unequivocal, a trend towards higher phase loading with increasing pore size is apparent.

C. Pore Size Effects

Although pore size and structure are important considerations in size-exclusion chromatography, these parameters have received little attention in the study of bonded-phase systems. Most solutes chromatographed under reversed-phase conditions are too small for size-exclusion mechanisms to have any significant effect. Perhaps the main concern about pore size with bonded phases has been that the

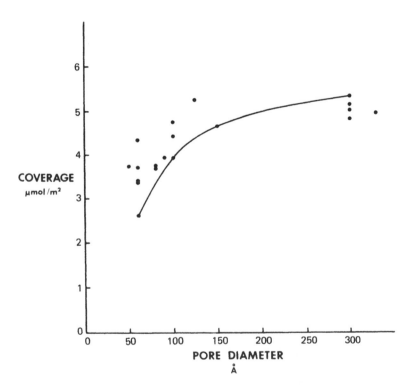

Figure 11 Bonded-phase surface coverage, plotted as a function of nominal pore diameter for polymeric phase syntheses in Table 4. The curve shown is for 60, 100, 150, and 300 Å pore-diameter Zorbax substrates. (From Ref. 30.)

pores be large enough to allow unhindered solute diffusion and thus aid solute mass transfer. Because of their high specific surface areas, most substrates currently used with bonded phases have pore diameters of about 100 Å or less. A pore diameter limit of about 60 Å is imposed for C_{18} phases because an extended C_{18} chain is about 24 Å long. Thus, a 60 Å diameter cylindrical pore would be nearly filled after modification with octadecylsilane. Berendsen, Pikaart, and de Galan [18] have suggested that surface coverage of long-chain monomeric phases should decrease for very small-pore substrates. They envisioned the curvature of small pores as a limiting factor in bonded phase synthesis because high curvature would increase steric hindrance of adjacent alkyl chains.

To examine selectivity differences that might occur as a function of pore size, a series of monomeric and polymeric C_{18} phases have been synthesized on 60, 100, 150, and 300 Å pore diameter Zorbax substrates [30]. The two PAH mixtures described earlier were used in the evaluation of the chromatographic characteristics of these bonded phases: SRM 1647 and the three-component mixture, BaP, PhPh, TBN (see Figs. 1 and 9). Representative separations of the two mixtures for a C_{18} monomeric phase are shown in Fig. 12. With the exception of the 60 Å substrate, retention decreases with increasing pore diameter as a consequence of the lower surface areas for the wide-pore substrates. Thus, even though the 100, 150, and 300 Å substrates have similar surface coverage values (2.76, 2.22, and 2.31 μmol/m^2, respectively), the quantity of phase in the column decreases with surface area. The lower retention observed for the 60 Å substrate is probably due to the low surface coverage value for that phase (1.26 μmol/m^2). Careful comparison of the separations reveals no significant differences in selectivity between the phases. The elution-order of the components is the same for each phase. Better separation of the 16-component mixture occurs for the 100 Å pore diameter substrate simply because retention is greatest for that phase.

More significant differences are observed for polymeric phases synthesized on these same Zorbax substrates (Fig. 13). Unlike the monomeric phases, the best separations are obtained on wide-pore substrates. All 16 components of SRM 1647 are separated on the 150 and 300 Å substrates, whereas only partial separation is obtained on the narrow-pore substrates. For example, components 9 and 10, benz[*a*]anthracene and chrysene, coelute on the 60 Å substrate, are partially resolved on the 100-Å substrate, and are fully resolved on 150 and 300 Å substrates. Similar changes occur for acenaphthene and fluorene (components 3 and 4) and benzo[*ghi*]perylene and indeno-[1,2,3-*cd*]pyrene (components 15 and 16) on the four phases. Differences in selectivity are also apparent for the separation of PhPh, BaP, and TBN. BaP elutes before TBN for the narrow-pore substrates, but for the 150- and 300 Å substrates, the elution order is PhPh, TBN, BaP. The atypical retention behavior observed for the

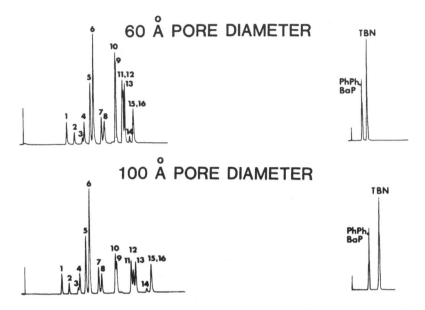

Figure 12 Separation of 16 PAH (SRM 1647) on representative monomeric phases, prepared on substrates with varying pore diameter. The separation was performed by using gradient elution, 40–100% acetonitrile in water over 45 min at 2 ml/min. The three-component mixture was chromatographed isocratically with acetonitrile-water (85:15). Peak assignments are as listed in Figs. 1 and 9. (From Ref. 30.)

narrow-pore polymeric phases is similar to that for oligomeric or monomeric phases. This suggests that the polymeric character of the phase decreases with pore diameter.

D. Polymeric Synthesis Model

The differences in retention behavior observed among polymeric C_{18} phases prepared on substrates of different pore size cannot adequately be explained by a size-exclusion mechanism. Not only are PAH molecules too small to undergo significant size differentiation in 60–300 Å pores, but the best separations occur for the wide-pore substrates— better separations would be expected for the narrow-pore substrates if a size-exclusion mechanism were in effect. Instead, differences in the retention behavior of the phases are probably caused by differences in the phases originating at the time of synthesis.

The differences in selectivity can be explained in terms of a size-exclusion mechanism that limits the extent of polymeric modification during bonded-phase synthesis [30]. For monomeric phases, the

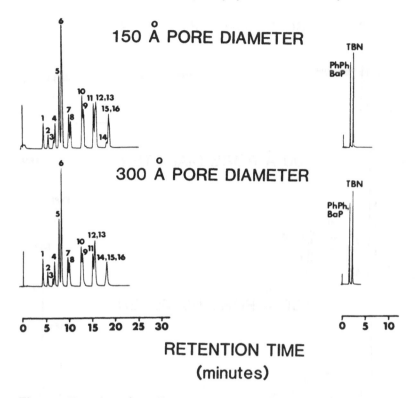

**RETENTION TIME
(minutes)**

Figure 12 (continued)

bonding reaction should be relatively insensitive to substrate pore size, as long as the pores are large enough to permit diffusion of the silane reagent. For water-slurry-type polymeric syntheses, a mixture of silane monomer and polymer exist together in solution in the reaction slurry. Both species are reactive and can bond to the silica surface. However, because the silane polymer is larger than the monomer, size differentiation is possible within the pore network. Unger [38] has reported that the diffusion and reaction kinetics of silane molecules may be affected by pore diameter. Thus, silane molecules of different molecular weight can be expected to react at different rates within the pore network.

A schematic representation of this size-exclusion mechanism is shown in Fig. 14. For small-pore substrates, a phase with high monomeric character would result if the silane polymer were excluded

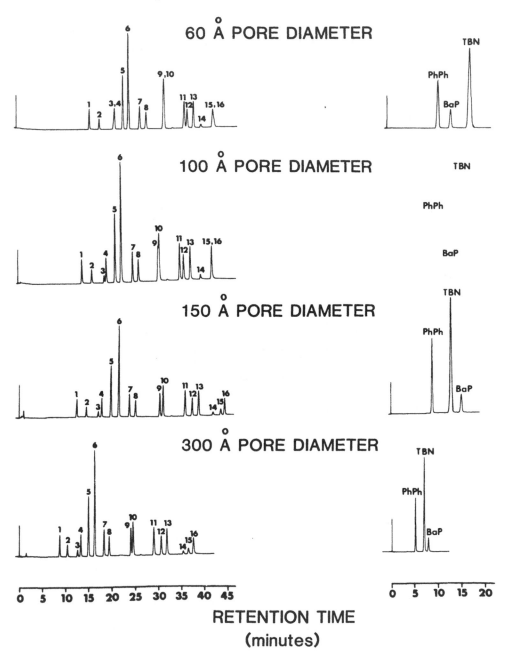

RETENTION TIME

(minutes)

Figure 13 Separation of SRM 1647 on polymeric phases, prepared on substrates with varying pore diameter. Chromatographic conditions and peak identification are as in Fig. 12. (From Ref. 30.)

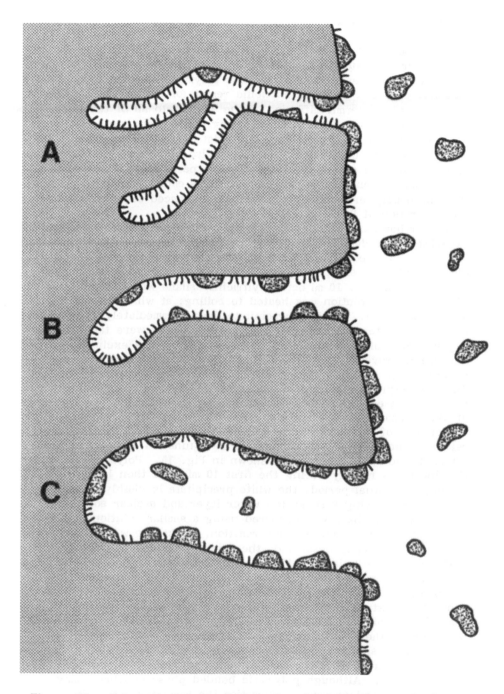

Figure 14   Schematic representation of a size-exclusion mechanism that may limit polymeric coverage during phase synthesis. (From Ref. 30.)

from the pores (Fig. 14a). Greater polymeric behavior would result with larger-pore substrates because more of the pore volume of the substrates is accessible to the reactive polymer molecules (Fig. 14b). Maximum polymeric modification would result for substrates with pores large enough to permit diffusion of the silane into all of the pore volume (Fig. 14c). Even in this limiting case, a mixture of monomeric and polymeric groups probably exist on the silica surface, since both are present in solution. It should also be noted that regardless of pore size, the exterior surface of silica particles will be modified with monomeric and polymeric groups. This could explain why even small-pore polymeric bonded phases have enhanced selectivity toward the separation of PAH.

To further examine the nature of polymeric phase syntheses, an experiment was designed to study the reactive silane polymer in solution [30]. The conditions for a polymeric phase synthesis were duplicated with one exception—no silica was added to the reaction mixture. Specifically, 10 ml octadecyltrichlorosilane was added to 100 ml of CCl_4. This solution was heated to boiling, at which time 0.5 ml water was added. A white precipitate formed immediately upon addition of the water. Aliquots of the reaction slurry were taken at regular intervals. Analysis of these samples by size-exclusion chromatography revealed the presence of a high-molecular-weight component. The elution volume for this component corresponded to that for a polystyrene standard of molecular weight of approximately 5000. Because size-exclusion chromatographic separations occur on the basis of hydrodynamic volume and not molecular weight, this value is only an estimate of the molecular weight of the silane polymer. A plot of relative concentration of this polymer as a function of reaction time (from the addition of water) is shown in Fig. 15. Polymer concentration increases rapidly during the first 10 min and then levels off. During this initial period, the white precipitate is visibly observed to dissolve, leaving a separated water layer and a clear solution.

A polymeric phase was prepared using a similar solution of the silane polymer in CCl_4. In this reaction, unlike the water-slurry synthesis previously described, the silica was added to the reaction mixture after refluxing for 15 min. At this point, effectively all of the white precipitate had dissolved. A column was prepared from the bonded silica and evaluated using the three-component "phase type" test mixture. Interestingly enough, the selectivity and retention of this column for PAH was nearly identical to monomeric phases prepared on the same substrate. It is thus clear that most or all of the polymeric surface modification occurs during the first few minutes of the reaction, during the time in which the silane polymer precipitate is visible. Although polymeric bonded-phase synthesis may seem a haphazard process, in practice the reactions are highly reproducible.

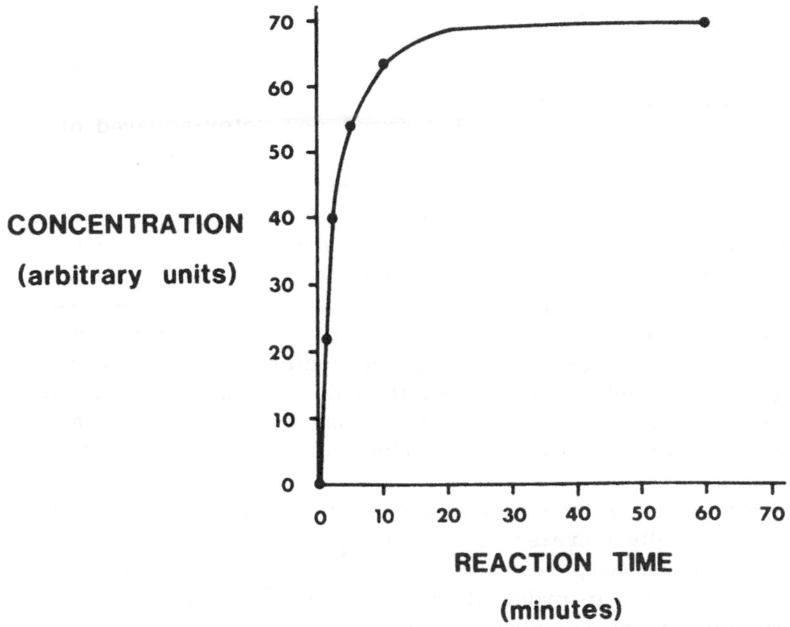

Figure 15 Concentration of silane polymer in solution plotted as a function of reaction time after the initial addition of water to the reaction mixture (*water-slurry synthesis*). (From Ref. 30.)

V. THE EFFECT OF SURFACE COVERAGE ON SELECTIVITY

A. Changes in Selectivity Versus Surface Coverage

To investigate the effect of C_{18} surface coverage on PAH selectivity, the C_{18} surface concentrations of the seven different commercial materials, which were previously evaluated for retention characteristics (Table 1), were determined [35]. The specific surface area, carbon content, and the calculated C_{18} surface concentrations are are summarized in Table 2. The surface concentrations were calculated from the measured carbon contents and the specific surface areas of the C_{18} modified silica (Eq. 5) as well as from the surface areas for the underivatized silica (values obtained from manufacturers' literature) (Eq. 3):

$$N^*(\mu mol/m^2) = 10^6 \frac{P_c}{1200 n_c S^*} \tag{5}$$

where P_C and n_C are as defined in Eq. (3), and S* is the surface
area of the substrate after bonding. Surface concentration as defined by Eq. (3) is based on micromoles of ligand per meter-squared
of underivatized silica, whereas Eq. (5) provides a surface concentration based on micromoles of bonded ligand per meter-squared of
chemically modified silica. Equation (3) is the most widely accepted
method for the determination of surface concentration. The physical
meaning of surface concentration as determined by Eq. (5) is not
completely understood, because of questions about the theoretical
justification of S*. However, surface concentration values from Eq.
(5) are often useful to observe trends when comparing different commercial phases where the manufacturers' nominal surface area values
for the underivatized silica may provide misleading results. In this
paper, surface area values determined from Eq. (5) will be followed
by an asterisk to differentiate them from those determined by the
more widely used Eq. (3) (N and N* values should not be directly
compared).

As shown in Table 2, the k' values for benzo[a]pyrene and benzo-
[b]chrysene generally increase as the surface concentration increases,
as expected. In the comparison of different columns in Table 1, the
columns were arranged in order of increasing surface coverage.
Some interesting trends are observed when these data are analyzed
in this manner. Of particular interest is the separation of the isomeric pair benz[a]anthracene and chrysene. For the monomeric materials, chrysene elutes prior to benz[a]anthracene with nearly baseline resolution on the material with lowest surface coverage. As the
surface coverage increases, these two compounds elute closer together
until they reverse elution order on the two polymeric materials and
are completely resolved on the polymeric material on the wide-pore
silica. This same trend can be observed in Fig. 10 for the comparison of monomeric, oligomeric, and polymeric phases. For the PAH
pairs of anthracene/phenanthrene, pyrene/fluoranthene, benzo[e]-
pyrene/benz[a]anthracene, the selectivity factors increase from low
to high surface coverage (see Table 1). The behavior of the pairs
benzo[ghi]perylene/dibenz[a,h]anthracene and indeno[1,2,3-cd]-
pyrene/benzo[b]chrysene are unique on the polymeric C_{18} on wide-
pore silica.

To evaluate the effect of surface coverage on selectivity, seven
different bonding lots of 5-μm wide-pore polymeric C_{18} materials from
one manufacturer were studied. The results of the percent carbon
analyses and the bonded-phase surface-area measurements for these
seven lots are summarized in Table 5. Lots 12, 15, and 16 were different bonding lots of the same batch of silica, whereas lots 10, 11,
13, and 14 represented different lots of silica as well as different
bonding lots [46]. The manufacturer indicated that lots 10 and 11
had low carbon loadings (based on incomplete resolution of benz[a]-
anthracene and chrysene), whereas lots 15 and 16 had high carbon

Table 5 Physical Characteristics for Different Lots of 5-μm Polymeric C_{18} Material

Column (lot)	Percent carbon	Surface area (m^2/g)	Surface coverage $(\mu mol/m^2)$ Eq. 3^a N	Eq. 5 N*	k' (BaP)
10	4.5 ± 0.3	48.5	2.7 ± 0.2	4.3 ± 0.3	2.7
11	7.5 ± 0.2	61.0	4.7 ± 0.1	5.7 ± 0.2	3.3
12	7.7 ± 0.4	60.0	4.8 ± 0.1	5.9 ± 0.3	3.5
13	8.0 ± 0.1	61.6	5.0 ± 0.1	6.0 ± 0.1	4.0
14	8.0 ± 0.3	50.1	5.0 ± 0.1	7.4 ± 0.3	5.0
15	9.3 ± 0.1	54.4	6.0 ± 0.1	8.2 ± 0.1	5.6
16	8.1 ± 0.3	48.3	5.1 ± 0.2	7.8 ± 0.3	6.3
17 (monomeric)	4.0 ± 0.1	53.4	2.1 ± 0.1	3.1 ± 0.1	0.9

[a] An average value of 84 m^2/g was assumed for the surface area of the underivatized silica.
Source: Refs. 34 and 35.

loadings. These four lots were rejected by the manufacturer's normal criteria for selectivity in PAH separations. Lot 14 was designated as an intermediate loading; the remaining two lots, 12 and 13, are representative of those currently commercially available as 5-μm polymeric C_{18} material from the manufacturer. These lots provide selectivity similar to that shown in Fig. 10 for the polymeric phase prepared in our laboratory as described earlier.

As shown in Table 2, the k' values for benzo[a]pyrene (BaP) generally increase with increasing surface coverage. The selectivity factors, α, for selected PAH relative to benzo[a]pyrene versus the k'_{BaP} on the various columns are shown in Figs. 16 and 17. The selectivity factors on the monomeric column are indicated on the far left side to illustrate the differences in selectivity on a monomeric phase. The position of the x-axis point for the monomeric phase does not represent the k'_{BaP} on the monomeric columns (which is actually similar to lot 14 of the polymeric C_{18} material in Table 3, i.e., k' = 5.5), but is only representative of a high-surface-coverage monomeric material (Zorbax ODS). The monomeric and polymeric columns used to obtain the selectivity data in Figs. 16 and 17 had comparable efficiencies (i.e., 20,000 plates/column). A difference of approximately 0.07 in α values provides baseline resolution of the two components.

Some interesting trends are observed in Fig. 16, particularly between monomeric and polymeric phases, and these results clarify some

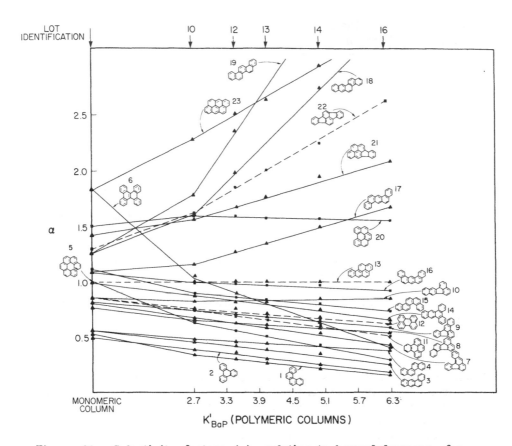

Figure 16 Selectivity factors (α), relative to benzo[a]pyrene, for PAH on five different polymeric C_{18} phases and a monomeric C_{18} phase. (From Ref. 35.)

conflicting literature reports of PAH elution sequence on various C_{18} columns. The resolution of benz[a]anthracene (no. 3) and chrysene (no. 4) has often been viewed as a difficult separation on many columns. As illustrated in Fig. 16, these isomers are unresolved on the monomeric column and α values increase as the surface concentration values (which are related to k'_{BaP}) increase on the polymeric C_{18} materials. As shown earlier by the data in Table 1, these two compounds can be separated with a reversed elution order on a low-surface-coverage column (see column E in Table 1). Benzo[c]phenanthrene (no. 1) and triphenylene (no. 2), isomers of molecular weight 228, have reversed elution order on the monomeric phase as compared to the polymeric phase. Of particular interest in the determination of PAH in environmental mixtures, is the separation of

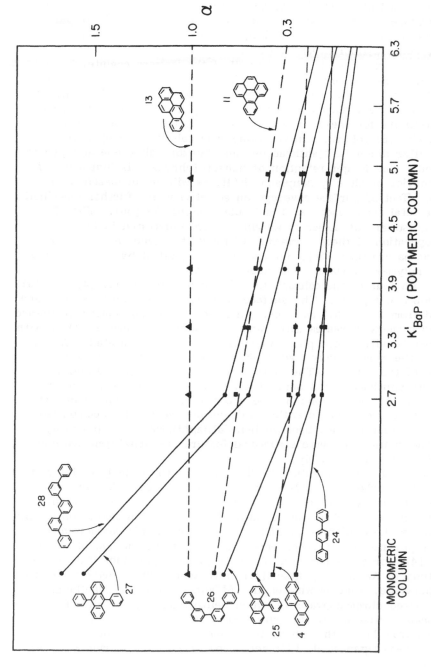

Figure 17 Selectivity factors (α), relative to benzo[a]pyrene, for polyphenyl arenes on five different polymeric C$_{18}$ phases and a monomeric C$_{18}$ phase. (From Ref. 35.)

isomers of molecular weight 252 (compounds 7 through 13). On the high-surface-coverage material, all seven isomers are at least partially resolved. On the low loading (lot 10) and the monomeric phase, benzo[a]fluoranthene (no. 7) and benzo[j]fluoranthene (no. 8) reverse elution sequence, and several of the isomers coelute.

Other isomers of interest are those of molecular weight 278. Selectivity data for six of these isomers are shown in Fig. 16. All six isomers were resolved on the polymeric C_{18} columns, and the greatest selectivity for these six isomers was achieved on the heavily loaded (lot 16) column. On the monomeric column, benzo[c]chrysene elutes after dibenz[a,h]anthracene; and benzo[b]chrysene and picene coeluted. The longer retention of benzo[c]chrysene is indicative of the behavior of slightly nonplanar PAH, as will be discussed later.

The effect of surface coverage on selectivity was further examined for the total set of isomers of molecular weight 278 [36]. The 12 possible isomers of molecular weight 278 are illustrated in Fig. 18. The separation of these isomers is difficult if conventional monomeric C_{18} phases are used [36], but can be accomplished by using heavily loaded polymeric phases. The separation of 11 of the 12 possible isomers on four of the columns from Table 5 (lots 10, 12, 14, and 16) are shown in Fig. 19. The surface-coverage values for these columns ranged from 2.7 to 5.1 $\mu mol/m^2$. In general, column selectivity toward the 278 isomers increases with increasing surface coverage. Only with the most heavily loaded phase were all 11 isomers separated. As expected, the more heavily loaded phases had the highest absolute retention of the solutes. The elution order of components 9 and 10, benzo[a]naphthacene and benzo[b]chrysene, changes with surface coverage. A similar trend has been observed for the four-ring catacondensed isomers benz[a]anthracene and chrysene. This comparison is of particular interest because benz[a]naphthacene and benzo[b]-chrysene contain the same structural features as benz[a]anthracene and chrysene, respectively.

In Fig. 16, the separation of dibenz[a,h]anthracene [no. 17] and benzo[ghi]perylene [no. 20] is of particular interest. These compounds are easily resolved on monomeric C_{18} phases, as shown in Fig. 16 and reported in the literature [5,8—11]. However, conflicting reports on the elution order of this PAH pair on this commercial polymeric C_{18} material have been reported. In the LC method of Ogan et al. [47] for the separation of the 16 priority pollutant PAH and in the study of Amos [11], dibenz[a,h]anthracene eluted prior to benzo-[ghi]perylene, whereas Wise et al. [5] and Colmsjö and MacDonald [10] showed chromatograms in which benzo[ghi]perylene eluted first. Presumably, Wise et al. [5], and Colmsjö and MacDonald [11] used columns from lots with high surface coverage (such as lot 16 in Fig. 16), whereas Ogan et al. [47] and Amos [11] used material with low surface coverage.

PEAK ID	STRUCTURE	IDENTIFICATION	L/B
(1)	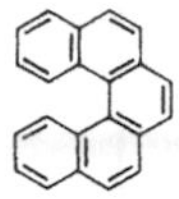	Dibenzo[c,g]phenanthrene	1.12
(4)	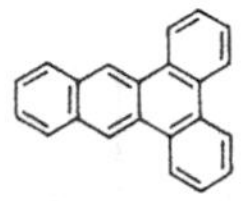	Dibenz[a,c]anthracene	1.24
(2)	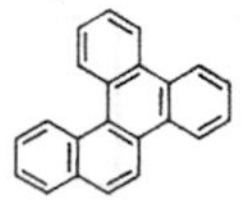	Benzo[g]chrysene	1.32
(3)	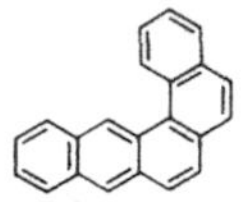	Dibenzo[b,g]phenanthrene	1.33
(5)	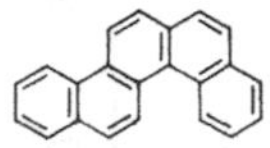	Benzo[c]chrysene	1.47
(6)	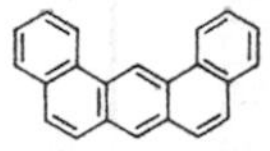	Dibenz[a,j]anthracene	1.47
(7)	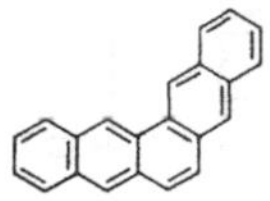	Pentaphene	1.73
(9)	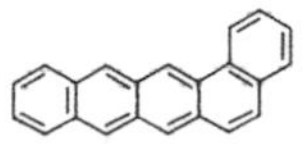	Benzo[a]naphthacene	1.77
(8)	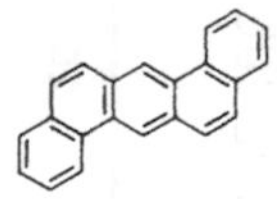	Dibenz[a,h]anthracene	1.79
(10)	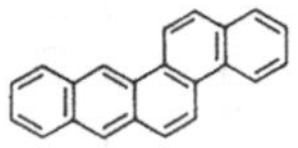	Benzo[b]chrysene	1.84
(11)	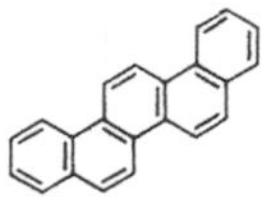	Picene	1.99
(12)	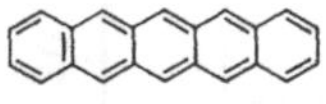	Pentacene	2.22

Figure 18   PAH isomers of molecular weight 278 and their corresponding length-to-breadth ratios.   (From Ref. 36.)

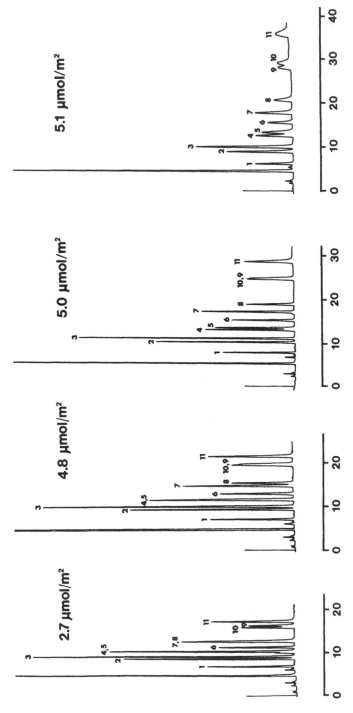

Figure 19 Reversed-phase separation of PAH isomers of molecular weight 278 on four polymeric C18 columns (columns 10, 12, 14, and 16 in Table 5) with increasing C18 loading (see Fig. 18 for peak identifications). Mobile phase: linear gradient from 85% acetonitrile in water to 100% acetonitrile at 1%/min. (From Ref. 36.)

Differences in selectivity for PAH on different lots of this polymeric C_{18} material have been reported by Katz and Ogan [9]. They compared the selectivity factors, α, for several PAH on different lots of HC-ODS material and several other C_{18} materials. HC-ODS and Vydac 201TP are similar materials made by the same manufacturer. Their work showed that HC-ODS lots with low surface coverage (4.8* $\mu mol/m^2$) provided very short retention times for PAH and exhibited selectivity factors for several PAH pairs which were similar to other C_{18} columns (primarily monomeric materials); whereas the HC-ODS materials with medium surface coverage (7.3* $\mu mol/m^2$) exhibited suitable selectivity characteristics for their PAH separations [9,48]. One lot of HC-ODS material with high surface coverage (8.2* $\mu mol/m^2$) exhibited very long retention times and a reversal of the elution order of dibenz[a,h]anthracene and benzo[ghi]perylene. As a result of their studies, lots of HC-ODS material that have certain selectivity characteristics are selected for columns designed specifically for the separation of the 16 priority pollutant PAH [9,49]. However, earlier columns of 10 μm HC-ODS and Vydac 201TP were not screened for selectivity; thus these reports of different selectivity have appeared in the literature. At present, 5-μm Vydac 201TP columns are screened to provide columns that have selectivity similar to that of lots 12 and 13 in Table 2. In general, the selectivity criteria used to screen these materials is based on the separation of two pairs of PAH, i.e., (1) benz[a]anthracene and chrysene and (2) benzo[ghi]perylene and dibenz[a,h]anthracene.

The unusual behavior of nonplanar PAH such as phenanthro[3,4-c]phenanthrene (hexahelicene) (no 5) and 1,2:3,4:5,6:7,8-tetrabenzonaphthalene (no. 6) is illustrated in Fig. 16. For these two compounds, the α values (relative to the planar benzo[a]pyrene) increase as the surface coverage decreases on the polymeric phases, and the α values increase significantly on the monomeric phase. Because the α values for these compounds are very sensitive to changes in the phase type and the surface coverage, these compounds were selected as part of a chromatographic test mixture to characterize C_{18} phases as described earlier.

The effect of solute planarity is also illustrated in Fig. 17 for several polyphenyl arenes. As the surface coverage decreases, the α values for the polyphenyl arenes increase and the retention (as indicated by α) on the monomeric column is significantly longer, as compared to the fused-ring PAH. The retention characteristics of 13 polyphenyl arenes on a monomeric and a polymeric C_{18} material have been reported [4]. The polyphenyl arenes are generally retained longer, relative to the fused-ring PAH, on the monomeric than on the polymeric C_{18} materials, e.g., m-quaterphenyl, 1,2'-binaphthyl, 2,2'-binaphthyl 9,9'-bianthryl, 9-phenylanthracene, 1-phenylanthracene, and 9-phenylphenanthrene all elute prior to benz[a]anthracene on the polymeric C_{18} phase, but after it on the monomeric C_{18} phase.

Reversed-phase LC on C_{18} phases provides excellent selectivity for the separation of methyl-substituted PAH. The selectivity factors for the methyl-substituted chrysene isomers on the polymeric columns of varying surface coverage are shown in Fig. 20. As with the parent PAH, the longer isomers have increasing α values with the increasing surface coverage. Polymeric C_{18} materials generally provide more resolution of individual methyl isomers than do the monomeric materials [4]. However, as illustrated in Fig. 20, even though the resolution of individual methyl isomers improves, the parent compound elutes closer to the methyl isomers as the surface coverage increase. These same trends have also been observed for methylbenzo[a]pyrene and methylbenzo[b]naphtho[2,1-d]thiophene isomers.

B. Mixed C_{18} Phases for Modification of Selectivity

Because column selectivity is dependent upon bonded-phase surface coverage, significantly different separations of PAH can result not only among columns from different manufacturers, but also among

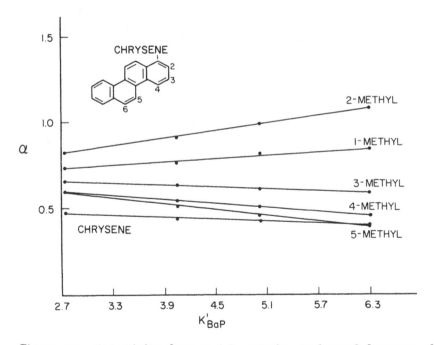

Figure 20 Selectivity factors (α), relative to benzo[a]pyrene, for chrysene and five methylchrysene isomers on four different polymeric C_{18} phases. (From Ref. 35.)

columns from different lots from a single manufacturer. Numerous examples of such selectivity differences, including shifts in relative retention and changes in elution order, have been discussed [5,8—11, 49]. It is clear that no single column will be suitable for the separation of all PAH. In general, heavily loaded polymeric phases give the best overall separation of PAH, but it is possible that any specific solute pair may be better resolved on a phase with lower surface coverage.

The feasibility of controlling selectivity by combining two or more dissimilar bonded phases has recently been investigated [34]. Two approaches were taken. In the first case, columns were prepared by blending 5-μm C_{18} polymeric phases from different production lots. High- and low-surface-coverage phases were blended in the proportions 30/70, 50/50, and 70/30 (w/w). Columns of each individual phase were also prepared. In the second set of experiments, a series of short columns of appropriate lengths were prepared that contained the (unblended) polymeric phases. The columns were coupled together in combinations so as to produce phase proportions similar to the blended phases.

The reversed-phase separation of 7 PAH on the blended C_{18} phases is shown in Fig. 21. The lot designations are those assigned in Table 5. Lot 11 is a low-loaded polymeric phase (4.5* μmol/m^2) prepared on a wide-pore low-surface-area silica substrate. Lot 15 refers to a polymeric phase with high loading (5.9* μmol/m^2), and lot 17 is a monomeric phase (2.1* μmol/m^2), also prepared on the same silica substrate. As might be expected, the worst separation of the PAH mixture occurs with the monomeric C_{18} phase (lot 17). The best separation, however, does not occur for the high-loaded polymeric phase (lot 15). As can be observed from Fig. 21, components 1 and 4 (chrysene and tetrabenzonaphthalene) and 6 and 7 (dibenz[a,h]-anthracene and benzo[ghi]perylene) are incompletely resolved on the high-loaded phase. Complete separation of all seven components is obtained on the column prepared from a 70/30 blend of lots 11 and 15. The two other blended-phase columns (30/70 and 50/50) did not completely resolve all the components.

The elution order of several PAH can be observed to vary as a function of overall phase loading. As discussed earlier, the selectivity factors of components 2 and 4 relative to planar PAH (e.g., benzo[a]pyrene) can be used as an indicator of phase coverage. Thus, the blended phases exhibit selectivities toward the planar and nonplanar PAH that are similar to comparably loaded homogeneous polymeric phases.

The selectivities of "blended" and "coupled" columns were compared for a series of selected PAH. Two short columns (12.5 cm) were prepared, one containing the low-loaded polymeric phase (lot 11) and the other the high-loaded phase (lot 15). The blended-phase

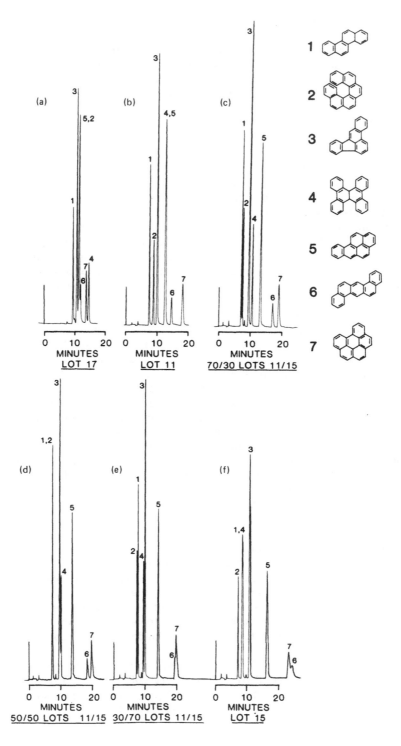

Table 6 Selectivity Factors (α) for Selected PAH on
Mixed versus Coupled Columns

Compound	50/50 Lots 11 & 15	
	Blended	Coupled
m-Tetraphenyl	0.288	0.291
Phenanthro[3,4-c]phenanthrene	0.436	0.438
9,10-Diphenylanthracene	0.454	0.459
m-Quinquephenyl	0.557	0.564
Tetrabenzonaphthalene	0.676	0.676
Dibenz[a,c]anthracene	0.728	0.730
Dibenz[a,j]anthracene	0.939	0.941
Dibenz[a,h]anthracene	1.42	1.43

Source: Ref. 34.

column was prepared from a 50/50 mixture of lots 11 and 15. Selectivity factors for the compounds are listed in Table 6. The results indicate that column selectivity for "combined phases" does not depend on how the phases are mixed—either by blending the phases or by coupling short columns. The use of coupled columns provides an easy method for adjusting column selectivity for chromatographers that do not prepare their own bonded phases. A collection of short columns with different bonded-phase surface coverages could be coupled together to give overall column selectivities that would be appropriate for a wide range of PAH mixtures.

VI. THE EFFECT OF SOLUTE CHARACTERISTICS

A. Shape of Solute

In addition to investigations of the influence of stationary-phase parameters on retention and selectivity in reversed-phase LC, other studies have examined the correlation of reversed-phase retention data for PAH with various molecular structure descriptors. Sleight [50] studied structure-retention relationships of PAH on a C_{18} column

Figure 21 Reversed-phase LC separation of selected PAH on a monomeric C_{18} column from lot 17 (a) and on polymeric C_{18} columns from lot 11 (b) and lot 15 (f) (see Table 5 for information on different lots) and from mixtures of these two polymeric lots [70/30 (c), 50/50 (d), and 30/70 lot 11/15 (e); (w/w)]. (From Ref. 34.)

and derived a simple expression relating the retention to the number of carbons for unsubstituted PAH. Recently, we described a relationship between the shape of PAH solutes, defined as the length-to-breadth ratio (L/B), and reversed-phase LC retention [4].

In gas chromatography, Janini, et al. [51–53] had noted that the retention of PAH on liquid-crystal stationary phases correlated with the shape, specifically the length-to-breadth ratio, of the molecule. They found that for isomeric PAH, those solutes having greater length-to-breadth ratios were retained longer on the liquid crystal phases. Radecki et al. [54] discussed the relationship between GC retention indexes on liquid crystal phases and the shape of the PAH. These workers used two parameters, the connectivity index, χ, and a shape parameter, η, to develop this relationship. The connectivity index is a topological parameter that describes the size of the molecule. The shape parameter, η, was defined as the ratio of the longer side to the shorter side of a rectangle having a minimum area which enclosed the molecule. Kaliszan and Lamparczyk [55] found a correlation between retention indexes on a nonpolar GC phase and the connectivity index.

1. Unsubstituted PAH

In comparing the retention and selectivity characteristics of various C_{18} phases, we found that the polymeric C_{18} phases on wide-pore silicas provided excellent selectivity for the separation of PAH isomers, including methyl-substituted isomers. To examine the effect of PAH shape on the selectivity, the length-to-breadth ratios (L/B) of a number of PAH were determined in a manner similar to that described by Radecki and co-workers [54]. The calculation of the L/B ratio is illustrated in Fig. 22 for 2-methylchrysene. The dimensions of the PAH were determined using the approximate bond lengths: (A) C-C aromatic (1.4 Å), (B) C-C aromatic to aliphatic (1.54 Å), and (C) C-H (1.1 Å). The van der Waals radius of the hydrogen atoms (r) was 1.2 Å. More accurate bond lengths from crystallographic data were used for the calculation of L/B of fluoranthene and benzo[a]pyrene. Radecki et al. [54] used a shape parameter based on the rectangle with minimum area that could envelope the molecule. In our study, the rectangle enclosing the PAH molecule which maximizes the length-to-breadth ratio, was used. The shape parameter, η, used by Radecki is equivalent to the maximized L/B ratio used in our study [4] except for several PAH. In all cases for the unsubstituted PAH studied, the predicted elution orders based on η and the maximum L/B ratio were identical. This was not the case, however, for the methyl-substituted PAH. Because the maximum length-to-breadth ratio was better in predicting the elution order for the methyl-substituted PAH, it was used in our studies instead of the Radecki shape parameter (see Ref. 4 for the comparison of the L/B ratios calculated by both methods).

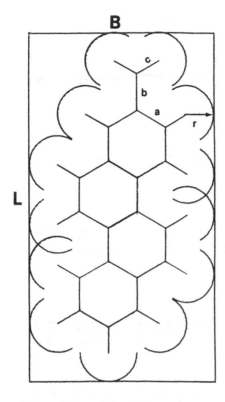

Figure 22 Different bond types used in the calculation of the length-to-breadth ratio for 2-methylchrysene. a = 1.4 Å (C-C aromatic), b = 1.54 Å (C-C aromatic to aliphatic), c = 1.1 Å (C-H single bond), and r = 1.2 Å. (From Ref. 4.)

The L/B ratios and the LC retention on both a polymeric and monomeric C_{18} phase for a large number of unsubstituted PAH are summarized in Table 7. In general, when isomeric PAH containing only catacondensed six-membered aromatic rings are compared, the LC retention increases with increasing L/B ratios. The exceptions to this trend, e.g., benzo[c]phenanthrene, benzo[g]chrysene, and dibenzo[b,g]phenanthrene, are slightly nonplanar. The effect of this nonplanarity will be described later.

The PAH containing a fluoranthenic ring (e.g., the benzofluoranthenes, indeno[1,2,3-cd]pyrene, and the dibenzofluoranthenes) have elution orders consistent with the L/B values when compared to each other. The elution order of the nonfluoranthenic PAH of equivalent molecular weight is also consistent with the L/B ratios. However, when the two groups are combined (i.e., fluoranthenic and

Table 7 Summary of Length-to-Breadth Ratios and Reversed-Phase LC Retention for Unsubstituted PAH

Compound	MW	L/B	LC retention (log I)		GC retention		
			Polymeric	Monomeric	BMBT[a]	BBBT[b]	PS[c]
Fluorene	166	1.57	2.70	2.75	2456		
Phenanthrene	178	1.46	3.00	3.00	2800		
Anthracene	178	1.57	3.20	3.14	2874		
Fluoranthene	202	1.22	3.37	3.43	3204		
Pyrene	202	1.27	3.58	3.65	3301		
Benzo[c]fluorene	216	1.34	3.49	3.64			
Benzo[a]fluorene	216	1.68	3.72	3.76			
Benzo[b]fluorene	216	1.78	3.84	3.77			
Triphenylene	228	1.12	3.70	3.83	4017		2
Benzo[c]phenanthrene	228	1.22	3.64	3.91			1
Benz[a]anthracene	228	1.58	4.00	4.00	4169		3
Chrysene	228	1.72	4.10	3.99	4198		4
Naphthacene	228	1.89	4.51	—			5
Benzo[e]pyrene	252	1.12	4.28	4.48	4650	1.00	
Perylene	252	1.27	4.33	4.50	4739	1.20	
Benzo[a]pyrene	252	1.50	4.53	4.58	4834	1.42	
Benzo[a]fluoranthene	252	1.16	4.22	4.45		0.88	
Benzo[j]fluoranthene	252	1.39	4.24	4.37			

Compound							Ref.
Benzo[b]fluoranthene	252	1.40	4.29	4.46		0.88	1
Benzo[k]fluoranthene	252	1.48	4.42	4.52	4629	0.96	6
Benzo[ghi]perylene	276	1.12	4.76	5.36	5262	3.7	3
Anthanthrene	276	1.35	5.08	5.61	5439		2
Indeno[1,2,3-cd]pyrene	276	1.40	4.84	5.23			4
Indeno[1,2,3-cd]fluoranthene	276	1.62	4.93	5.05			5
Dibenzo[c,g]phenanthrene	278	1.12	4.07	4.51			7
Dibenz[a,c]anthracene	278	1.24	4.40	4.73	5099	2.5	11
Benzo[g]chrysene	278	1.32	4.27	4.71			
Dibenzo[b,g]phenanthrene	278	1.33	4.33	4.80			
Benzo[c]chrysene	278	1.47	4.45	4.85			
Dibenz[a,j]anthracene	278	1.47	4.56	4.84			
Pentaphene	278	1.73	4.67	4.96			
Benzo[a]naphthacene	278	1.77	4.99	4.50			
Dibenz[a,h]anthracene	278	1.79	4.73	4.86	5325	5.5	8
Benzo[b]chrysene	278	1.84	5.00	5.00	5448	5.9	9
Picene	278	1.99	5.18	5.02			10
Dibenzo[j,l]fluoranthene	302	1.14	4.78	5.35			
Dibenzo[a,e]fluoranthene	302	1.14	4.90	5.50			
Dibenzo[b,e]fluoranthene	302	1.15	4.80	5.48			
Dibenzo[def,p]chrysene	302	1.18	4.89	5.50	1d		
Naphthol[1,2,3,4-def]chrysene	302	1.24	4.97	5.56	2d		
Dibenzo[a,k]fluoranthene	302	1.25	4.90	5.51			
Dibenzo[de,qr]naphthacene	302	1.28	4.92	5.53			
Benzo[b]perylene	302	1.38	5.04	5.55			
Dibenzo[b,k]fluoranthene	302	1.58	5.27	5.64			
Dibenz[e,k]acephenanthrylene	302	1.61	5.27	5.59			
Naphtho[1,2-k]fluoranthene	302	1.62	5.00	5.34			

Table 7 (continued)

Compound	MW	L/B	LC retention (log I)		GC retention		
			Polymeric	Monomeric	BMBT[a]	BBBT[b]	PS[c]
Naphtho[2,1,8-qra]naphthacene	302	1.69	5.87	5.92			
Dibenzo[a,i]pyrene	302	1.73	5.73	5.93	3[d]		
Dibenzo[b,def]chrysene	302	1.73	6.00	6.00	4[d]		
Naphtho[2,3-k]fluoranthene	302	1.74	5.92	5.79			

[a]Kovats retention indexes from Ref. 53.
[b]GC retention relative to benzo[e]pyrene from Ref. 4.
[c]GC elution orders on liquid crystal polysiloxane phase from Refs. 59 and 60.
[d]GC elution orders on BPhBT from Ref. 52.
Source: Refs. 4 and 22.

nonfluoranthenic with the same molecular weight), the elution order does not always follow L/B ratios. Fluoranthenic PAH are slightly nonplanar [56], which would tend to reduce the L/B ratios. An empirical observation can be made that if the L/B ratios for the benzofluoranthenes were all decreased by approximately 0.28, the ratios of the seven isomers of molecular weight 252 would be consistent with the LC elution order.

The largest group of nonsubstituted PAH isomers studied contained 11 isomers of molecular weight 278. The structures of the 12 possible catacondensed PAH isomers of molecular weight 278 are shown in Fig. 18. The separation of 11 of these 12 possible isomers on a polymeric C_{18} phase with a high surface coverage is shown in Fig. 23. The remaining isomer, pentacene, was not included in the mixture, due to limited solubility in the mobile phase. However, pentacene does elute significantly later than picene on a polymeric C_{18} column [57], as would be expected on the basis of the large L/B value.

In general, the elution order for these 11 isomers follows increasing L/B values. One isomer with an apparent anomalous behavior is dibenz[a,c]anthracene (L/B = 1.24). However, it might also be argued that benzo[g]chrysene (L/B = 1.32) and dibenzo[b,g]phenanthrene (L/B = 1.33), which both contain the nonplanar benzo-[c]phenanthrene structure, are eluting earlier than expected because of the solute nonplanarity. The same argument would explain the separation of benzo[c]chrysene and dibenz[a,j]anthracene, which both have L/B values of 1.47. Benzo[a]naphthacene (L/B = 1.77) has greater retention than would be predicted by the L/B value. This may indicate that the naphthacene structure (i.e., four rings with linear annulation) has greater influence on retention than the overall L/B value of the PAH.

2. Methyl-substituted PAH Isomers

The relationship between PAH shape and C_{18} retention is also illustrated by the elution characteristics of methyl-substituted isomers. In fact, it was this observation of the excellent selectivity of polymeric C_{18} phases for the separation of five of the six methylchrysene isomers illustrated in Fig. 24 that initiated our investigations of the effect of solute shape on LC retention. The position of the methyl group on the PAH can significantly change the length-to-breadth ratio of the solute. The LC retention data and L/B values for a large number of methyl-substituted PAH are summarized in Tables 8–10.

In Tables 8 and 9, the elution orders on the polymeric C_{18} column for the methyl-substituted fluorenes, anthracenes, phenanthrenes,

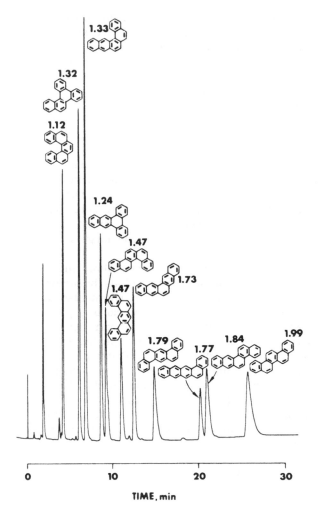

Figure 23 Reversed-phase separation of PAH isomers of molecular weight 278 on a polymeric C_{18} column (column 16 in Table 5). Mobile phase: linear gradient 85% acetonitrile in water to 100% acetonitrile at 1%/min. (From Ref. 36.)

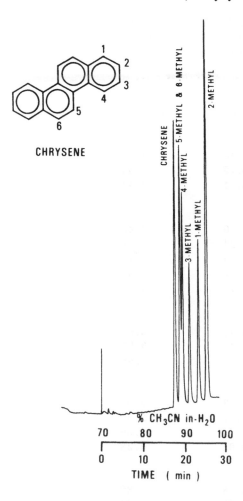

Figure 24 Reversed-phase LC separation of chrysene and methyl-chrysene isomers on polymeric C_{18} column. (From Ref. 5.)

Table 8 Comparison of Shape and LC Retention for Methyl-substituted Isomers of Fluorene, Phenanthrene, Anthracene, Fluoranthene, and Pyrene

Compound	L/B	Retention (log I)	
		Polymeric	Monomeric
Fluorene			
4-Methyl	1.35	3.10	3.34
1-Methyl	1.40	3.15	3.33
2-Methyl	1.73	3.24	3.42
Phenanthrene			
4-Methyl	1.25	3.26	3.40
9-Methyl	1.25	3.34	3.51
3-Methyl	1.37	3.29	3.46
1-Methyl	1.45	3.39	3.51
2-Methyl	1.58	3.71	3.71
Anthracene			
9-Methyl	1.38	3.41	3.53
1-Methyl	1.41	3.41	3.55
2-Methyl	1.71	3.71	3.69
Phenanthrene			
9,10-Dimethyl	1.23	3.63	3.90
3,6-Dimethyl	1.26	3.57	3.91
1,8-Dimethyl	1.46	3.79	3.99
2,7-Dimethyl	1.71	4.01	4.23
Fluoranthene			
1-Methyl	1.13	3.68	3.89
7-Methyl	1.22	3.76	3.94
3-Methyl	1.33	3.84	3.95
8-Methyl	1.36	3.80	3.99
Pyrene			
4-Methyl	1.10	3.96	4.12
1-Methyl	1.27	3.98	4.15
2-Methyl	1.38	4.05	4.19

Source: Ref. 4.

Table 9 Comparison of Shape and Chromatographic Retention of Methyl-substituted Isomers of Benzo[c]phenanthrene, Benz[a]anthracene, and Chrysene

| Compound | L/B | LC retention (log I) | | GC retention[a] | | |
		Polymeric	Monomeric	BBBT	BABT[b]	PS[c]
Benzo[c]phenanthrene						
6-Methyl	1.12	3.94	4.36			
2-Methyl	1.17	3.83	4.27			
1-Methyl	1.21	3.62	4.09			
5-Methyl	1.22	3.97	4.36			
3-Methyl	1.36	4.01	4.41			
4-Methyl	1.36	4.04	4.37			
Benz[a]anthracene						
6-Methyl	1.38	4.10	4.39	1.19	1.15	3
5-Methyl	1.43	4.28	4.48	1.43	0.91	6
11-Methyl	1.45	4.13	4.41	1.15	1.08	1
1-Methyl	1.47	4.14	4.38	1.40	0.89	4
2-Methyl	1.50	4.09	4.40	1.15	1.37	2
7-Methyl	1.50	4.14	4.35	1.62	1.43	8
12-Methyl	1.51	4.10	4.35	1.54	1.22	7
8-Methyl	1.57	4.19	4.39	1.39	1.24	5
10-Methyl	1.59	4.24	4.47		1.82	8
4-Methyl	1.64	4.33	4.46			
9-Methyl	1.71	4.39	4.53	1.78	1.97	9
3-Methyl	1.71	4.39	4.53	1.78	1.97	10

Table 9 (continued)

| Compound | L/B | LC retention (log I) | | GC retention[a] | | |
		Polymeric	Monomeric	BBBT	BABT[b]	PS[c]
Chrysene						
6-Methyl	1.48	4.14	4.36	1.37		1
5-Methyl	1.48	4.14	4.35	1.44		2
4-Methyl	1.51	4.18	4.35	1.62		3
3-Methyl	1.63	4.29	4.42	1.55		4
1-Methyl	1.71	4.43	4.46	1.92		5
2-Methyl	1.85	4.52	4.52	2.17		6

[a]Retention relative to benz[a]anthracene.
[b]From Ref. 62.
[c]Elution order on smectic biphenylcarboxylate ester liquid crystalline polysiloxane phase from Refs. 59 and 61.

pyrenes, fluoranthenes, benzo[c]phenanthrenes, and chrysenes are predicted from the L/B ratios, except for 3-methyl- and 9-methyl-phenanthrene, 8-methyl- and 3-methylfluoranthene, and 1-methyl-, 2-methyl-, and 6-methylbenzo[c]phenanthrene. The anomalous behavior of the 1-methylbenzo[c]phenanthrene may be explained by its significant deviation from planarity due to steric crowding of the methyl group in this position.

The linear correlation between the LC retention (polymeric phase) and the L/B ratios for 23 methyl-substituted isomers of benzo[c]-phenanthrene, benz[a]anthracene, and chrysene is illustrated in Fig. 25 (correlation coefficient = 0.936). For the six methylchrysenes, the correlation coefficient for the LC retention (polymeric phase) and the L/B ratios is excellent (correlation coefficient = 0.991). The correlation coefficient for the eight isomeric methyl-substituted phenanthrenes and anthracenes is 0.913. For the methylbenz[a]anthracenes the correlation between retention and L/B ratios is less than the other PAH studied (correlation coefficient = 0.805); however, the five isomers with the largest length-to-breadth ratios do have the greatest retention. Only the elution of the 5-methyl isomer differs significantly from that predicted by the L/B ratio.

The LC retention and L/B ratios for the 12 isomeric methylbenzo-[a]pyrenes are summarized in Table 10. The separation of 10 of these isomers on the polymeric C_{18} phase is illustrated in Fig. 26. The correlation coefficient for the LC retention indexes versus the L/B ratio for the methylbenzo[a]pyrenes is 0.914. As was the case with the unsubstituted PAH, the polymeric C_{18} phases provided greater selectivity and a greater correlation of retention and L/B ratios for all the isomeric sets than did the monomeric C_{18} phase.

3. Polyphenyl Arenes

The LC retention of polyphenyl arenes was also investigated in relation to the length-to-breadth ratios. These data are summarized in Table 11. The length-to-breadth ratios for the polyphenyl arenes were determined as for the other PAH, even though these compounds deviate from planarity because of rotation about the biaryl carbon-carbon bond. However this nonplanarity does not alter the ratios enough to change the predicted elution orders. As shown in Table 11, the elution order within a group of isomeric polyphenyl arenes on the polymeric C_{18} phase was correctly predicted by the increasing L/B ratio in all cases. The significant selectivity differences for these compounds on the monomeric and polymeric C_{18} phase are discussed in detail later.

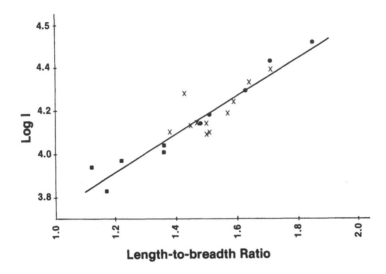

Figure 25 Linear correlation of length-to-breadth ratio versus LC retention on a polymeric C_{18} column for methyl-substituted benzo[c]-phenanthrenes (■), benz[a]anthracenes (×), and chrysenes (●). (From Ref. 4.)

Table 10 Comparison of Length-to-Breadth Ratios and Reversed-Phase Liquid Chromatographic Retention of Methyl-substituted Benzo[a]pyrenes

Compound	L/B	Retention (log I) Polymeric	Retention (log I) Monomeric	GC retention (BPhBT)[a]
Benzo[a]pyrene				
12-Methyl	1.30	4.59	5.04	
5-Methyl	1.30	4.61	5.04	1
11-Methyl	1.33	4.63	5.04	2
10-Methyl	1.41	4.69	5.06	3
6-Methyl	1.48	4.68	5.00	6
7-Methyl	1.50	4.72	5.04	4
4-Methyl	1.50	4.81	5.16	
3-Methyl	1.50	4.89	5.15	8
1-Methyl	1.51	4.82	5.12	5
9-Methyl	1.51	4.77	5.12	4
2-Methyl	1.60	4.94	5.23	7
8-Methyl	1.62	4.95	5.22	7

[a]GC elution order from Ref. 63.
Source: Ref. 4.

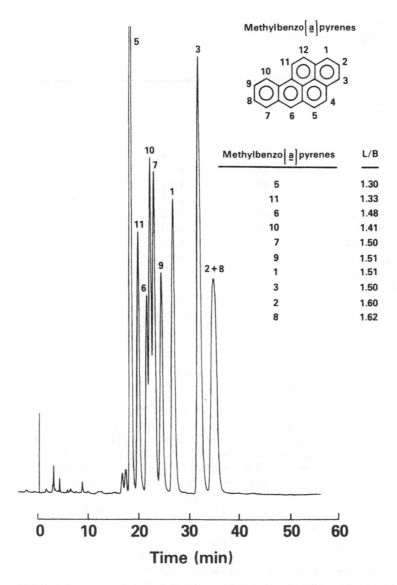

Figure 26 Reversed-phase LC separation of methyl-substituted benzo-[a]pyrene isomers on a polymeric C_{18} column. Column: Vydac 201TP 5μm; mobile phase: 90% acetonitrile in water at 1 ml/min. (From Ref. 4.)

Table 11   Comparison of Length-to-Breadth Ratios and LC
Retention for Polyphenyl Arenes

Compound	L/B	Retention (log I)	
		Polymeric	Monomeric
o-Terphenyl	1.11	2.97	3.31
m-Terphenyl	1.47	3.16	3.56
p-Terphenyl	2.34	3.74	3.75
1,1'-Binaphthyl	1.34	3.35	3.92
1,2'-Binaphthyl	1.49	3.56	4.13
2,2'-Binaphthyl	2.02	4.05	4.23
9-Phenylanthracene	1.00	3.55	4.21
9-Phenylphenanthrene	1.33	3.58	4.20
1-Phenylphenanthrene	1.47	3.65	4.23
9,9'-Bianthryl	1.07	3.88	4.44
9,9'-Biphenanthryl	1.51	4.19	5.30
m-Quaterphenyl	1.86	3.81	4.49
p-Quaterphenyl	2.97	5.04	5.48

Source: Ref. 4.

B.   Comparison of Reversed-Phase LC and GC on Liquid Crystal Phases

Most workers agree that GC separations on liquid crystal phases are
based to some degree on the shape of the molecule.   Zielinski and
Janini [53] and Radecki et al. [54] correlated the relative length-to-
breadth ratio for PAH isomers with the relative GC retention on
nematic liquid crystal phases.   More recently, Lee and co-workers
[58—61] have reported GC separations of various PAH on several new
nematic and smectic liquid crystalling phases.

Relative GC retention data (or elution orders) on several liquid
crystalline phases are listed in Table 7 for a number of the unsubstituted PAH.   In general, the GC elution orders for the various
isomers are in agreement with the reversed-phase LC elution orders.
This is of particular interest in cases where the reversed-phase LC
elution order did not strictly follow increasing L/B values.   For example, in the case of the five isomers of molecular weight 228, the
LC elution order of benzo[c]phenanthrene and triphenylene does not
follow L/B values.   However, the same anomalous elution order is observed in GC on liquid crystalline phases.

Of particular interest are the GC separations of the PAH isomers
of molecular weight 278 on a conventional SE-54 phase and a smectic
liquid crystalline phase, as described by Lee and co-workers [60]

and shown in Fig. 27. The elution order is generally in agreement
with that obtained in reversed-phase LC on a polymeric C_{18} phase
(see Table 7 and Fig. 23). As was observed in LC, the GC retention of dibenz[a,c]anthracene and benz[a]naphthacene on the liquid
crystal phase was also greater than would be predicted on the basis
of L/B values. It is interesting to note that capillary GC on a conventional stationary phase was unsuccessful in separating all of these
isomers.

Liquid-crystalline-phase GC retention data and chromatograms for
methyl-substituted PAH have been reported [4,61—63] and are summarized in Tables 8, 9, and 10. The separation of the six methyl-
chrysene isomers is an excellent example of the correlation of L/B
and GC retention on a liquid crystalline phase. Retention on a
nematic liquid crystal phase (BBBT in Table 9) does not strictly
follow increasing L/B, i.e., 3-methylchrysene (L/B = 1.63) elutes
before 4-methylchrysene (L/B = 1.51) with only partial separation,
whereas the LC elution order on the polymeric C_{18} phase does follow
increasing L/B. Recently however, Markides et al. [59] reported
the complete resolution of all six isomers on a new smectic liquid
crystal phase, with the elution order in agreement with increasing
L/B values.

The separation of the methylbenz[a]anthracene isomers has been
reported in several papers [4,61,62], and the data are given in
Table 9. In GC, the methylbenz[a]anthracene isomers generally elute
in a similar elution order as with reversed-phase LC; the isomers that
do not elute in agreement with L/B in LC also show similar behavior
in GC with liquid crystalline phases, e.g., 5-methylbenz[a]anthracene,
which elutes later than would be predicted by L/B, also elutes late
in GC. The correlation coefficient for GC and L/B for two of the
data sets in Table 9 were 0.766 [4] and 0.881 [62] compared to
0.805 for the LC retention.

Issaq et al. [63] compared reversed-phase LC separation of 10
isomers of methylbenzo[a]pyrene with GC separation on two nematic
liquid crystal stationary phases. The GC elution order of these 10
isomers of methylbenzo[a]pyrene was similar to the elution order in
LC (see Table 10), with the exception of the 6-methyl isomer eluting
after the 1-methyl isomer and the 3-methyl isomer eluting after the
8-methyl isomer. In both cases, these differences in the GC elution
order versus the LC elution order provide less correlation between
elution order and L/B values.

C. Solute Planarity

Previous studies [4,36] have indicated that the retention behavior of planar and nonplanar solutes differs considerably on monomeric and polymeric phases. The property of planarity (or nonplanarity) can be considered an extension of the molecular descriptor L/B to include a thickness

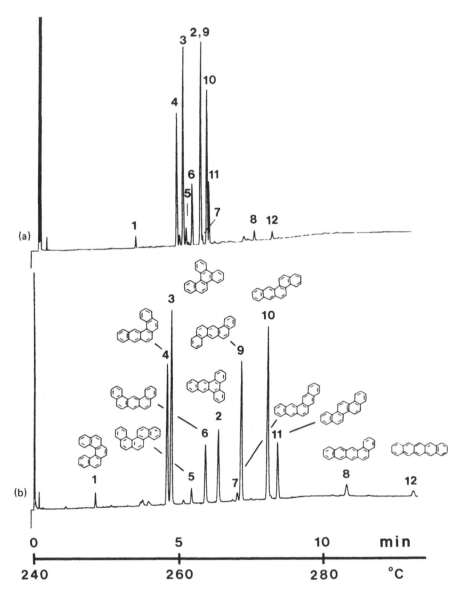

Figure 27   Gas chromatographic separation of 12 PAH isomers of molecular weight 278 on (a) SE-54 and (b) a smectic liquid crystalline phase.   (From Ref. 60.)

parameter. Planar and nonplanar PAH of similar molecular weights and with similar L/B ratios can often be separated on reversed-phase C_{18} columns. The retention mechanisms that control such separations are not well understood, but several explanations have been proposed and will be discussed later.

To study the effect of nonplanarity on solute retention, a series of planar and nonplanar PAH of similar size and shape were selected (see Fig. 28). For example, phenanthro[3,4-c]phenanthrene has a nonplanar, helical conformation due to steric hindrance of hydrogens on opposing rings of the molecule. Coronene, on the other hand, has no opposing hydrogens and is planar. The retention behavior of linear and nonlinear molecules was also studied using positional isomers of quaterphenyl (tetraphenyl) and terphenyl (Fig. 28).

Four columns were used in these retention studies: a monomeric C_{18} phase prepared on narrow-pore Polygosil silica, and three polymeric C_{18} phases with different phase loadings, prepared on wide-pore Vydac TP silica. Percent carbon loadings, surface areas, and surface coverage values for the columns are listed in Table 12. Each of the solutes was chromatographed individually under mobile-phase conditions of 85% or 100% acetonitrile in water. Selectivity factors for the planar/nonplanar and linear/nonlinear solute pairs are summarized in Table 13. In every instance, planar and linear compounds were retained longer than their nonplanar and nonlinear analogues. Selectivity factors were generally small for separations performed on the monomeric phase at both mobile-phase compositions. Column selectivity increased significantly for the polymeric phases, with further increases occurring as a function of phase loading. As an example of how differently the monomeric and (heavy) polymeric phases behave, at 100% acetonitrile, the selectivity factors for coronene/phenanthro[3,4-c]phenanthrene are 3.4 and 17.1, respectively. For p-quaterphenyl/m-tetraphenyl, the values are 1.4 and 34.5, a factor of about 25 difference!

Because the retention of planar and nonplanar compounds depends strongly on phase type and surface coverage, mixtures of these compounds make excellent probes for the evaluation of phase properties. The three-component mixture, benzo[a]pyrene (BaP), phenanthro[3,4-c]phenanthrene (PhPh), and tetrabenzonaphthalene (TBN), described earlier, consists of planar and nonplanar compounds. BaP is planar, but PhPh and TBN are nonplanar. TBN is thought to have a saddle conformation, with opposing rings above and below the plane of the molecule. For monomeric phases, retention of the planar solute BaP is low, and it elutes before either of the larger nonplanar solutes PhPh or TBN. With increasing phase coverage, the elution order of the mixture shifts first to PhPh, BaP, TBN (typically for very low-loaded polymeric

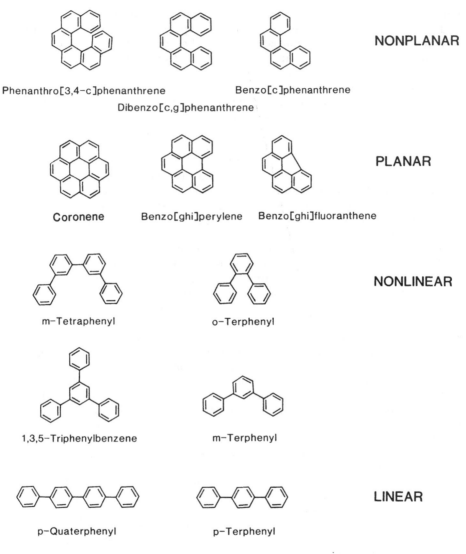

NONPLANAR

Phenanthro[3,4-c]phenanthrene

Dibenzo[c,g]phenanthrene

Benzo[c]phenanthrene

PLANAR

Coronene

Benzo[ghi]perylene

Benzo[ghi]fluoranthene

NONLINEAR

m-Tetraphenyl

o-Terphenyl

1,3,5-Triphenylbenzene

m-Terphenyl

LINEAR

p-Quaterphenyl

p-Terphenyl

Figure 28 Structures of planar/nonplanar and linear/nonlinear PAH pairs. (From Ref. 36.)

Table 12   Percent Carbon, Surface Area, and C_{18} Surface Coverage
Values for C_{18} Phases Synthesized for the Planar/Nonplanar Retention
Study

C_{18} phase	Percent carbon	Surface area (m^2/g)	Surface coverage[a] $(\mu mol/m^2)$
Monomeric	14.4	245	3.01
Polymeric (low)	5.87	84	3.55
Polymeric (normal)	7.50	84	4.67
Polymeric (heavy)	7.84	84	4.91

[a]Calculated using Eq. (3).
Source: Ref. 36.

phases or oligomeric phases) and finally to PhPh, TBN, BaP (for
higher-loaded polymeric phases. This mixture has been used for
phase characterization simply because of the clear changes in elution
order that are observed with phase type. Other mixtures of planar
and nonplanar compounds could also be used, perhaps with phase
type being correlated with the planar/nonplanar selectivity factors.

D.   Other Studies on Shape Versus LC Retention

Hasan and Jurs [64] used the reversed-phase retention data set from
the study of Wise et al. [4] to predict LC retention based on PAH
structural features. They generated numerical descriptors from the
PAH molecular structures (e.g., fragments, molecular connectivity,
substructure environment, geometric and calculated physical property
descriptors) which were used as predictor variables in multiple linear
regression analysis. Two linear equations (i.e., one for the monomeric phase and one for the polymeric phase) were developed that
expressed retention as a function of these variables. The variables
included the following descriptors: number of aromatic rings, molecular volume, cluster-three molecular connectivity, largest and smallest
principal axes, smallest principal moment, and X/Y principal moments.
   The type of descriptors selected to represent the PAH structures
in the equations of Hasan and Jurs [64] supports the results of our
studies on the influence of solute structure on LC retention. The
number of aromatic rings and the molecular volume are directly related to the size of the molecules. The cluster-three connectivity
index indicates the degree of branching and the presence and position
of a methyl group in the molecule. The presence of the principal axis
and the principal moment in the equations strongly supports the

Table 13 Planar/Nonplanar Selectivity Coefficients[a]

Compounds	85% Acetonitrile/water				100% Acetonitrile			
	Monomeric	Polymeric			Monomeric	Polymeric		
		Low	Normal	Heavy		Low	Normal	Heavy
p/m-Terphenyl	1.2	1.3	2.1	3.0	1.2	1.3	2.3	3.5
p/o-Terphenyl	1.3	1.7	2.6	3.8	1.4	1.5	2.9	4.3
p/m-Tetraphenyl	1.4				1.4	4.4	11.7	34.5
p-Tetraphenyl/1,3,5-Triphenylbenzene	1.5				1.5	3.0	14.2	46.0
Benzo[ghi]fluoranthene/Benzo[c]phenanthrene	1.1	1.2	1.4	1.6	1.2	1.4	1.7	1.8
Benzo[ghi]perylene/Dibenzo[c,g]phenanthrene	1.8	2.8	4.5	4.1	2.1	3.3	7.8	7.5
Coronene/Phenanthro[3,4-c]phenanthrene	2.7	5.1	9.6		3.4	6.2	12.1	17.1

[a]Selectivity coefficient = k' ratio of planar/nonplanar or linear/nonlinear solutes.
Source: Ref. 36.

importance of shape, because the magnitude of the three principal axes can be associated with the length, breadth, and thickness of the PAH. The third principal axis and the principal moment descriptors (i.e., the thickness) serve as correction factors for the nonplanar PAH. As discussed previously, the nonplanar PAH have shorter retention on polymeric than monomeric C_{18} phases. This observation was supported by the larger negative correlation for this descriptor in the retention equation for the polymeric C_{18} column compared to the monomeric C_{18} column. Hasan and Jurs [64] demonstrated that these equations could be used to predict the retention of several PAH that were not included in the original data set.

Jinno and Kawasaki [65] correlated reversed-phase LC retention for 26 PAH on several different stationary phases with five descriptors: a topological descriptor (molecular connectivity, χ), a geometric descriptor (van der Waals volume, V_W), a calculated physical property descriptor (logarithm of the partition coefficient in n-octanol/water), a correlation factor (F) as defined by Hurtubise et al. [66], and a shape parameter (L/B). They found that each of these parameters, except L/B, showed high correlation with LC retention. This is not surprising, since all of these parameters, except L/B, are related to the molecular weight of the PAH. However, these parameters provide little or no differentiation among PAH isomers. The length-to-breadth ratio, on the other hand, does differentiate between isomers because it provides a measure of the shape of the PAH, but it has little meaning when nonisomeric PAH are compared, as attempted in the study of Jinno and Kawasaki [65].

VII. RETENTION MODELS

A. Retention Behavior

To summarize the differences in retention behavior between monomeric and polymeric C_{18} phases, better separations of PAH result, in general, from the latter phase type. Selectivity towards PAH increases as a function of bonded-phase surface coverage. Because polymeric surface modification results in surface coverages about twice those of monomeric syntheses, polymeric phases exhibit enhanced selectivity toward PAH. Solute retention in reversed-phase systems is dependent on a number of factors, the most important of which is probably hydrophobic retention. The usual trends of increasing retention as a function of size (also molecular weight and contact area) are often observed for PAH. Exceptions to these trends, however, are not uncommon, and it is clear that solute shape also influences retention behavior. For isomeric PAH, retention is closely correlated with L/B ratios. Long, narrow PAH are retained to a greater degree than square PAH of the same molecular weight. Solute nonplanarity further affects retention, in that planar PAH are retained longer than their nonplanar analogues. This trend is also observed for linear and nonlinear PAH, where linear solutes are retained longer than their nonlinear analogues. All of these trends are observed on monomeric and polymeric C_{18} phases alike; however, the effects are much greater with polymeric phases.

B. Slot Model

The influence of solute shape on the retention of PAH has been described in terms of a schematic representation of the bonded phase referred to as the *slot model* [36]. This model does not attempt to describe in detail the physical structure of the bonded phase or specific solute - bonded-phase interactions. Instead it models retention in terms of a hypothetical phase consisting of a number of narrow slots into which solute molecules can penetrate (see Fig. 29). Planar molecules would be able to fit more easily into the slots than nonplanar molecules, and in so doing would be able to interact strongly with the stationary phase. Nonplanar molecules, on the other hand, would not penetrate as far into the slots, and would be retained less. The planar/nonplanar PAH pairs shown in Fig. 28 have similar molecular weights and L/B values, and the differences in retention are attributed to the increased "thickness" of the nonplanar solute, which would hinder penetration of the solute into the slots. For the retention of planar PAH isomers, the long, narrow solutes would fit into the bonded phase more readily and be more retained than square solutes. This is the trend observed with increasing L/B. Viewed in this way, the property of nonplanarity is simply an extension of L/B to include a thickness parameter. Penetration of "one-dimensional" (linear) solutes into the bonded phase would follow a similar mechanism based on the concept that long narrow solutes would be retained longer than their nonlinear analogues.

   The relationship between phase selectivity, solute size, and bonded-phase length has been examined by a number of workers. Lochmüller and Wilder [67] observed differences in retention behavior between large and small PAH. They attributed these differences to simultaneous changes in thermodynamic and phase-ratio parameters that occur as a function of PAH size. Tchapla and co-workers [68] found discontinuities in the retention of homologues, also as a function of size. They proposed that the effect was due to penetration of the solute into a finite stationary-phase thickness. As solute size approached the chain length of the stationary phase, retention leveled off.

   Martire and Boehm [69], in their "unified theory of retention and selectivity in liquid chromatography," concluded that chemically bonded phases exhibit shape selectivity which increases as the chains become more fully extended, and that rigid-rod solutes have greater retention than globular solutes. It is clear that phase thickness, phase density,

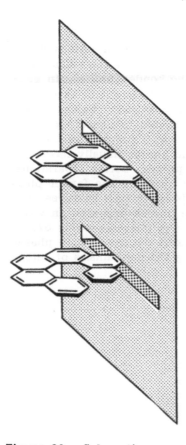

Figure 29 Schematic representation of the "slot model" for retention of PAH on polymeric C_{18} stationary phases. (From Ref. 36.)

or both increase with loading for the polymeric C_{18} phases. The fact that polymeric phases exhibit a greater selectivity for molecular shape than monomeric phases and that shape selectivity increases with increasing loading for polymeric phases suggests that polymeric phases may be more "extended" or more rigid than the monomeric phases. In addition, the shape selectivity observed for polymeric C_{18} phases is similar to that observed for liquid crystalline phases used in gas chromatography. This similarity would also suggest that polymeric phases are more "ordered" than monomeric phases.

VIII. CONCLUSIONS

Column selectivity for the reversed-phase separation of PAH is dependent upon the physical and chemical properties of the silica substrate and the conditions of synthesis of the bonded phase. In general, PAH retention is dependent upon the total amount of carbon (from the bonded phase) contained within the column. Column selectivity, on the other hand, is more closely related to phase surface coverage. Selectivity toward PAH increases as a function of bonded-phase surface coverage. Because polymeric phases usually have greater surface coverages than monomeric phases, better separations of PAH, particularly isomers, are achieved by using the former phase type. In general the best separations of isomeric PAH mixtures are obtained by using heavily loaded polymeric phases prepared on wide-pore silica substrates. Because polymeric C_{18} phases provide excellent selectivity for PAH separations, the exact nature of such phases is of considerable interest in understanding the mechanisms of retention in reversed-phase LC.

Appendix   LC Retention Data for PAH[a]

Compound	Parent PAH MW	Polymeric C18 column[b]				Monomeric C18 column[c]			
		70%	80%	90%	Ave.	70%	80%	90%	Ave.
Benzene	78							1.00	1.00
Methylbenzene	78							1.64	1.64
Ethylbenzene	78						2.14		2.14
Propylbenzene	78						2.74		2.74
Butylbenzene	78						3.27		3.27
Pentylbenzene	78						3.78		3.78
Hexylbenzene (phenylhexane)	78						4.27		4.27
Heptylbenzene (phenylheptane)	78						4.75		4.75
Octylbenzene (phenyloctane)	78						5.23	5.11	5.17
Nonylbenzene (phenylnonane)	78							5.60	5.60
Decylbenzene (phenyldecane)	78							6.08	6.08
1-Methylfluorene	166					3.27			3.27
2-Methylfluorene	166					3.35			3.35
4-Methylfluorene	166					3.26			3.26
9-Methylfluorene	166					3.08			3.08
Anthracene	178	3.16	3.16		3.16	3.11	3.10		3.11
1-Methylanthracene	178	3.44	3.42		3.43	3.57	3.57		3.57
2-Methylanthracene	178	3.68	3.69		3.68	3.68	3.69		3.69
9-Methylanthracene	178	3.42	3.40		3.41	3.52	3.53		3.52
9,10-Dimethylanthracene	178	3.68	3.68		3.68	3.89	3.92		3.91
Phenanthrene	178	3.00	3.00	3.00	3.00	3.00	3.00		3.00
1-Methylphenanthrene	178	3.40	3.40		3.40	3.49	3.51		3.50
2-Methylphenanthrene	178	3.68	3.67		3.68	3.70	3.72		3.71
3-Methylphenanthrene	178	3.34	3.33		3.34	3.47	3.47		3.47

Appendix (continued)

Compound	PAH MW	Polymeric C18 column[b]				Monomeric C18 column[c]			
		70%	80%	90%	Ave.	70%	80%	90%	Ave.
9-Methylphenanthrene	178	3.38	3.37		3.38	3.51	3.51		3.51
1,8-Dimethylphenanthrene	178	3.80	3.81		3.80	3.96	3.96		3.96
3,6-Dimethylphenanthrene	178	3.63	3.60		3.61	3.93	3.93		3.93
9-Methyl-10-ethylphenanthrene	178	3.80	3.77		3.78		4.17		4.17
9-Ethylphenanthrene	178	3.60	3.56		3.58	3.87	3.88		3.88
9-Isopropylphenanthrene	178	3.70	3.64		3.67		4.10		4.10
9-Propylphenanthrene	178	3.87	3.82		3.84		4.27		4.27
4,5-Methylenephenanthrene	190	3.22	3.22		3.22	3.32	3.33		3.32
Aceanthrylene	202	3.38	3.39		3.39	3.38	3.40		3.39
Acephenanthrylene	202	3.36	3.37		3.36	3.38	3.39		3.38
Fluoranthene	202	3.38	3.40		3.39	3.42	3.44		3.43
1-Methylfluoranthene	202	3.72	3.73		3.73	3.86	3.88		3.87
3-Methylfluoranthene	202	3.84	3.87		3.85	3.92	3.90		3.91
7-Methylfluoranthene	202	3.78	3.81		3.79	3.91	3.91		3.91
8-Methylfluoranthene	202	3.84	3.85		3.84	3.96	3.93		3.95
1-Methyl-7-isopropylphenanthrene (retene)	202		4.11		4.11		4.65		4.65
Pyrene	202	3.55			3.55	3.61	3.65		3.63
1-Methylpyrene	202	3.96	3.99		3.97		4.15		4.15
2-Methylpyrene	202		4.04		4.04		4.21		4.21
4-Methylpyrene	202	3.96	3.99		3.98		4.13		4.13
4,9-Dimethylpyrene	202		4.40		4.40		4.77		4.77
1-Ethylpyrene	202		4.09		4.09		4.45		4.45
1-Butylpyrene	202		4.46		4.46			5.32	5.32
11H-Benzo[a]fluorene	216	3.81	3.80		3.80	3.77	3.73		3.75

	MW								
9-Methyl-11H-benzo[a]fluorene	216	3.75	3.71	3.73		3.92	3.87		3.89
11H-Benzo[b]fluorene	216	3.82	3.81	3.82		3.80	3.75		3.78
Benzo[ghi]fluoranthene	226	3.93	3.96	3.95			4.07		4.07
Cyclopenta[cd]pyrene	226	3.94	3.94	3.94		3.95	3.94		3.95
Benz[a]anthracene	228	4.00	4.00	4.00	4.00	4.00	4.00		4.00
1-Methylbenz[a]anthracene	228		4.18	4.18			4.39		4.39
2-Methylbenz[a]anthracene	228		4.14	4.14			4.43		4.43
3-Methylbenz[a]anthracene	228		4.39	4.39			4.51		4.51
5-Methylbenz[a]anthracene	228		4.28	4.28			4.48		4.48
6-Methylbenz[a]anthracene	228		4.15	4.15			4.41		4.41
7-Methylbenz[a]anthracene	228		4.17	4.17			4.36		4.36
8-Methylbenz[a]anthracene	228		4.21	4.21			4.40		4.40
9-Methylbenz[a]anthracene	228		4.37	4.37			4.52		4.52
10-Methylbenz[a]anthracene	228		4.18	4.18			4.42		4.42
11-Methylbenz[a]anthracene	228		4.17	4.17			4.36		4.36
12-Methylbenz[a]anthracene	228		4.14	4.14			4.37		4.37
1,12-Dimethylbenz[a]anthracene	228		4.18	4.18			4.39		4.39
3,9-Dimethylbenz[a]anthracene	228		4.80	4.80				5.13	5.13
5,7-Dimethylbenz[a]anthracene	228		4.37	4.37			4.76		4.76
6,8-Dimethylbenz[a]anthracene	228		4.42	4.42			4.88		4.88
7,10-Dimethylbenz[a]anthracene	228		4.84	4.84				5.52	5.52
7,12-Dimethylbenz[a]anthracene	228		4.26	4.26			4.64		4.64
9,10-Dimethylbenz[a]anthracene	228		4.26	4.26			4.65		4.65
1,7,12-Trimethylbenz[a]anthracene	228		4.18	4.18			4.85		4.85
2,7-12-Trimethylbenz[a]anthracene	228		4.41	4.41			4.97		4.97
6,7,8-Trimethylbenz[a]anthracene	228		4.47	4.47					4.97
6,8,12-Trimethylbenz[a]anthracene	228							5.18	5.18
Benzo[c]phenanthrene	228	3.69	3.70	3.69		3.91	3.91		3.91
1-Methylbenzo[c]phenanthrene	228	3.75		3.72			4.10		4.10
2-Methylbenzo[c]phenanthrene	228	3.95	3.92	3.94			4.29		4.29

Appendix   (continued)

Compound	PAH MW	Polymeric C$_{18}$ column[b]				Monomeric C$_{18}$ column[c]			
		70%	80%	90%	Ave.	70%	80%	90%	Ave.
3-Methylbenzo[c]phenanthrene	228		4.09		4.09		4.41		4.41
4-Methylbenzo[c]phenanthrene	228		4.04		4.04		4.37		4.37
5-Methylbenzo[c]phenanthrene	228		4.04		4.04		4.37		4.37
6-Methylbenzo[c]phenanthrene	228		4.17		4.17		4.37		4.37
1,12-Dimethylbenzo[c]phenanthrene	228		3.90		3.90			4.34	4.34
5,8-Dimethylbenzo[c]phenanthrene	228		4.24		4.24		4.82		4.82
Chrysene	228		4.06		4.06		3.97		3.97
1-Methylchrysene	228		4.39		4.39		4.46		4.46
2-Methylchrysene	228		4.49		4.49		4.54		4.54
3-Methylchrysene	228		4.28		4.28		4.41		4.41
4-Methylchrysene	228		4.20		4.20		4.36		4.36
5-Methylchrysene	228		4.17		4.17		4.36		4.36
6-Methylchrysene	228		4.17		4.17		4.35		4.35
5-Ethylchrysene	228		4.19		4.19		4.61		4.61
Triphenylene	228	3.75	3.75		3.75	3.82	3.81		3.82
Benz[e]aceanthrylene	252		4.25		4.25		4.34		4.34
Benz[j]aceanthrylene	252		4.26		4.26		4.29		4.29
Benz[l]aceanthrylene	252		4.26		4.26		4.38		4.38
Benz[k]acephenanthrylene	252		4.39		4.39		4.43		4.43
Benzo[a]fluoranthene	252		4.24		4.24		4.45		4.45
Benzo[b]fluoranthene	252		4.29		4.29		4.44		4.44
Benzo[j]fluoranthene	252		4.26		4.26		4.37		4.37
Benzo[k]fluoranthene	252		4.38		4.38		4.50		4.50
Benzo[a]pyrene	252		4.51		4.51		4.68		4.68

Compound	MW				
1-Methylbenzo[a]pyrene	252	4.83		5.24	5.24
2-Methylbenzo[a]pyrene	252	4.93		5.33	5.33
3-Methylbenzo[a]pyrene	252	4.91		5.25	5.25
4-Methylbenzo[a]pyrene	252	4.81		5.26	5.26
5-Methylbenzo[a]pyrene	252	4.60		5.15	5.15
6-Methylbenzo[a]pyrene	252	4.72		5.11	5.11
7-Methylbenzo[a]pyrene	252	4.73		5.13	5.13
8-Methylbenzo[a]pyrene	252	4.89		5.33	5.33
9-Methylbenzo[a]pyrene	252	4.73		5.18	5.18
10-Methylbenzo[a]pyrene	252	4.66		5.16	5.16
11-Methylbenzo[a]pyrene	252	4.60		5.14	5.14
12-Methylbenzo[a]pyrene	252	4.59		5.14	5.14
1,2-Dimethylbenzo[a]pyrene	252		5.21	5.73	5.73
1,3-Dimethylbenzo[a]pyrene	252		5.26	5.75	5.75
1,4-Dimethylbenzo[a]pyrene	252		5.08	5.77	5.77
1,6-Dimethylbenzo[a]pyrene	252		5.10	5.65	5.65
2,3-Dimethylbenzo[a]pyrene	252		5.37	5.74	5.74
3,6-Dimethylbenzo[a]pyrene	252		5.09	5.63	5.63
3,11-Dimethylbenzo[a]pyrene	252		5.01	5.68	5.68
3,12-Dimethylbenzo[a]pyrene	252	4.94	4.93	5.62	5.62
4,5-Dimethylbenzo[a]pyrene	252	4.89	4.90	5.54	5.54
7,10-Dimethylbenzo[a]pyrene	252	4.82	4.82	5.51	5.51
Benzo[e]pyrene	252	4.29	4.51		
Perylene	252	4.33	4.52		
Cholanthrene	254	4.43	4.69		
20-Methylcholanthrene	254	4.80			
13H-dibenzo[a,g]fluorene	266	4.53	4.62		
13H-dibenzo[a,h]fluorene	266	4.96		4.77	4.77
11H-indeno[2,1-a]phenanthrene	266	4.91		4.74	4.74
Anthanthrene	276	5.08		5.61	5.61

Appendix (continued)

Compound	PAH MW	Polymeric C18 column[b]				Monomeric C18 column[c]			
		70%	80%	90%	Ave.	70%	80%	90%	Ave.
Benzo[ghi]perylene	276		4.76	4.76	4.76			5.36	5.36
Indeno[1,2,3-cd]fluoranthene	276		4.93	4.93	4.93			5.05	5.05
Indeno[1,2,3-cd]pyrene	276		4.84	4.84	4.84			5.23	5.23
Benzo[b]chrysene	278	5.00	5.00		5.00			5.00	5.00
Benzo[c]chrysene	278		4.45		4.45		4.85		4.85
Benzo[g]chrysene	278		4.27		4.27		4.71		4.71
Benzo[a]naphthacene	278		4.99		4.99			4.50	4.50
Dibenz[a,j]anthracene	278		4.56		4.56		4.84		4.84
Dibenzo[a,h]anthracene	278		4.73		4.73		4.86		4.86
Dibenzo[b,g]phenanthrene	278		4.33		4.33		4.80		4.80
Dibenzo[c,g]phenanthrene	278		4.07		4.07		4.51		4.51
Pentaphene	278		4.67		4.67		4.96		4.96

Compound	MW						
Picene	278			5.18	5.18	5.02	5.02
13-Methylpicene	278		4.87	4.82	4.84	5.22	5.22
Dibenz[e,k]acephenanthrylene	302			5.31	5.31	5.56	5.56
Dibenzo[b,def]chrysene	302	6.00	6.00	6.00	6.00	6.00	6.00
Dibenzo[def,p]chrysene	302		4.94	4.93	4.93	5.46	5.46
Dibenzo[a,e]fluoranthene	302		4.95	4.94	4.94	5.47	5.47
Dibenzo[a,k]fluoranthene	302		4.95	4.94	4.95	5.50	5.50
Dibenzo[b,k]fluoranthene	302			5.28	5.28	5.61	5.61
Dibenzo[j,l]fluoranthene	302		4.82	4.83	4.83	5.32	5.32
Dibenzo[a,i]pyrene	302			5.76	5.76	5.93	5.93
Naphtho[1,2,3,4-def]chrysene	302			5.00	5.00		
Phenanthro[3,4-c]phenanthrene	328		4.21	4.21	4.77	4.77	4.77
3-Methylphenanthro[3,4-c]phenanthrene	328		4.40	4.40	4.40	5.16	5.16
1,2:3,4:5,6:7,8-Tetrabenzonaphthalene	328		4.45	4.45	4.45	5.53	5.53

[a]Retention data reported as log I, as described in Refs. 4 and 5.
[b]Vydac 201TP.
[c]Zorbax ODS.

REFERENCES

1. J. A. Schmit, R. A. Henry, R. C. Williams, and J. F. Dieckman, J. Chromatogr. Sci. *9*, 645 (1971).
2. M. L. Lee, D. L. Vassilaros, C. M. White, and M. Novotny, Anal. Chem. *51*, 768 (1979).
3. Fed. Regist. 44(No. 233), 69514 (1979).
4. S. A. Wise, W. J. Bonnett, F. R. Guenther, and W. E. May, J. Chromatogr. Sci. *19*, 457 (1981).
5. S. A. Wise, W. J. Bonnett, and W. E. May, in *Polynuclear Aromatic Hydrocarbons: Chemistry and Biological Effects* (A. Bjørseth and A. J. Dennis, eds.), Battelle Press, Columbus, Ohio, 1980, p. 791.
6. M. Popl, V. Dolansky, and J. Mostecky, J. Chromatogr. *117*, 117 (1976).
7. D. L. Vassilaros, R. C. Kong, D. W. Later, and M. L. Lee, J. Chromatogr. *252*, 1 (1982).
8. K. Ogan and E. Katz, J. Chromatogr. *188*, 115 (1980).
9. E. Katz and K. Ogan, J. Liq. Chromatogr. *3*, 1151 (1980).
10. A. Colmsjö and J. C. MacDonald, Chromatographia *13*, 350 (1980).
11. R. Amos, J. Chromatogr. *204*, 469 (1981).
12. S. A. Wise, in *Handbook of Polycyclic Aromatic Hydrocarbons* (A. Bjørseth, ed.), Marcel Dekker, New York, 1983, p. 183.
13. H. Engelhardt and G. Ahr, Chromatographia *14*, 227 (1981).
14. M. C. Hennion, C. Picard, and M. Caude, J. Chromatogr. *166*, 21 (1978).
15. K. K. Unger, N. Becker, and P. Roumeliotis, J. Chromatogr. *125*, 115 (1976).
16. R. Karch, I. Sebestian, and I. Halasz, J. Chromatogr. *122*, 3 (1976).
17. G. E. Berendsen, L. de Galan, J. Liq. Chromatogr. *1*, 561 (1978).
18. G. E. Berendsen, K. A. Pikaart, and L. de Galan, J. Liq. Chromatogr. *3*, 1437 (1980).
19. H. Hemetsberger, W. Maasfeld, and H. Ricken, Chromatographia *9*, 303 (1976).
20. H. Hemetsberger, P. Behrensmeyer, J. Henning, and H. Ricken, Chromatographia *12*, 71 (1979).
21. P. Roumeliotis and K. K. Unger, J. Chromatogr. *149*, 211 (1978).
22. S. A. Wise and L. C. Sander, unpublished data, 1985.
23. M. B. Evans, A. D. Dale, and C. J. Little, Chromatographia *13*, 5 (1980).
24. H. Colin and G. Guiochon, J. Chromatogr. *141*, 289 (1979).
25. L. Boksanyi, O. Liardon, and E. Kovats, Adv. Colloid Interphase Sci. *6*, 95 (1976).

26. G. E. Berendsen and L. de Galan, J. Liq. Chromatogr. *1*, 403 (1978).
27. L. C. Sander and S. A. Wise, Anal. Chem. *56*, 504 (1984).
28. M. Verzele and P. Mussche, J. Chromatogr. *254*, 117 (1983).
29. R. E. Majors, J. Chromatogr. Sci. *12*, 767 (1974).
30. L. C. Sander and S. A. Wise, J. Chromatogr. *316*, 163 (1984).
31. C. H. Lochmüller and D. B. Marshall, Anal. Chim. Acta *142*, 63 (1982).
32. C. J. Little, J. A. Whatley, A. D. Dale, and M. B. Evans, J. Chromatogr. *171*, 435 (1979).
33. B. I. Diamondstone, S. A. Wise, and L. C. Sander, J. Chromatogr. *321*, 318 (1985).
34. S. A. Wise, L. C. Sander, and W. E. May, J. Liq. Chromatogr. *6*, 2709 (1983).
35. S. A. Wise and W. E. May, Anal. Chem. *55*, 1479 (1983).
36. S. A. Wise and L. C. Sander, J. High Resol. Chromatogr. Chromatogr. Commun. *8*, 248 (1985).
37. L. C. Sander and S. A. Wise, in *Polynuclear Aromatic Hydrocarbons: Mechanisms, Methods and Metabolism* (M. W. Cook and A. J. Dennis, eds.), Battelle Press, Columbus, Ohio, 1984, pp. 1133–1144.
38. K. K. Unger, in *Porous Silica*, Elsevier, Amsterdam–New York, pp. 1–53, (1979).
39. H. Engelhardt and H. Müller, J. Chromatogr. *218*, 395 (1981).
40. H. Müller and H. Engelhardt, in *Practical Aspects of Modern High Performance Liquid Chromatography* (I. Molnar, ed.), Walter de Gruyter, New York, 1983, pp. 25–39.
41. M. Verzele, J. Van Dijck, P. Mussche, and C. Dewaele, J. Liq. Chromatogr. *5*, 1431 (1982).
42. D. E. Damagolska and C. R. Loscombe, Chromatographia *15*, 657 (1982).
43. C. Dewaele and M. Verzele, J. Chromatogr. *260*, 13 (1983).
44. M. Verzele, M. de Potter, and J. Bhysels, J. High Resol. Chromatogr. Chromatogr. Commun. *2*, 151 (1979).
45. M. Verzele and C. Dewaele, J. Chromatogr. *217*, 399 (1981).
46. K. Harrison, The Separations Group, personal communication, 1983.
47. K. Ogan, E. Katz, and W. Slavin, Anal. Chem. *51*, 1315 (1979).
48. K. Ogan and E. Katz, Perkin-Elmer Corp., unpublished work, 1979.
49. J. G. Atwood and J. Goldstein, J. Chromatogr. Sci. *18*, 650 (1980).
50. R. B. Sleight, J. Chromatogr. *83*, 31 (1973).
51. G. M. Janini, K. Johnston, and W. L. Zielinski, Jr., Anal. Chem. *47*, 670 (1975).
52. G. M. Janini, G. M. Muschik, J. A. Schroer, and W. L. Zielinski, Jr., Anal. Chem. *48*, 1879 (1976).

53. W. L. Zielinski and G. M. Janini, J. Chromatogr. *186*, 237 (1979).

54. A. Radecki, H. Lamparczyk, and R. Kaliszan, Chromatographia *12*, 595 (1979).

55. R. Kaliszan and H. Lamparczyk, J. Chromatogr. Sci. *16*, 246 (1978).

56. A. C. Hazell, D. W. Jones, and J. M. Sowden, Acta Crystallogr. *1333*, 1516 (1977).

57. J. C. Fetzer, Chevron Research Co., personal communication, 1985.

58. R. C. Kong, M. L. Lee, Y. Tominaga, R. Pratap, M. Iwao, and R. C. Castle, Anal. Chem. *54*, 1802 (1982).

59. K. E. Markides, M. Nishioka, B. J. Tarbet, J. S. Bradshaw, and M. L. Lee, Anal. Chem. *57*, 1296 (1985).

60. M. L. Lee, S. R. Goates, K. Markides, and S. A. Wise, in *Polynuclear Aromatic Hydrocarbons: Chemistry, Characterization, and Carcinogenesis*, Battelle Press, Columbus, Ohio, in press.

61. M. Nishioka, B. A. Jones, B. J. Tarbet, J. S. Bradshaw, and M. L. Lee, *J. Chromatogr.*, in press.

62. G. M. Janini, R. I. Sato, and G. M. Muschik, Anal. Chem. *52*, 2417 (1980).

63. H. J. Issaq, G. M. Janini, B. Poehland, R. Shipe, and G. M. Muschik, Chromatographia *14*, 655 (1981).

64. M. N. Hasan and P. C. Jurs, Anal. Chem. *55*, 263 (1983).

65. K. Jinno and K. Kawasaki, Chromatographia *17*, 445 (1983).

66. R. J. Hurtubise, T. W. Allen, and H. F. Silver, J. Chromatogr. *235*, 517 (1982).

67. C. H. Lochmüller and D. R. Wilder, J. Chromatogr. Sci. *17*, 574 (1979).

68. A. Tchapla, H. Colin, and G. Guiochon, Anal. Chem. *56*, 621 (1984).

69. D. E. Martire and R. E. Boehm, J. Phys. Chem. *87*, 1045 (1983).

5

Liquid Chromatographic Analysis of the Oxo Acids of Phosphorus

Roswitha S. Ramsey   *Oak Ridge National Laboratory, Oak Ridge, Tennessee*

I. INTRODUCTION

The oxo acids of phosphorus are unique among inorganic oxo acids in that they comprise a class of polymeric, polyelectrolytic compounds of which the individual members form a homologous series. The acids can be further categorized on the basis of the number of phosphorus atoms in the compound, the oxidation numbers of the phosphorus atoms, the skeleton formation or structure, and the ratio of metal oxide to phosphorus pentoxide (M_2O/P_2O_5) in corresponding salts of the acids [1–3]. The more common acids can be classified as lower oxo acids, linear condensed acids, or cyclic condensed acids. Lower oxo compounds contain one or more phosphorus atoms with an oxidation number of less than five and are characterized by P-H or P-P bonds. Condensed phosphates contain only phosphorus atoms with an oxidation number of five and have the general formula $H_{(n+2)}P_nO_{(3n+1)}$. Orthophosphoric acid is often included in this group as the first member of the series of linear polymers. The cyclic compounds are also called metaphosphates and have the general formula $H_nP_nO_{3n}$, with phosphorus atoms of oxidation number five. The separation of the compounds in these three classes, some of which are shown with their structural formulas in Table 1, is the subject of this review. The abbreviated notation shown in the table for these compounds [4] indicates the oxidation number of the phosphorus atoms with a superscript, and the degree of polymerization or cyclic formation with a subscript.

All of the phosphorus oxo acids contain ionizable POH groups and show considerable ionic character in aqueous solution. This feature and the structural properties are the principal factors that have been used to influence the separation of mixtures. The methods most frequently used are ion-exchange and gel-permeation chromatography. As early as the 1950s, ion exchange was used to determine the composition of strong phosphoric acids (i.e., aqueous solutions containing greater than $\sim 70\%$ P_2O_5) [5]. Since then, it has primarily been used to separate the lower members of the condensed polymers and lower oxo acids. Gel chromatography has been used to fractionate the higher condensed phosphates ($>P_{10}$), cyclic phosphates, or separate various classes (monomeric, dimeric, and trimeric oxo acids). The most recent advances in separations involve the use of more efficient packing material and automated detection methods. Because there are several excellent reviews describing the early work in phosphate separations [2,3,6–8], only the studies conducted in approximately the past ten years will be considered here.

A factor that must be taken into consideration in order to achieve the separation of the phosphates is their hydrolytic stability. The lower oxo acids are readily hydrolyzed in acidic and basic solutions but are relatively stable at neutral pH. The condensed acids, both linear and cyclic, are stable in basic and neutral solution but are rapidly hydrolyzed at acidic pH. Hydrolytic stability is also influenced by temperature and the presence of certain cations. Procedures that include special precautionary measures to preserve sample integrity, and investigations that have examined the effects of various chromatographic conditions on stability will be noted in this article.

Table 1   Oxo Acids of Phosphorus

Classification	Trivial name (anion)	Structural formula	Abbreviated notation
Lower oxo acids	Hypophosphorous acid (Phosphinate) $H_2PO_2^{-1}$	$H-\overset{\displaystyle OH}{\underset{\displaystyle O}{\overset{\vert}{\underset{\Vert}{P}}}}-H$	$\overset{1}{P}$
	Phosphorous acid (Phosphonate) HPO_3^{-2}	$H-\overset{\displaystyle OH}{\underset{\displaystyle O}{\overset{\vert}{\underset{\Vert}{P}}}}-OH$	$\overset{3}{P}$
	Diphosphorous acid	$H-\overset{\displaystyle OH}{\underset{\displaystyle O}{\overset{\vert}{\underset{\Vert}{P}}}}-\overset{\displaystyle OH}{\underset{\displaystyle O}{\overset{\vert}{\underset{\Vert}{P}}}}-OH$	$\overset{2}{P}-\overset{4}{P}$
	Hypophosphoric acid	$HO-\overset{\displaystyle OH}{\underset{\displaystyle O}{\overset{\vert}{\underset{\Vert}{P}}}}-\overset{\displaystyle OH}{\underset{\displaystyle O}{\overset{\vert}{\underset{\Vert}{P}}}}-OH$	$\overset{4}{P}-\overset{4}{P}$

Table 1 (continued)

Classification	Trivial name (anion)	Structural formula	Abbreviated notation
Lower oxo acids	Diphosphorous acid (diphosphonate) $H_2P_2O_5^{-2}$	$\begin{array}{c} \text{OH} \quad\;\; \text{OH} \\ \mid \qquad\;\; \mid \\ H-P-O-P-H \\ \parallel \qquad\;\; \parallel \\ O \qquad\quad\; O \end{array}$	$\overset{3}{P}-O-\overset{3}{P}$
	Isohypophosphoric acid (isohypophosphonate) $HP_2O_6^{-3}$	$\begin{array}{c} \text{OH} \quad\;\; \text{OH} \\ \mid \qquad\;\; \mid \\ H-P-O-P-OH \\ \parallel \qquad\;\; \parallel \\ O \qquad\quad\; O \end{array}$	$\overset{3}{P}-O-\overset{5}{P}$
		$\begin{array}{c} \qquad\quad \text{OH} \\ \text{OH} \quad\; \mid \quad\; \text{OH} \\ \mid \qquad \mid \qquad \mid \\ H-P-O-P-O-P-OH \\ \parallel \qquad \parallel \qquad \parallel \\ O \qquad O \qquad O \end{array}$	$\overset{3}{P}-O-\overset{5}{P}-O-\overset{5}{P}$
Cyclic condensed phosphoric acids	Trimetaphosphoric acid (trimetaphosphate) $P_3O_9^{-3}$		P_{3m}

Cyclic condensed phosphoric acids

Tetrametaphosphoric acid (tetrametaphosphate)

$P_4O_{12}^{-4}$ P_{4m}

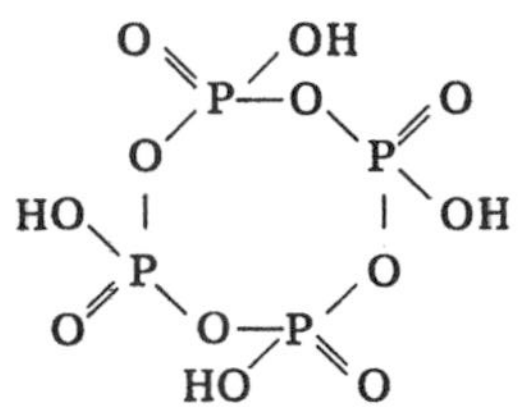

Cyclic polymeric phosphoric acid (_ metaphosphate)

$P_nO_{3n}^{-n}$ P_{nm}

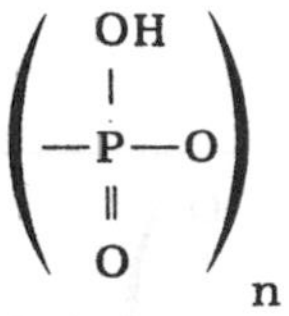

Linear condensed phosphoric acids

Orthophosphoric acid (orthophosphate)

PO_4^{-3} P_1

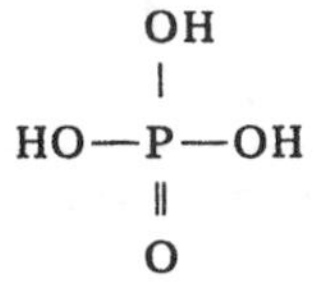

Pyrophosphoric acid, diphosphoric acid (pyrophosphate)

$P_2O_7^{-4}$ P_2

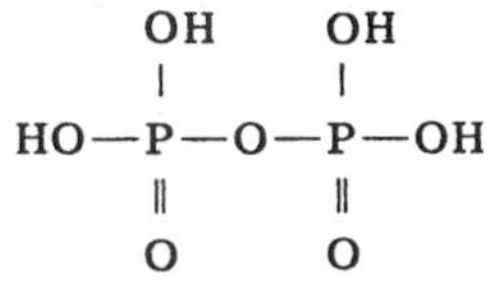

Tripolyphosphoric acid (tripolyphosphate)

$P_3O_{10}^{-5}$ P_3

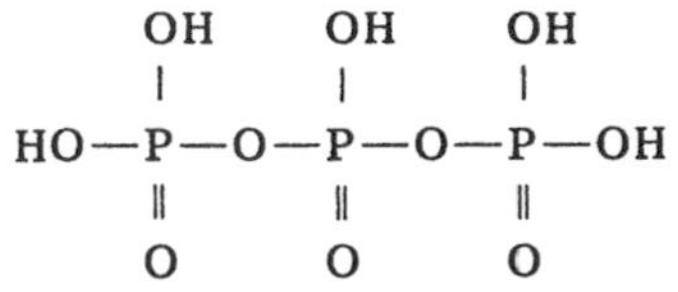

Table 1 (continued)

Classification	Trivial name (anion)	Structural formula	Abbreviated notation
	Long-chain condensed phosphoric acid $P_n O_{3n+1}^{(n+2)-}$	$\left(\begin{array}{c} OH \\ | \\ OH\!-\!P\!-\!O\!-\!H \\ \| \\ O \end{array} \right)$	P_n

II. DETECTION METHODS

The phosphorus oxo acids are most frequently determined by means of complexation reactions that are specific for the orthophosphate ion [2]. Condensed phosphates are converted to orthophosphate by hydrolysis, whereas the lower oxo acids are oxidized. The conditions necessary for quantitative hydrolysis may vary slightly, according to the species and the degree of polymerization. In general, however, hydrolysis proceeds rapidly in an acidic medium at elevated temperatures. Oxidation can be accomplished by means of a variety of reagents, although the selection is partially species dependent because some acids are oxidized only in acidic solution whereas others require neutral conditions. Sodium hydrogen sulfite in acid solution has been used as an oxidizing agent for phosphonate, phosphinate, and other lower oxo acids [9–12]. To quantitatively convert anions containing P-P bonds to orthophosphate, a sodium hypochlorite reagent can be used [13]. Other reagents including those that allow selective oxidation of some species have been discussed elsewhere [3]. Following the conversion of the oxo acids, the resulting orthophosphate is reacted with molybdenum reagents to form a colored complex which can be detected spectrophotometrically. Ammonium molybdate and antimony potassium tartrate react with orthophosphate to form an antimony-phosphate-molybdate complex, which is reduced with ascorbic acid to form a molybdenum blue complex [14]. Orthophosphate also combines with ammonium molybdate in an acidic medium to form a molybdophosphate, which can be reduced to form a heteropolyblue complex [1–3,15]. Ascorbic acid [2], 1-amino-2-naphthol-4-sulfonic acid [14], stannous chloride [2,3], hydrogen sulfide [2], hydrazine hydrochloride [3], and hydrazine sulfate [13] have been used as reducing agents. The formation of the heteropolyacids is dependent upon the concentration of the reacting substances, the temperature, and the acidity.

Prior to reduction, the phosphomolybdate is yellow. The change in color to blue is attributed to the reduction of peripheral molybdenum atoms in the heteropoly complex [1]. A mixed reagent containing molybdenum (V) and molybdenum (VI) can also be used to react with orthophosphate to form the blue complex [16,17]. With this reagent, the hydrolysis of the polyphosphates and the color reaction proceed simultaneously. Equations that describe the various reactions are given:

$$P_n \longrightarrow P_1 + Mo(V)\text{-}Mo(VI)$$

$$P_n \longrightarrow P_1 + Mo(VI) \quad P_1\text{-}Mo(VI) \xrightarrow{\text{Reducing agent}} P_1\text{-}Mo(V)\text{-}Mo(VI)$$

$$P^1, P^3 \xrightarrow{\text{Oxidizing agent}} P_1 + Mo(V)\text{-}Mo(VI)$$

A mixture of ammonium molybdate and ammonium metavanadate also reacts with orthophosphoric acid to form molybdovanadophosphoric acid, which can be determined spectrophotometrically [14,18]. In all of these complexes, the color intensity is proportional to the orthophosphate concentration in the sample. Absorbance measurements are usually made at 830 nm for the blue complex and approximately 420 nm for the yellow complex (molybdovanadophosphate method). Hypophosphoric acid, however, reacts with the molybdenum (V) - molybdenum (VI) reagent to form a complex with absorption maxima at 590 and 670 nm [3].

In the past, most of the analyses used to monitor the effluents from separation columns or to determine phosphate anions on thin layers were conducted in a batch mode (i.e., fractionation or isolation followed by stepwise manual determinations). Automated methods using air-segmented autoanalyzers (Technicon) and flow injection systems greatly reduce analysis time and provide on-line visualization of detected compounds for column separation methods. Flow injection methods are discussed in greater detail in Sec. III.D. Lundgren and Loeb were the first to describe the use of an autoanalyzer for the determination of condensed phosphates following separation by ion-exchange chromatography [19]. In this system, the column effluent is acidified and pumped through a heated coil to hydrolyze the polyphosphates to orthophosphate. The resulting solution is then passed through a dialyzer and mixed with ammonium molybdate and reducing agent to form the molybdophosphate complex. Since this method was introduced, it has been modified and optimized to provide maximum sensitivity [20,21] and to extend the applications to include the lower oxo acids [13]. A diagram of the AutoAnalyzer II (Technicon), used as an HPLC postcolumn reaction detector for condensed phosphates, is shown in Fig. 1. Air segmentation is used to prevent sample carry-over and assist in reagent-sample mixing. Resampling following the hydrolysis reaction assures a uniform flow rate and uniform dilution with reagents added downstream. Additional pumps and reaction coils are used to introduce and mix the effluent with an oxidizing reagent when lower oxo acids are determined.

Nakamura et al. have examined the factors affecting the quantitative analysis on an autoanalyzer, of condensed phosphates with average degrees of polymerization up to 100 [20]. Changes in the eluent composition on the response of the detector were also investigated. Although most autoanalyzer systems are based on the formation of a heteropolyblue complex, Deelder et al. have described a system based on the formation of the molybdovanadophosphate complex [22].

Other methods for continuous monitoring of the phosphorus acids include atomic absorption [23—25] and flame emission [26]. Atomic absorption coupled with gel chromatography has been used to determine pyrophosphate, tripolyphosphate, tetrapolyphosphate, and Kurrol's salt

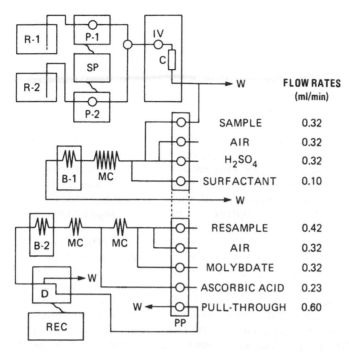

	FLOW RATES (ml/min)
SAMPLE	0.32
AIR	0.32
H₂SO₄	0.32
SURFACTANT	0.10
RESAMPLE	0.42
AIR	0.32
MOLYBDATE	0.32
ASCORBIC ACID	0.23
PULL-THROUGH	0.60

Figure 1 Diagram of Technicon AutoAnalyzer II. R-1 and R-2, solvent reservoirs; P-1 and P-2, dual-piston high-pressure pumps; SP, solvent programmer, M, mixer; IV, injection valve; C, analytical column; W, waste; MC, mixing coil; B-1, heating bath for hydrolysis reaction; B-2, heating bath for color development; D, absorbance detector; PP, proportionating pump; REC, recorder. (Adapted from Ref. 56.)

a highly polymeric phosphate [23]. The method is based on the formation and detection of a magnesium phosphate complex, produced by the elution of a phosphate anion on a gel column preequilibrated and eluted with a solution of magnesium chloride of known concentration. By monitoring the absorption due to magnesium in the effluent, the concentration of a phosphate anion can be determined. The flame emission detector has a burner assembly which can accept the total effluent from an HPLC column and can be used for the low-level detection of inorganic phosphates [26]. Voltametry has also been used in a flow injection procedure to determine low concentrations of phosphate by measuring the reduction current of molybdophosphate at a glassy carbon electrode held at a positive potential relative to a saturated calomel electrode [27]. Although this method has only been used for the determination of total phosphate, it could also be useful as a postcolumn reaction detector.

III. SEPARATION METHODS

A. Paper and Thin-Layer Chromatography

These analytical techniques were among the first used to separate
mixtures of the oxo acids. Because they have largely been replaced
by advanced separation methods that provide greater accuracy, faster
analysis time, and improved resolution, they will not be discussed in
detail. Instead, reviews of the two techniques covering the periods
up to 1972 [2,3] and 1975 [6] are recommended. R_f values for some
of the condensed phosphates, determined in various solvent systems
by ascending paper and thin-layer chromatography, have been com-
piled in the review by Ohashi [3]. More recent applications of the
two techniques have utilized variations of these earlier methods,
either to improve quantitation or optimize a particular separation
[28—32].

Investigations that have systematically examined the effects of
various ionic constituents on the separations, however, should be
noted. Metallic cations influence the separations of condensed phosphates and distort both paper and thin-layer chromatograms so that
the spots may have long leading or tailing edges. Multivalent metals,
as compared to the alkali metals, have been reported to have considerable effect on thin-layer separations, with the strongest observed for
Al (III) and Fe (III) [33]. Ammonium ions and the anions accompany-
ing the metal cations do not influence the separations. Interfering
effects can be eliminated by removing the cations in the sample solutions prior to chromatography by extraction with ion-exchange resins,
the addition of EDTA, or complexation with hydroxyquinoline, followed
by nonaqueous solvent extraction of the complex. The formation of
soluble metal phosphate complexes is thought to be responsible for
changes in the chromatographic patterns for components in solutions
containing relatively high concentrations of cations. The effects of
different metal ions on the separation of tripolyphosphate, trimetaphosphate, and tetrametaphosphate from their respective hydrolysis
products have also been demonstrated by paper chromatography [34].

B. Ion-Exchange Column Chromatography

Column liquid-chromatographic techniques have in general provided
more accurate and precise analysis of the phosphates than either thin-
layer or paper chromatography, where quantitation has ordinarily relied on the extraction of a spot or band from a cellulose filter or thin-
layer plate. Chromatographic conditions can be more closely controlled
and generally greater resolution can be obtained by LC. Strong basic
anion exchangers such as Dowex 1-X8, BioRad AG 1-X8, and Amberlite IRA-400 have primarily been used for phosphate separations by
ion-exchange with selectivity provided by stepwise or gradient elution
techniques. Solutions of sodium or potassium chloride, increasing in

ionic strength and buffered at a pH ranging from 5 to 10, have been used as eluents. Detection of the separated phosphates is ordinarily accomplished with spectrophotometry by means of the various complexation reactions. A review covering the period up to 1972 for the separation of condensed phosphates by anion exchange is given by Zar-Ayan [7].

More recently, the hydrolytic stability of the phosphates on ion-exchange columns was investigated. Fukuda, Nakamura, and Ohashi examined the chromatographic behavior of diphosphates [35]. A leading peak edge was observed when disodium dihydrogen diphosphate or tetrasodium diphosphate was eluted from a BioRad AG 1-X8 column with sodium chloride solutions. The anomalous peak shape was obtained both in buffered and unbuffered eluents. The addition of 5×10^{-3} M EDTA to the eluent provided a symmetrical peak shape. Collection and analysis of the shoulder revealed that it was composed substantially of orthophosphate anions, supporting the view that diphosphate hydrolyzes to form orthophosphate during separation. Trace metal ions present in the chromatographic system are believed to accelerate the process. The hydrolytic effect of metal ions has also been reported by other investigators [36].

The separation of condensed phosphates (glassy sodium polyphosphate with an average chain length of 4.5) was subsequently examined, using eluents containing chelating agents (i.e., sodium salts of citric acid or EDTA) [37]. The linear phosphates from ortho to P_8 were well resolved when sodium chloride/EDTA was used. It was noted that the retention volumes of the anions were reduced in the presence of this chelating agent. This reduction restricts the final concentration that can be added to the eluent before the peaks begin to overlap. Nakamura, Yano, Nunokawa, and Ohashi also examined the anion-exchange separation of a mixture of lower oxo acids and condensed acids on BioRad AG 1-X8, using sodium chloride with EDTA as the eluent [12]. The results are shown in Fig. 2. Although some of the peaks overlap, the analytical conditions described are useful for examining a complex mixture of oxo anions with different oxidation states and structures.

C. Gel Chromatography

Size-exclusion chromatography has been used to analyze phosphate anions to determine molecular weights, polydispersion, and molecular dimensions as well as to isolate and purify individual compounds. Separations by this method occur principally by a molecular sieve mechanism, which relates the size or molecular weight of the analyte to the elution volume. In the absence of secondary effects resulting from interaction of the gel with the analyte or eluent, the elution volume is a linear function of the log of the molecular weight. In achieving separations by gel filtration, ion-exclusion and adsorption

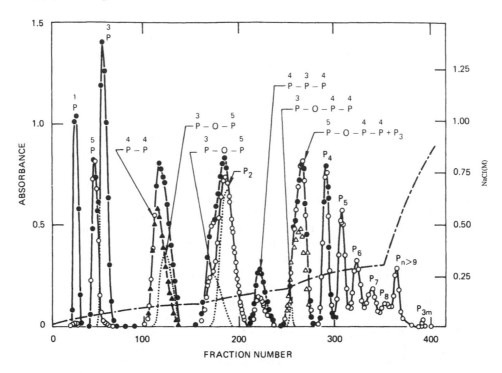

Figure 2 Chromatogram of phosphorus oxo anions separated by anion exchange on a BioRad AG1-X8 column. The eluents contained 5 mM EDTA and were as follows for the various fractions: fractions 1–150, 0.02 M NaCl in the mixing bottle and 0.12 M NaCl in the reservoir; fractions 151–250, 0.20 M NaCl; fractions 251–350, 0.35 M NaCl; fractions 351–400, 1.50 M NaCl. Concentration of NaCl in the eluent (dashed line); calculated curve based on absorbance data (dotted line). Conditions for color development and absorbance measurement: (○) Mo(V) - Mo(VI) reagent, λ = 810 nm; (●) Mo(V) - Mo(VI) reagent and 1 M NaHSO$_3$, λ = 810 nm; (△) Mo(V) - Mo(VI) reagent, λ = 610 nm; (▲) Mo(V) - Mo(VI) reagent + 1 M NaHSO$_3$, λ = 610 nm. (Adapted from Ref. 12.)

should be minimized. Ion-exclusion effects have been attributed to the presence of a small number of fixed charges in the gel matrix which repel the analyte, restricting penetration in the gel phase and reducing elution volumes [38]. An eluent that contains a supporting electrolyte such as sodium chloride or potassium chloride is used to eliminate exclusion effects in phosphate analysis [38,39]. Alternatively, a cooperating effect between a molecular sieve and ion-exchange mechanism has been used to separate phosphates by ion-exchange gels.

Ohashi, Yoza, and Ueno were the first to report on the separation of ortho-, pyro-, and tripolyphosphate on a gel column (Sephadex G-25) [40]. Since then, the gel chromatographic behavior of linear and cyclic polyphosphates and of the lower oxo acids have been investigated [38—49]. The dextran gels (e.g., Sephadex) have been primarily used for separations. Partition coefficients, elution order, and peak shapes have been examined as a function of gel porosity [38,39], particle size [38], eluent concentration [39], pH [39,41], and various other column parameters [41,38]. Pechkovskii, Cherches, and Kuz'menkov have reviewed the principles of gel chromatography as applied to the separation of condensed phosphates [8]. The partition coefficients of some of the phosphate anions, including those containing phosphorus atoms with different oxidation states, on a Sephadex G-25 column are shown in Table 2.

Yoza et al. have used gel chromatography to separate and purify the reaction products between diphosphate and diphosphonate [10]. On a Sephadex G-25 column, $\overset{3}{P}-O-\overset{5}{P}-O-\overset{5}{P}$ was separated from the starting products. An additional peak observed in the elution profile was believed to correspond to a tetramer ($\overset{3}{P}-O-\overset{5}{P}-O-\overset{5}{P}-O-\overset{3}{P}$). A Sephadex G-10 column was used for preparative purification.

The cyclic phosphates from trimeta- to octametaphosphate contained in Graham's salt have been separated on an anion-exchange gel, QAE Sephadex A-25, which contains tertiary alkyl ammonium groups in a dextran matrix [42]. Two separate chromatographic runs were required for complete resolution of all six compounds. When 0.30 M potassium chloride was used as the eluent, P_{3m}, P_{4m}, P_{5m}, and P_{7m} were separated from each other and from P_{6m} and P_{8m}. The latter two cyclic phosphates, which coeluted with the above eluent, were separated with 0.25 M potassium chloride. Greater resolution was obtained on QAE Sephadex A-25, as compared to Dowex I-X4. The preparative-scale isolation and purification of trimetaphosphate through nonametaphosphate produced from the condensation of orthophosphoric acid with carbodiimides, have been carried out on a DEAE cellulose column [43]. NMR shifts for the isolated cyclic compounds, as well as some long-chain polyphosphates, were determined. Penta- and heptametaphosphate have also been isolated from Graham's salt on a

Table 2 Partition Coefficients of Various Oxo Acids of Phosphorus on a Sephadex G-25 Column[a]

Phosphate sample		Partition coefficients with eluent:			
Compound	Acid	Pure water	0.1 M KCl	0.5 M KCl	1.0 M KCl
$NaPH_2O_2 \cdot H_2O$	$\overset{1}{P}$		0.80	0.83	0.83
$Na_2PHO_3 \cdot 5H_2O$	$\overset{3}{P}$		0.78	0.83	0.84
$NaH_2PO_4 \cdot 2H_2O$	P_1	0.36	0.78	0.83	0.84
$Na_2P_2HO_5 \cdot 12H_2O$	$\overset{2}{P}\text{--}\overset{4}{P}$		0.64	0.74	0.76
$Na_2H_2P_2O_6 \cdot 6H_2O$	$\overset{4}{P}\text{--}\overset{4}{P}$		0.67	0.77	0.79
$Na_2P_2H_2O_5$	$\overset{3}{P}\text{--O--}\overset{3}{P}$		0.59	0.66	0.69
$Na_3P_2HO_6 \cdot 4H_2O$	$\overset{3}{P}\text{--O--}\overset{5}{P}$		0.61	0.70	0.73
$Na_4P_2O_7 \cdot 10H_2O$	P_2	0.30	0.62	0.71	0.75
$Na_5P_3O_8 \cdot 14H_2O$	$\overset{4}{P}\text{--}\overset{3}{P}\text{--}\overset{4}{P}$		0.61	0.73	0.76
$Na_4P_2HO_8 \cdot H_2O$	$\overset{5}{P}\text{--O--}\overset{4}{P}\text{--}\overset{4}{P}$		0.51	0.67	0.70
$(NH_4)_5P_3O_9 \cdot \times H_2O$	$\overset{5}{P}\text{--O--}\overset{4}{P}\text{--}\overset{4}{P}$		0.50	0.62	0.70
$Na_5P_3O_{12} \cdot 6H_2O$	P_3	0.21	0.52	0.62	0.70
$Na_3P_3O_9 \cdot 6H_2O$	P_{3m}		0.52	0.67	0.70
$Na_6P_4O_{11} \cdot \times 4H_2O$	$\overset{4}{P}\text{--}\overset{4}{P}\text{--O--}\overset{4}{P}\text{--}\overset{4}{P}$		0.46	0.62	0.64
$Na_4P_4O_{10} \cdot 4H_2O$	$(\text{--}\overset{4}{P}\text{--}\overset{4}{P}\text{--O--})$		0.48	0.62	0.65
$Na_4P_4O_{12} \cdot 4H_2O$	P_{4m}		0.46	0.60	0.65

[a]Bed volume, 150 ml, sample concentration, 2×10^{-3} to 5×10^{-3} g-atom P/liter.
Source: Data from Ref. 39.

preparative scale via a QAE Sephadex A-25 gel column [44]. Cyclic
and long-chain linear phosphates were initially separated from each
other on a Sephadex G-25 column. Individual cyclic compounds were
then separated on the ion-exchange gel. The distribution of the
cyclic phosphates through octametaphosphate in Graham's salt was
determined. IR spectra for these compounds were also measured,
and characteristic adsorption bands identified.

Sodium phosphate glasses have been fractionated by gel chromatography to determine molecular weight composition. For this application, the degree of polymerization in the eluate fractions is usually
determined by potentiometric tritration. The relationship between
elution volume (Ve) and the degree of polymerization has been examined by several investigators [45-48]. Pechkovskii et al. have
fractionated a phosphate glass with an average chain length ($\bar{n}$) of
36 on Molselekt G-25 (a dextran gel, Hungary) [45]. The elution
volumes of the isolated fractions were directly proportional to the
log of their degree of polymerization. Cherches et al. have examined
the chromatographic behavior of polydispersed samples of alkali metal
phosphates, having average chain lengths ranging from 2.9 to 55,
on Molselekt G-50 [46]. Elution volumes were dependent on both the
degree of polymerization in a given fraction and the average degree
of polymerization of the test polymer. The cations associated with
the phosphates were also found to have an effect on Ve. Miyajima,
Yamauchi, and Ohashi have fractionated glasses with $\bar{n}$ values of 113
and 19 on Sephadex G-100 and G-50, respectively, to examine the
possibility of using gel chromatography for obtaining samples of polyphosphates with narrow chain-length distribution [47]. The samples
were initially fractionated on preparative-scale columns. Portions of
each fraction were then analyzed on an analytical-scale column, and
the total amount of phosphate in the effluent was monitored with a
Technicon AutoAnalyzer detector. A linear relationship was found for
Ve values and the log of $\bar{n}$ for the fractions. These investigators
have also characterized other glasses with nominal $\bar{n}$ values ranging
from 4 to 75 on a Sephadex G-100 column coupled to an automated
phosphate analyzer [48].

Although the soft gels (cross-linked dextrans) have primarily been
used for gel filtration, Miyajima and Ohashi have reported using a
semirigid hydrophilic porous vinyl polymer (Toyopearl HW-55, Toyo
Soda, Japan) to characterize various glasses ($\bar{n}$ values from 9 to 75)
[49]. They obtained the molecular-weight distribution profiles of
long-chain polyphosphate mixtures in less than 30 min on this polymeric column material, using a relatively high mobile-phase velocity.

D. High-Performance Liquid Chromatography

With the advent of small-particle-size and narrow-distribution packing materials for LC, high-performance and high-efficiency separations
for the phosphates have become possible. The advanced column technology coupled with automated methods of analysis has, in fact, been

the most significant in terms of facilitating phosphate determinations. Improved resolution, faster analysis time, and reduced requirements for operator assistance are some of the advantages HPLC has over some of the other methods that have been discussed. Detection is accomplished by continuous monitoring of the effluent from the column as it passes a flow-through detector. Although different methods can be used for monitoring the column effluent (e.g., voltametry and conductivity), the most commonly used are based on absorption measurements. The column effluent is reacted with various reagents to form a colored complex, the intensity of which is measured at a specific wavelength. Automated air-segmented analyzers such as the Technicon AutoAnalyzers have already been discussed in Sec. II. Flow injection systems in which the sample or column effluent is introduced into an unsegmented stream of reagent, which is pumped at a constant flow rate, have been recently described for phosphate determinations [11,17,50]. They differ from air-segmented analyzers in that they do not require complete mixing of the sample and reagent prior to product measurement. Mixing in flow injection analysis (FIA) occurs by diffusion and convection. In general, the advantages of such a system are reduced analysis time, reduced sample and reagent consumption, and improved precision [51]. Because mixing induces band broadening, improved resolution may also be obtained.

A diagram for a typical HPLC/FIA system, which is based on the formation and the absorption of the molybdenum-blue complex for the determination of linear and cyclic acids, is shown in Fig. 3. The carrier reagent is introduced at a constant flow rate by a reciprocating pump, and the column eluent is added downstream through a low-dead-volume tee. The polyphosphates are hydrolyzed by the acidic reagent at an elevated temperature in the reaction coil to form orthophosphate, which then condenses with the molybdenum (V) and molybdenum (VI) reagent, to form the colored complex. Maximum response is obtained when the hydrolysis, which is influenced by the temperature of the reaction coil and the residence time of the sample within the coil, is complete. The color development proceeds rapidly, once the orthophosphate has been formed. The solution is then cooled before it passes through the detector. A back-pressure coil attached to the detector outlet is used to prevent bubble formation caused by heating of the reaction coil. This system can also be used for the rapid analysis of total phosphate by injecting samples at I_2.

Hirai, Yoza, and Ohashi were the first to examine the applicability of a flow injection system as a postcolumn reaction detector for the separation of inorganic polyphosphates [50]. Ortho-, pyro-, and tripolyphosphate were resolved on an anion-exchange column, TSK-GEL,IEX-220 SA (Toyo Soda, Japan). The system was compared to an air-segmented autoanalyzer and was found to have greater sensitivity and to provide greater resolution than the latter. This was attributed to decreased sample-residence time, the absence of extracolumn

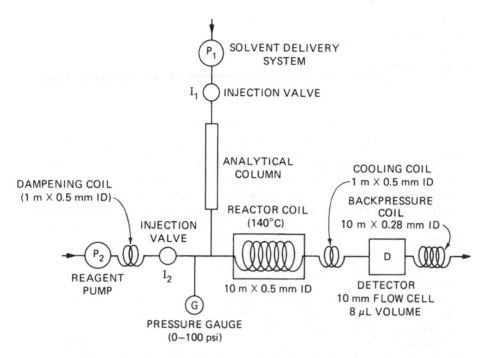

Figure 3   Schematic diagram of HPLC/FIA system.   (From Ref. 54.)

contributions to band broadening, and reduced sample-reagent dilution.   Total amounts of phosphates were also determined at a rate of
45 samples per hour with a relative standard deviation of <1%.

The detection system has been modified to analyze the lower oxo
acids of phosphorus as well [11].   Phosphinate, phosphonate, and
orthophosphate were separated on a TSK-GEL, IEX-220 SA anion-
exchange column and detected in a system containing two pumping
channels; one to introduce an oxidizing agent, and another to deliver
the molybdenum reagent.   Sodium hydrogen sulfite was used as the
oxidizing agent and was introduced with a four-way loop injection
valve into an inert carrier stream.   The column effluent and molybdenum reagent were continuously added to the carrier.   This arrangement allows the detection of all the acids in the presence of the oxidizing agent or the selective detection of orthophosphate in its absence.   This feature should also be useful for determining the chemical composition of lower oxo acids containing P^3 and P^5 units.   The
maximum response for the compounds considered in this study was

obtained at 140°C reaction coil temperature and 1.0 M sodium hydrogen sulfite concentration.

Applications of HPLC/FIA have included the analysis of detergents [52], wastewaters [53], and phosphorus smokes [54]. In the latter case, the aerosol formed from the combustion of red and white phosphorus was examined and found to contain a complex mixture of condensed polyphosphates. Orthophosphate through the P_{13} polymer, P_{3m}, and P_{4m} were resolved on an Aminex A-27, 8% cross-linked quaternary ammonium anion-exchange column, via a linear gradient of sodium chloride solutions (0.28–0.53 M NaCl containing 5 mM tetrasodium EDTA). Polyphosphates up to the P_{18} polymer could be resolved on an Aminex A-14 column, 4% cross-linked resin. The analysis time, however, was comparatively long, and the efficiency was less than with the Aminex A-27 column. A study was also conducted to examine the hydrolytic stability of the polyphosphates, following dissolution of the aerosol samples in water. It was determined that as the extracted sample hydrolyzed, the concentration of long-chain polymers decreased and the concentrations of P_{3m} and P_{4m} increased. The changes in concentration of various species as a function of time are shown in Table 3. Samples could be preserved for a minimum of 72 hr, if they were extracted in NaCl/EDTA solution.

HPLC/FIA has also been used to examine the hydrolysis of trimetaphosphate in aqueous solution [55]. Linear tripolyphosphate is formed initially, and then degrades to pyro- and orthophosphate. From the hydrolytic rate constants at various temperatures, the activation energy for scission of a P–O–P linkage was estimated.

HPLC has also been used with air-segmented autoanalyzers for the separation and detection of various phosphates. Yamaguchi et al. resolved orthophosphate through the P_{30} polymer on a Hitachi 2630 ion-exchange resin column [56]. The effects of eluent composition, flow rates, solvent program modes (linear or convex), the degree of resin cross-linking (4 or 8%) were examined. Orthophosphate through the P_{10} polymer were well resolved on an 8% cross-linked resin column (Hitachi 2632), whereas the 4% column was more effective for the separation of condensed phosphates with a higher degree of polymerization (up to the P_{30} polymer). Cyclic phosphates were eluted later than condensed phosphates with the same number of phosphorus atoms, and their elution order was the reverse of the order of increasing ionic charge. This result has been observed by other investigators using ion-exchange or ion-exchange gel columns [42,57]. The addition of EDTA was also found to be necessary to obtain reproducible profiles. Chromatograms of the polyphosphates separated on the Hitachi 2630 column with and without EDTA in the eluent are shown in Figs. 4 and 5, respectively.

The practical utility of an air-segmented system for the routine analysis of ortho-, pyro-, tripoly-, and tetrapolyphosphate has been considered [58]. Quantitative monitoring requires a residence time

Table 3 Compositional Changes for the Phosphoric Acids in Phosphorus Smokes as a Function of Sample Analysis Time[a]

Sample analysis time following collection (hr)	Composition (%)							
	Ortho-phosphate	Pyro-phosphate	Tripoly-phosphate	Tetrapoly-phosphate	P_5-P_{13}	Higher polyphosphates	Tetrameta-phosphate	Trimeta-phosphate
0[b]	9.5	3.6	3.9	4.5	24.3	53.3	—	—
4	10.0	3.5	3.8	4.9	27.2	49.5	0.7	0.4
24	12.0	4.8	6.5	9.8	42.3	16.5	0.9	7.2
48	15.2	6.5	9.8	13.9	36.3	—	1.4	16.9
72	16.8	7.5	12.3	16.2	26.2	—	2.3	18.7

[a]Total phosphate concentration in aerosol sample, 1.5 mg PO_4^{-3}/liter.
[b]Unaged aerosol (analyzed immediately following collection).
Source: Adapted from Ref. 54.

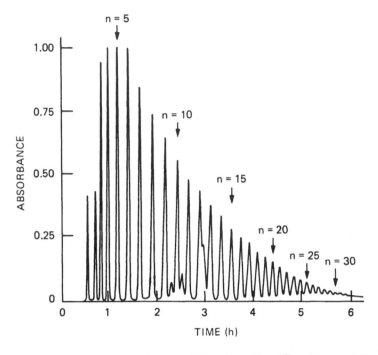

Figure 4   Chromatogram of P_1, P_2, P_n ($\bar{n} = 5$), and P_n ($\bar{n} = 10$) separated on Hitachi 2630 anion-exchange resin and eluted with a convex gradient of NaCl solution (0.22 M to 0.52 M) containing 5 mM Na_4EDTA. (From Ref. 56.)

in the autoanalyzer of approximately 18 min and thus restricts the overall analysis rate. Via an isocratic elution system and separation on a TSK-GEL, IEX 220 SA column, six samples containing P_1, P_2, and P_3 could be analyzed per hour.

E. Miscellaneous Methods

In size-exclusion chromatography, the electrostatic repulsion between sample ions and fixed ionic groups of an ion-exchange resin has been used to separate various phosphate oxo anions [59–65]. Conventionally this type of chromatography has been used to separate electrolytes and nonelectrolytes. Ions having the same charge as the functional groups of a resin are eluted rapidly because of ionic repulsive forces which exclude the species from the resin phase. Neutral compounds are free to occupy the interstitial volume of the resin and are eluted more slowly. Various acids have also been separated by ion-exclusion, and retention volumes have been related to their first

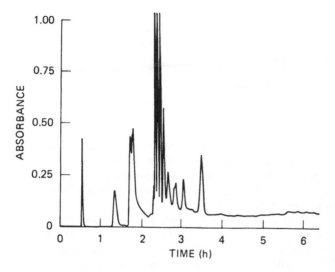

Figure 5 Chromatogram of P_1, P_2, P_n ($\bar{n}$ = 5), and P_n ($\bar{n}$ = 10) separated on Hitachi 2630 anion-exchange resin and eluted with a convex gradient of NaCl solution (0.25 to 0.52 M) without EDTA. (From Ref. 56.)

dissociation constants (pK_1) [59]. The mechanism has been attributed to the Donnan membrane equilibrium between hydrogen ions of the solute and hydrogen ions attached to the resin.

The chromatographic conditions required for the separation of phosphates by ion exclusion have been investigated by Tanaka and Ishizuka [60]. Using a strong cation-exchange column (Hitachi 2613) and an eluent containing a mixture of water and an organic solvent, these investigators resolved ortho-, pyro-, tripoly-, and trimetaphosphate. The column effluent was monitored with a flow-coulometric and conductometric detector. The best separation was obtained with 80% dioxane-water. Because the retention volumes were found to increase as the concentration of the solvent in the eluent increased, both partition and ion-exclusion are believed to influence separations in water-organic solvent systems. Phosphate, phosphonate, and phosphinate have been separated in a similar manner with 70% acetone-water [61].

Waki and Tokunaga have used a sulfonated cross-linked dextran cation-exchanger, SP-Sephadex C-25, and tetramethylammonium chloride solution (1.0 M) as an eluent to resolve various condensed oxo acids and lower oxo acids [62]. Complexation with a molybdenum (V) molybdenum (VI) reagent was used for detection. The separation of highly charged anions such as P_3, P_4, and P_5 was facilitated by reducing their charge either through protonation or ion-pair formation by use of an eluent of lower pH or different composition.

Steric factors have been found to influence the separation of larger anions (P_{4m} and P_{8m}) [63]. The term *Donnan exclusion chromatography* was coined for the separations based only on ion-exclusion effects, and a structural size parameter was introduced for large anions influenced by steric factors in theoretical models explaining the separations. The advantages of Donnan exclusion chromatography, as compared to ordinary ion-exchange separations, are reduced analysis time and elution volumes; the ions are eluted in one column liquid volume.

Applications of exclusion chromatography include the evaluation of stability constants for trimetaphosphate complexes of magnesium and calcium ions [64]. A review of the theory and applications, including discussions on the separation of phosphorus oxo anions, has been presented by Waki and Tokunaga [65].

Nonsuppressed ion chromatography has been used to separate pyro- and tripolyphosphate. A column containing a surface agglomerated resin prepared by coating a 40-m-diameter cation exchanger with a 0.6-m-diameter colloidal anion exchanger has been used with sodium trimesate (10^{-2} M, pH 8) as the eluent to resolve these two phosphates [66]. The anions were detected indirectly by monitoring the background absorbance of the eluent at 294 nm. These phosphates have been determined in a similar manner in detergent formulations [67]. They were separated on an RSil-AN anion column (Alltech Assoc.) and detected by monitoring the quantitative attenuation of the high background absorbance of the eluent (4×10^{-3} M trimesate at pH 8, 285 nm) as the analyte eluted. The sensitivity of this indirect photometric method was greater than that of determinations by conductivity.

IV. CONCLUSIONS

In the past several years, significant advances have been made in the separation methods used for the analysis of the oxo acids of phosphorus. The time-consuming paper and thin-layer methods have largely been replaced with the more efficient automated HPLC methods. Analyses that previously required several days can presently be performed in several hours. The lower members of the condensed phosphates (orthophosphate through the P_{13} polymer) can be resolved by gradient elution techniques on HPLC anion-exchange columns in approximately 1 hr [54]. Routine determinations of ortho-, pyro-, and tripolyphosphate can be conducted by isocratic elution every 10 min [58]. Determinations by automated postcolumn reactions, especially with flow injection techniques, also provide very precise measurements.

The separation of the longer-chain polyphosphates (up to the P_{30} polymer) still requires several hours, and cyclic structures may be incompletely resolved from the condensed forms in samples containing complex mixtures. Although gel chromatography has been very useful for the characterization of various phosphate glasses, the semi-rigid polymer packings as compared to the soft gels may improve efficiency. Further research in ion-exclusion chromatography may provide a comprehensive understanding of the separations mechanisms and lead to methods that provide an even faster analysis time.

One additional area requiring some attention in phosphate analysis is detection methods. In order to differentiate lower oxo acids from condensed forms, two separate postcolumn analyses are required. In the presence of an oxidizing reagent all of the acids can be determined, whereas in the absence of the reagent, only the condensed species or P segments can be determined. A method that could provide some structural information would greatly simplify the analyses of mixtures of oxo anions with different oxidation states and structural complexities.

ACKNOWLEDGMENTS

This work was supported under Army Project Order No. 9600 by the U. S. Army Medical Research and Development Command, Fort Detrick, Frederick, MD 21701, performed at Oak Ridge National Laboratory under U. S. Department of Energy Contract DE-AC05-840R21400 with Martin Marietta Energy Systems, Inc.

REFERENCES

1. J. R. Van Wazer, *Phosphorus and Its Compounds*, Vol. 1, Interscience, New York, 1958.
2. Y. Kiso, M. Kobayoshi, and Y. Kitaoka, in *Analytical Chemistry of Phosphorus Compounds* (M. Halmann, ed.), Wiley-Interscience, New York, 1972, p. 93.
3. S. Ohashi, in *Analytical Chemistry of Phosphorus Compounds* (M. Halmann, ed.), Wiley-Interscience, New York, 1972, p. 409.
4. B. Blaser and K. H. Worms, Z. Anorg. Allgem. Chem. *300*, 225 (1959).
5. R. F. Jameson, J. Chem. Soc. Part I, 752–759 (1959).
6. S. Greenfield and M. Clift, *Analytical Chemistry of the Condensed Phosphates*, Pergamon Press, New York, 1975.
7. M. Zar-Ayan, in *Chem., Phys. Chem. Anwendungstech. Grenzflaechenaktivin Stoffe*, Ber. Int. Kongr., 6th, 1972, Carl Hansen Verlag, Munich, 1973, p. 563.
8. V. V. Pechkovskii, G. Kh. Cherches, and M. I. Kuz'menkov, Rus. Chem. Rev. *44*, 86 (1975).

9. N. Yoza and S. Ohashi, Bull. Chem. Soc. Jpn. *37*, 37 (1964).

10. N. Yoza, K. Ishibashi, and S. Ohashi, J. Chromatogr. *134*, 497 (1977).

11. Y. Hirai, N. Yoza, and S. Ohashi, J. Chromatogr. *206*, 501 (1981).

12. T. Nakamura, T. Yano, T. Nunokawa, and S. Ohashi, J. Chromatogr. *161*, 421 (1978).

13. F. H. Pollard, G. Nickless, D. E. Rogers, and M. T. Rothwell, J. Chromatogr. *17*, 157 (1965).

14. ASTM D515-82, 1982 Annual Book of ASTM Standards, p. 575.

15. D. P. Lundgren, Anal. Chem. *32*, 842 (1960).

16. F. Lucenda—Conde and L. Prat, Anal. Chim. Acta *16*, 473 (1957).

17. Y. Hirai, N. Yoza, and S. Ohashi, Chem. Lett, No. 5, 499 (1980).

18. J. M. Kolthoff, P. J. Elving, and E. S. Sandell, *Treatise on Analytical Chemistry*, Part II, Vol. 5, Interscience, New York, 1961, p. 351.

19. D. P. Lundgren and N. P. Loeb, Anal. Chem. *33*, 366 (1961).

20. T. Nakamura, H. Yamaguchi, and S. Ohashi, J. UOEH *2*, 199 (1980).

21. J. Hirai, N. Yoza, and S. Ohashi, J. Liq. Chromatogr. *2*, 677 (1979).

22. R. S. Deelder, M. G. F. Kroll, A. J. B. Buren, and J. H. M. Van Den Berg, J. Chromatogr. *149*, 669 (1978).

23. N. Yoza, K. Kouchiyama, T. Miyajima, and S. Ohashi, Anal. Lett. *8*, 641 (1975).

24. K. Kouchiyama, N. Yoza, and S. Ohashi, J. Chromatogr. *147*, 271 (1978).

25. N. Yoza, K. Kouchiyama, and S. Ohashi, At. Absorption Newsl. *18*, 39 (1979).

26. B. G. Julin, H. W. Vandenborn, and J. J. Kirkland, J. Chromatogr. *112*, 443 (1975).

27. A. G. Fogg and N. K. Bsebu, Analyst *107*, 566 (1982).

28. Y. Tonogai and M. Iwaida, J. Food Protection *44*, 835 (1981).

29. T. N. Galkova, L. I. Petrovskaya, I. L. Shashkova, and E. A. Prodan, Zh. Anal. Khim. *38*, 1640 (1983).

30. E. A. Prodan and I. L. Shashkova, Vesti Akad. Navrik BSSR, Ser. Khim. Navuk *4*, 79 (1978).

31. G. N. Sondor, B. Janos, F. Terez, and S. Andras, Magy. Kem. Lapja *32*, 57 (1977).

32. Kh. O. Vil'bok, M. Poldme, J. Poldme, and P. Raudsepp, Eesti NSV Tead. Akad. Torm., Keem; Geol. *24*, 44 (1975).

33. E. A. Prodan, I. L. Shashkova, and T. N. Galkova, Zh. Anal. Khim. *33*, 2304 (1977).

34. M. Watanabe and T. Yamada, Chubu Kogyo Daigaku Kiyo. [A] *16*, 111 (1980).

35. T. Fukuda, T. Nakamura, and S. Ohashi, J. Chromatogr. *128*, 212 (1976).

36. R. K. Osterheld, in *Topics in Phosphorus Chemistry* (E. J. Griffith and M. Grayson, eds.), Interscience, New York, 1972, p. 103.

37. T. Nakamura, T. Yano, A. Fujita, and S. Ohashi, J. Chromatogr. *130*, 384 (1977).

38. P. A. Neddermeyer and L. B. Robers, Anal. Chem. *41*, 94 (1969).

39. Y. Ueno, N. Yoza, and S. Ohashi, J. Chromatogr. *52*, 469 (1970).

40. S. Ohashi, N. Yoza, and Y. Ueno, J. Chromatogr. *24*, 300 (1966).

41. Y. Ueno, N. Yoza, and S. Ohashi, J. Chromatogr. *52*, 481 (1970).

42. G. Kura and S. Ohashi, J. Chromatogr. *56*, 111 (1971).

43. T. Glonek, J. R. Van Wazer, M. Mudgett, and T. C. Myers, Inorg. Chem. *11*, 567 (1972).

44. S. Ohashi, G. Kura, Y. Shimada, and M. Hara, J. Inorg. Nucl. Chem. *39*, 1513 (1977).

45. V. V. Peckkovskii, M. I. Kuz'menkov, and G. Kh. Cherches, Inorg. Mat. *9*, 97 (1973).

46. G. Kh. Cherches, V. V. Pechkovskii, and M. I. Kuz'menkov, Zh. Anal. Khim. *32*, 33 (1977).

47. T. Miyajima, K. Yamauchi, and S. Ohashi, J. Liq. Chromatogr. *5*, 265 (1982).

48. T. Miyajima, K. Yamauchi, and S. Ohashi, J. Liq. Chromatogr. *4*, 1891 (1981).

49. T. Miyajima and S. Ohashi, J. Chromatogr. *242*, 181 (1982).

50. Y. Hirai, N. Yoza, and S. Ohashi, Anal. Chim. Acta *115*, 269 (1980).

51. J. Ruzicka and E. H. Hansen, Anal. Chim. Acta *114*, 19 (1980).

52. A. Nakae, K. Furuya, and M. Yamanaka, Nippon Kagaku Kaishi *5*, 708 (1978).

53. H. Kabeya, Nippon Kaguku Kaishi *1*, 65 (1983).

54. R. S. Brazell, R. W. Holmberg, and J. H. Moneyhun, J. Chromatogr. *290*, 163 (1984).

55. G. Kura, T. Nakashima, and F. Oshima, J. Chromatogr. *219*, 335 (1981).

56. H. Yamaguchi, T. Nakamura, Y. Hirai, and S. Ohashi, J. Chromatogr. *172*, 131 (1979).

57. M. Tominaga, T. Nakamura, and S. Ohashi, J. Inorg. Nucl. Chem. *34*, 1409 (1972).

58. N. Yoza, K. Ito, Y. Hirai, and S. Ohashi, J. Chromatogr. *196*, 471 (1980).

59. K. Tanaka and T. Ishizuka, J. Chromatogr. *174*, 153 (1979).

60. K. Tanaka and T. Ishizuka, J. Chromatogr. *190*, 77 (1950).
61. K. Tanaka and H. Sunahara, Bunseki Kagaku *27*, 95 (1978).
62. H. Waki and Y. Tokunaga, J. Chromatogr. *201*, 259 (1980).
63. Y. Tokunaga, H. Waki, and S. Ohashi, J. Liq. Chromatogr. *5*, 1855 (1982).
64. Y. Tokunaga and H. Waki, J. Liq. Chromatogr. *5*, 2169 (1982).
65. H. Waki and Y. Tokunaga, J. Liq. Chromatogr. *5*, 105 (1982).
66. H. Small and T. E. Miller, Jr., Anal. Chem. *54*, 462 (1980).
67. W. D. MacMillan, J. High Resol. Chromatogr. and Chromatogr. Comm. *7*, 102 (1984).

6

HPLC Analysis of Oxypurines and Related Compounds

Katsuyuki Nakano   *PL Medical Data Center, Tondabayashi, Osaka, Japan*

I. INTRODUCTION

Since oxypurines, e.g., xanthines and hypoxanthine (see Fig. 1), are intermediate metabolites in purine metabolism, the determination of these compounds in physiological samples provides important information on the malfunctioning of purine metabolism. In addition, some of the methylated xanthines are also important in pharmacological studies.

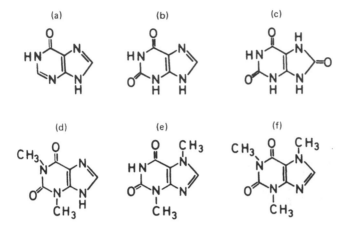

Figure 1   The structure of common oxypurines are as follows: (a) hypoxanthine, (b) xanthine, (c) uric acid, (d) 1,3-dimethylxanthine (theophylline), (e) 3,7-dimethylxanthine (theobromine), and (f) 1,3,7-trimethylxanthine (caffeine).

Several methods for the determination of oxypurines in biological samples have been developed. Enzymatic spectrophotometric [1,2] and spectrofluorometric [3,4] methods have been used. An enzymatic method with oxygen electrodes [5] and open-column chromatographic methods [6,7] have also been reported. However, these methods have a major disadvantage: They do not separate hypoxanthine from xanthine. Thin-layer chromatography [8,9] and two-dimensional paper chromatography [10] could separate out oxypurines, but required a comparatively long time for quantitation.

The identification of these compounds was achieved by the method of direct introduction of a lyophilized sample of blood and muscle [11] into a mass spectrometer. Gas-liquid chromatography coupled with mass spectrometry (GC-MS) [12,13] has high sensitivity and the ability to identify these compounds, but this technique requires derivatization of compounds prior to the analysis.

With the development of high-performance liquid chromatography (HPLC), it has become possible to determine accurately and with high resolution the naturally occurring nucleotides, nucleosides, and their bases in very small samples of tissue, blood, urine, and other physiological fluids.

A. Purine Metabolism

Hypoxanthine and xanthine are intermediate metabolites in purine nucleotide degradation. In purine metabolism (Fig. 2), hypoxanthine and guanine are degraded from inosine and guanosine by purine nucleoside

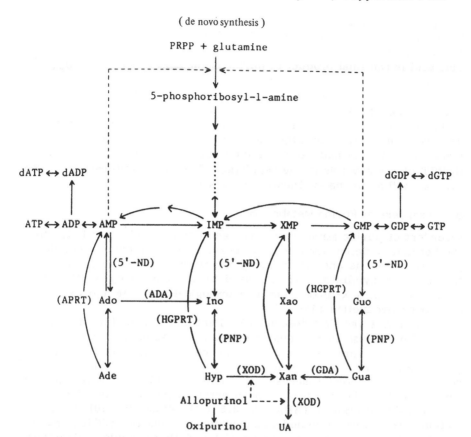

Figure 2 Pathway for purine metabolism. Abbreviations used: PRPP, 5-phosphoribosyl-1-pyrophosphate; ATP, adenosine triphosphate; ADP, adenosine diphosphate; AMP, adenosine monophosphate; Ado, adenosine; Ade, adenine; GTP, guanosine triphosphate; GDP, guanosine diphosphate; GMP, guanosine monophosphate; Guo, guanosine; Gua, guanine; IMP, inosine monophosphate; Ino, inosine; Hyp, hypoxanthine; XMP, xanthosine monophosphate; Xao, xanthosine; Xan, xanthine; UA, uric acid; HGPRT, hypoxanthine-guanine phosphoribosyltransferase; 5'-ND, 5'-nucleotidase; APRT, adenine phosphoribosyltransferase; ADA, adenosine deaminase; PNP, purine nucleoside phosphorylase; XOD, xanthine oxidase; GDA, guanine deaminase. Broken lines show the inhibition reactions.

phosphorylases (PNP, EC 2.4.2.1). Hypoxanthine is oxidized to xanthine by xanthine oxidase (XOD, EC 1.2.3.2), and xanthine to uric acid by the same enzyme, mainly in the liver and intestinal mucosa. Uric acid is the final product of purine metabolism in humans. Hypoxanthine and guanine are also converted to inosine monophosphate (IMP) and guanosine monophosphate (GMP) by hypoxanthine-guanine phosphoribosyltransferase (HGPRT, EC 2.4.2.8.).

Purine nucleotides are produced in the body either directly through de novo synthesis, or through the salvage pathway in which purines formed from the breakdown of the nucleotides are utilized. PNP and HGPRT play a key role in the regulation of de novo synthesis of nucleotides as well as in the synthesis via the salvage pathway.

B. Disorders of Purine Metabolism

Disorders of purine metabolism have been known mainly in relation to the hereditary deficiency of certain enzymes in the purine metabolic pathway. Gout is characterized by hyperuricemia aciduria. The Lesch-Nyhan syndrome [14] is caused by the deficiency of HGPRT associated with a X-linked human neurological disorder and excessive oxypurine production [15].

Xanthinuria was first described by Dent and Philpot [16] as a disorder of purine metabolism characterized by an abnormally high concentration of oxypurines and a very low concentration of uric acid in plasma and urine. A deficiency of XOD has been shown in the jejunal mucosa [17–19], rectal mucosa [19], and liver [17] of these patients. Thus, xanthinuria is a rare, inborn error in purine metabolism.

Adenosine deaminase (ADA, EC 3.5.4.4) deficiency [20] in association with severe combined immunodeficiency disease (SCID) involving disorders of thymus-dependent (T) and thymus-independent (B) lymphocyte function was found in children. PNP deficiency [21] was found in a child with severely defective T-cell immunity and normal B-cell immunity. The reduced activity of 5'-nucleotidase (5'-ND, EC 3.1.3.5) on the external surface of the lymphocyte was observed in a heterogeneous group of patients with hypogammaglobulinemia [22]. Since these discoveries, other inborn errors have been reported.

C. Other Clinical Implications of Purine Metabolites

During and after ischemia or hypoxia, excessive purine nucleotide (especially ATP) breakdown has been observed in the heart, as well as in other muscles. These breakdown products accumulate in the tissue and are released into the blood and urine. Because of high ADA and PNP activities and low XOD activity in the heart and blood, oxypurines seem to be a good marker for myocardial ischemia [23,24], and hypoxia of the fetus and in newborn infants [25,26].

The elevation of oxypurine concentration in the cerebrospinal fluids (CSF) from children with neurological signs or after cerebral convulsions has also been observed [27].

D. Pharmacological Implications of Methylxanthine

Some of the methylated xanthines are used as therapeutic drugs. For example, caffeine (1,3,7-trimethylxanthine) is a stimulator of the central nervous system and a drug used in the prevention of apnea in premature infants. Theophylline (1,3-dimethylxanthine) and aminophylline (theophylline ethylendiamine) are powerful bronchodilators and useful antiasthmatic drugs. Aminophylline and various theobromine (3,7-dimethylxanthine) salts are mild diuretic agents but are not often prescribed in edematous states at present. Clinically it is very important to monitor the narrow therapeutic range of theophylline in the blood during the treatment of asthma, and it has been found that HPLC can be used effectively to determine routinely serum levels of theophylline [28–32]. Early analyses of theophylline that employed conventional 25-cm RPLC columns required at least 10 min [28–31], but with very short columns the analysis can now be carried out in less than a minute [32].

II. ANALYSIS OF XANTHINE AND RELATED COMPOUNDS

In the late 1960s, the high-resolution ultraviolet (UV) analyzer with microreticular anion-exchangers was developed to separate the UV-absorbing constituents in physiological fluids [33]. A number of purine and pyrimidine metabolites, including uric acid, xanthine, and hypoxanthine, were found by this study and subsequent investigations [34,35]. Because of developments in HPLC, especially in packing materials, both separation and analysis time have been improved. Reviews are available for recent HPLC methods for the analysis of nucleotides, nucleosides, and bases [36,37], and methylxanthine drugs [38].

A. Sample Preparation

In general, there are four methods of sample preparation that were available prior to HPLC analysis. These include the use of solvent extraction, precolumns, deproteinization, and direct injection. Solvent extraction and the use of a reversed-phase precolumn are both advantageous for the isolation of methylxanthine drugs. The major advantages of the precolumn are that potential interferences by endogenous compounds are minimized and the drugs are concentrated.

Deproteinization (protein removal) is also a useful technique to stop enzymatic degradation of metabolites and to prevent irreversible adsorption of the proteins onto the packing materials (for a review, see Ref. 39). However, the technique has some limitations, such as the dilution of the sample (in the use of acetonitrile or methanol) and the loss of part of the protein-bound compounds (in the use of trichloroacetic acid or perchloric acid). Ultrafiltration (membrane filtration) is now a commonly used method and is easy to perform. In addition, the sample is not diluted. However, in this method, the concentrations of most of the protein-bound compounds are not determined.

Direct sample injection is not widely accepted. This procedure requires the use of a guard column to protect the analytical column from macromolecules in the sample matrix, especially proteins. After approximatley 30 injections, the guard column must be replaced. Moreover, even with the guard column, the lifetime of the analytical column is significantly shorter with direct sample injection.

B. Chromatography

1. Ion Exchange

In 1973, Singhal and Cohn [40] used anion-exchange columns (Dowex 1—8) and investigated optimal conditions for the separation of nucleosides and bases. Pfadenhauer [41] also reported an ion-exchange HPLC method for the rapid determination of some plasma oxypurines.

Using the anion-exchange packings (Aminex A-28), P. R. Brown et al. [42] initially obtained the separation of the major naturally occurring free nucleosides and bases, including hypoxanthine and xanthine, that were present in blood fluids. However, this method required an analysis time of 2 hr, and resolution of some of the compounds of interest were inadequate.

Comparatively rapid separation of bases by means of a cation exchanger (Aminex A-7) or an anion exchanger (Aminex A-28) was reported by Demushkin and Plyashkevich [43]. Eksteen et al. [44] also used an Aminex A-28 anion exchanger and examined the retention behavior of bases and nucleosides in a system in which the mobile phase consisted of buffer-ethanol mixtures. However, some of the compounds (Hyp/Guo, Xan/Xao) were not completely resolved in this system.

On the other hand, Thompson et al. [45] and Weinberger and Chidsey [46] separated oxypurines via a cation-exchange (Aminex A-5) HPLC method developed for the assay of theophylline and its metabolites.

In addition, ion-exchange HPLC methods have been used for the simultaneous separation of bases, nucleosides, and nucleotides in cell extracts. Floridi et al. [47] reported the HPLC method for these compounds, including hypoxanthine and xanthine, using an anion exchanger (Aminex A-14), and applied the method to the analysis of perchloric acid extracts of yeast and rat liver. Using an Aminex A-25 anion-exchange resin, Bakay et al. [48] and Nissinen [49] reported the simultaneous separation of the three groups of compounds, and applied their method to the quantitation of acid-soluble metabolites in fibroblast cells.

The chromatographic conditions for these analyses are listed in Table 1. In general, as is seen in Table 1, these ion-exchange methods required the use of low flow-rate, elevated temperatures, and a relatively long analysis time.

2. Reversed Phase

Because there are no ionized groups in nucleosides and bases, including oxypurines and methylxanthines, the reversed-phase partition mode of HPLC is particularly suited for the separation of these componds. Comprehensive studies for the naturally occurring nucleosides and bases have been performed by Brown and her collaborators. They first established the conditions for the separation of some standards, using microparticle, chemically bonded, reversed-phase packing (μ-Bondapak C-18) and a methanol gradient in phosphate buffer [50].

As shown in Fig. 3, Krstulovic et al. [51,52], using the reversed-phase mode of HPLC (RPLC), identified nucleic acid metabolites, such as uric acid, xanthine, guanine, xanthosine, inosine, guanosine, and other UV-absorbing constituents, in serum and plasma by the following techniques: peak height ratios (UV, 254 nm/280 nm), co-injection of standards, enzymatic peak shift, and stopped-flow UV scanning.

Hartwick et al. [53] carried this investigation further and made a study of a RPLC analysis of nucleosides and bases, including oxypurines and methylxanthines. Serum was also analyzed quantitatively for the low-molecular-weight, UV-absorbing constituents; the normal levels of 12 compounds including oxypurines were determined [54]. In addition, Hartwick et al. utilized the relationship between ln k' and the composition of an isocratic mobile phase to predict the retention times of nucleosides and bases when gradient elution was used [55].

Anderson and Murphy [56] examined the isocratic RPLC separation of some purine nucleotides, nucleosides, and bases including oxypurines or methylxanthines. Using GC and RPLC, Putterman et al. [57] reported procedures for the analysis of oxypurines, and of allopurinol and its metabolite, oxipurinol. In this study, the HPLC method had 50- to 100-fold greater sensitivity than the GC method. The simultaneous determination of 5-fluorouracil, allopurinol, oxipurinol, and of endogenous compounds such as uridine, hypoxanthine, xanthine, and uric acid was also carried out in plasma samples by Wung and Howell [58], using an RPLC technique with isocratic elution.

An RPLC analysis with gradient elution for plasma purines and pyrimidines, which was developed by McBurney and Gibson [59], was rather poor for the simultaneous determination of all these compounds. However, their method was suitable for hypoxanthine and oxipurinol levels, and it was used for the analyses of these compounds in plasma

Table 1 Chromatographic Conditions for Analyses of Xanthines and Related Compounds via Ion Exchange

Compounds resolved[a]	Stationary phase[b]	Mobile phase	Flow rate (ml/min)	Temp. (°C)	Analysis time (min)	Ref.
Nucleosides and bases including Hyp and Xan	AE: Aminex A-28 (0.2 × 25 cm, Varian)	A: 0.05 M ammonium acetate B: 1.00 M ammonium acetate (both adjusted to pH 9.7), linear gradient (A to B)	12 ml/hr	70	120	42
Bases including Hyp and Xan	AE: Aminex A-28 (0.2 × 30 cm, Bio-Rad)	0.2 M H_3PO_4/0.2 M H_3BO_3 (pH 8.25)	25 ml/hr	45	130	43
Bases including Hyp	CE: Aminex A-7 (0.2 × 30 cm, Bio-Rad)	1 M H_3PO_4 (pH 2.85)	12 ml/hr	60	20	43
Nucleosides and bases including Hyp and Xan	AE: Aminex A-28 (8–12 μm, 0.3 × 25 cm, Bio-Rad)	0.005 M citrate/0.05 M phosphate buffer (pH 9.25)/55% ethanol	—	70	30	44
Methylxanthines, Hyp, Xan	CE: Aminex A-5 (13 μm, 0.18 × 66.5 cm, Bio-Rad)	0.45 M $NH_4H_2PO_4$ (pH 3.65)	0.17	55	25	45

Methylxanthines, Hyp, Xan	CE: Aminex A-5 (13 μm, 0.16 × 85 cm, Bio-Rad)	0.45 M $NH_4H_2PO_4$ (pH 3.65)	0.4	55	30	46
Nucleotides, nucleosides, and bases including Hyp and Xan	AE: Aminex A-14 (20 μm, 0.6 × 50 cm, thick-walled glass column, Bio-Rad)	A: 0.1 M 2-methyl-2-amino-1-propanol (MAP), 0.1 M NaCl (pH 9.90) B: 0.1 M MAP, 0.4 M NaCl (pH 10.00), linear gradient from A to B	100 ml/hr	55	225	47
Nucleotides, nucleosides and bases including Hyp and Xan	AE: Aminex A-25 (17.5 μm, 0.18 × 70 cm, Bio-Rad)	A: 0.07 M $Na_2B_4O_7$/0.045 M NH_4Cl in 2.5% ethanol (pH 9.15) (for Ref. 49) B: 0.01 M $Na_2B_4O_7$/0.50 M NH_4Cl (pH 8.80), linear gradient from A to B	0.5	65	160	48 49

[a] Abbreviations used: Hyp, hypoxanthine; Xan, xanthine.
[b] AE, anion exchange; CE, cation exchange.

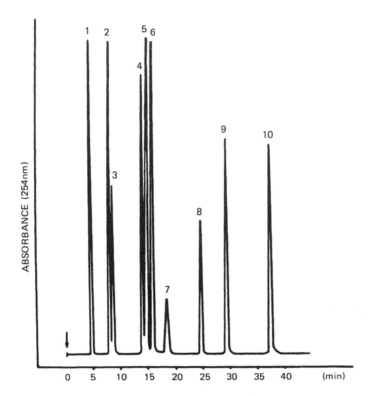

Figure 3 Separation of some reference compounds. Conditions: column, μ-Bondapak C-18; eluents, low concentration, 0.02 F KH_2PO_4, pH 5.5; high concentration, CH_3OH-H_2O (3:2); gradient, linear from 0 to 40% of the high-concentration eluent in 35 min; flow rate, 1.5 ml/min. Detector: UV at 254 nm. Temperature: ambient. Peaks: 1 = uric acid; 2 = hypoxanthine; 3 = xanthine; 4 = xanthosine; 5 = inosine; 6 = guanosine; 7 = tryptophan; 8 = theobromine; 9 = theophylline; 10 = caffeine. (From Ref. 52.)

from subjects with gout and renal failure and in urine from patients deficient in several enzymes related to purine metabolism.

Selective analytical procedures for plasma nucleosides and bases were examined by Taylor et al. [60]. They used an initial RPLC separation on deproteinized plasma and collected a number of fractions of the effluent. These fractions were then reapplied, and a second RPLC separation was carried out.

Simmonds and Harkness [61] developed an isocratic RPLC method for the determination of bases and nucleosides (Fig. 4). They applied this method to cells including erythrocytes, lymphocytes, and

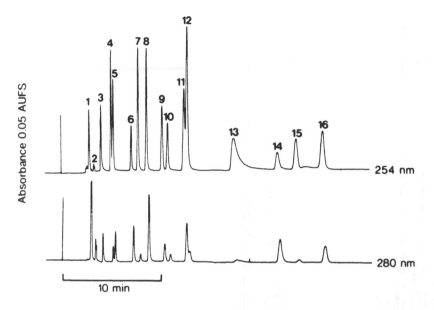

Figure 4 Chromatogram of standard compounds (30—100 pmol).
Peaks: 1 = orotic acid, 2 = uric acid, 3 = cytosine, 4 = uracil,
5 = pseudouridine, 6 = cytidine, 7 = hypoxanthine, 8 = xanthine,
9 = uridine, 10 = oxipurinol, 11 = thymine, 12 = allopurinol, 13 =
adenine, 14 = 7-methylguanine, 15 = inosine, 16 = guanosine. Conditions: 25 × 0.5 cm column of Shandon C-18 3-μm ODS; mobile
phase, 1 ml/min, 0.004 mol/liter KH_2PO_4 (pH 5.8) with 1% (v/v)
methanol at 30°C column temperature. N = 14,000. (From Ref. 61.)

polymorphonuclear neutrophil leukocytes. They also used this method
to determine the levels of hypoxanthine, xanthine, and uridine in
normal plasma samples.

Recently, several isocratic RPLC methods for the determination of
oxypurines in physiological fluids have been developed. De Abreu
et al. [62] applied their method to the analysis of compounds related
to purine and pyrimidine metabolism; however, this method was rather
poor for the separation of hypoxanthine and xanthine. Niklasson [63]
simultaneously determined hypoxanthine and xanthine, urate and creatinine in cerebral spinal fluid, by means of isocratic RPLC with direct injection. As shown in Fig. 5, the isocratic RPLC method developed by Boulieu et al. [64,65] is useful for the rapid determination of levels of hypoxanthine and xanthine in plasma, erythrocytes,
and urine.

Highly specific, simultaneous assays for hypoxanthine and xanthine
in biological fluids have been investigated by Kito et al., using an

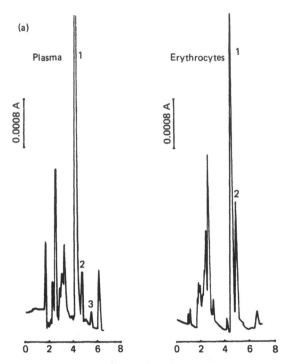

Figure 5 Chromatograms of plasma and erythrocyte samples from a healthy subject. (a) Plasma and erythrocytes were immediately separated by centrifugation after blood collection. (b) Plasma and erythrocytes were separated only 2 hr after blood collection (the whole blood was stored at room temperature). Chromatographic conditions: injection volume, 10 μl; column, Hypersil ODS, 3 μm; mobile phase, 0.02 M KH_2PO_4, pH 3.65; flow rate, 1.5 ml/min; detection, 254 nm. Peaks: 1 = uric acid, 2 = hypoxanthine, 3 = xanthine, 4 = 9-methylxanthine. (From Ref. 65.)

immobilized xanthine oxidase (IXO) enzyme reactor [66] and xanthine oxidase-peroxidase (IXO-IPO) reactors [67] set between the column and detector. In the former system, uric acid produced from hypoxanthine and xanthine by IXO was detected at 290 nm, and in the latter, fluorescent compounds produced from the reaction of p-hydroxyphenylacetic acid and H_2O_2 formed by IXO and IPO were determined using a fluorescence detector.

The effects of eluent pH [50,57,63,64,66,68] and organic solvents [55,66,68,69] on the RPLC retention of oxypurines and other compounds have been examined. Using these results, we can readily

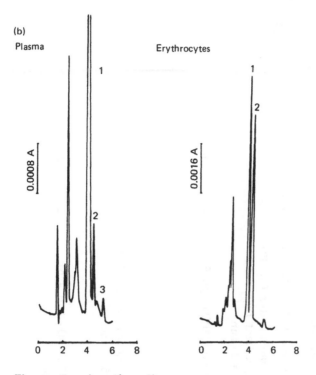

Figure 5 (continued)

obtain the separation of these compounds for specific analytical purposes. The chromatographic conditions for analyses using RPLC are listed in Table 2.

3. Ion Pairing

Since nucleosides and bases can readily be separated by RPLC, few analyses using ion-pair RPLC have been reported. N. D. Brown et al. [70] developed a method for the determination of hypoxanthine, xanthine, and uric acid in biological fluids using 1-heptane sulfonic acid as an ion-pairing agent. Voelter et al. [71] also reported the quantitative determination of hypoxanthine, xanthine, uric acid, and methylxanthines in physiological fluids, using tetrabutylammonium hydroxide as a pairing ion.

Kraak et al. [72] examined the retention behavior of nucleosides and bases, including oxypurines, using a mobile phase of water-ethanol mixtures containing small amounts of the ion-pairing agent, sodium dodecylsulfate (SDS). The chromatographic conditions for analyses using ion-pairing are listed in Table 3.

Table 2 Chromatographic Conditions for Analyses of Xanthines and Related Compounds via Reversed Phase

Compounds resolved[a]	Stationary phase	Mobile phase	Flow rate (ml/min)	Temp. (°C)	Analysis time (min)	Ref.
Nucleosides and bases including Hyp and Xan	μ-Bondapak C-18 (10 μm, 0.4 × 30 cm, Waters)	A: 0.010 F KH_2PO_4 (pH 5.5) B: methanol/water (80:20), linear gradient from A to 25% B in 30 min	1.5	amb.	30	50
Nucleosides and bases including Hyp and Xan	μ-Bondapak C-18 (10 μm, 0.4 × 30 cm, Waters)	A: 0.02 F KH_2PO_4 (pH 5.6) B: methanol/water (60:40), linear gradient from A to 40% B in 35 min	1.5	amb.	35	51 52 53 54
Purine nucleotides, nucleosides and bases including Hyp, Xan, methylxanthines	μ-Bondapak C-18 (0.4 × 30 cm, Waters)	0.05 M $NH_4H_2PO_4$ (pH 6.0), isocratic	2	amb.	25	56
UA, Hyp, Xan, allopurinol, oxipurinol	μ-Bondapak C-18 (0.4 × 30 cm, Waters)	0.05 M ammonium phosphate (pH 4.5), isocratic	1.5	—	14	57
5-FU, UA, Hyp, Urd, Xan, oxipurinol, allopurinol	μ-Bondapak C-18 (0.29 × 25 cm, Waters)	0.05 M KH_2PO_4 (pH 4.6), isocratic	1.5	25	15	58

Compounds	Column	Mobile phase	Flow	Temp.		Ref.
Nucleosides and bases including Hyp and Xan	μ-Bondapak C-18 (0.4 × 30 cm, Waters)	A: 0.073 mM KH_2PO_4 (pH 5.8) B: methanol, linear gradient (A to B)	1	amb.	30	59
Hyp, Thy, Thd, oxipurinol, allopurinol, Ino	μ-Bondapak C-18 (0.4 × 30 cm, Waters) LiChrosorb 10 RP-18	0.05 M ammonium phosphate (pH 5.0): 1st separation 0.025 M ammonium phosphate (pH 5.0): 2nd separation	2 2	23.5 23.5	25 10	60
Nucleosides and bases including Hyp and Xan	ODS-Hypersil (3 μm, 0.5 × 25 cm, Shandon Southern)	0.004 M KH_2PO_4 (pH 5.8) with 1% (v/v) methanol, isocratic	1	30	30	61
Bases, ribonucleosides, deoxyribonucleosides, cyclic ribonucleosides	Spherisorb 10-ODS (10 μm, 0.46 × 25 cm, Chrompack)	A: 0.05 M potassium phosphate buffer (pH 5.60) B: A/methanol/water (50:25:25, v/v) nonlinear ternary gradient (A to B)	1.5	40	35	62
UA, Creat, Hyp, Xan	μ-Bondapak C-18 (0.39 × 30 cm, Waters)	0.2 M potassium phosphate buffer (pH 6.6), isocratic	1	amb.	30	63
UA, Hyp, Xan, 9-M-Xan	Hypersil ODS (3 μm, 0.46 × 15 cm, Shandon)	0.02 M KH_2PO_4 (pH 3.65), isocratic	1.5	—	8	64 65

Table 2 (continued)

Compounds resolved[a]	Stationary phase	Mobile phase	Flow rate (ml/min)	Temp. (°C)	Analysis time (min)	Ref.
Hyp, Xan	Nucleosil 5 C-18 (5 μm, 0.4 × 20 cm, Macherey)	0.01 M phosphate buffer (pH 5.5)/acetonitrile (99:1 or 98:2, v/v)	0.7	—	30	66 or 67

[a]Abbreviations used: UA, uric acid; 5-FU, 5-fluorouracil; Urd, uridine; Thy, thymine; Thd, thymidine; Ino, inosine; Creat, creatinine; 9-M-Xan, 9-methylxanthine. See Table 1 for others.

Table 3 Chromatographic Conditions for Analyses of Xanthine and Related Compounds via Ion Pairing

Compounds resolved[a]	Stationary phase	Mobile phase	Flow rate (ml/min)	Temp. (°C)	Analysis time (min)	Ref.
UA, Hyp, Xan, allopurinol	μ-Bondapak C-18 (0.39 × 30 cm, Waters)	0.005 M 1-heptane sulfonic acid, 0.01 M sodium acetate (pH 4.00) (50:50, v/v)	1.5	amb.	6	70
Hyp, allopurinol, Xan, UA, orotic acid, oxipurinol	RP-18 (5 μm, 0.46 × 25 cm, Knauer)	0.01 M tetrabutylammonium hydroxide (pH 8.0)/CH$_3$CN (99.8:0.2, v/v)	2	35	25	71
Nucleosides, bases including Hyp and Xan	Hypersil ODS (5 μm, 0.45 × 15 cm, Shandon)	0.1 M HClO$_4$/ethanol (9:1, v/v) + 0.1% (w/w) SDS	v = 8 mm /sec	amb.	7	72

[a]For abbreviations, see Tables 1 and 2.

C. Detection and Peak Identification

Because xanthine and related compounds have chromophores with comparatively high extinction coefficients, detection has been carried out mainly by means of a UV detector.

Several techniques have been used for the peak identification of these compounds. Tentative identification was obtained from the retention behavior, cochromatography with standard compounds, ratios of absorbances detected at two different wavelengths (for example, UV 254 nm/280 nm) [50,51,53,54,57], and the stopped-flow UV scanning detectors [52,54,73]. Application of the enzymatic peak-shift techniques [50–54,63], and the use of xanthine oxidase for the determination of hypoxanthine and xanthine, purine nucleoside phosphorylase for inosine and guanosine, and uricase for uric acid, confirmed the identity and purity of the chromatographic peaks of these compounds.

The most powerful identification method is an LC-MS system. In 1983, Esmans et al. [74] reported on the analysis of a mixture of adenosine, thymidine, uridine, and inosine, and a mixture of hypoxanthine, xanthine, adenine, adenosine, uridine, uracil, and thymidine, via an RPLC-MS system and a gradient elution with methanol and ammonium formate. However, systems such as these are not readily available at present.

III. SELECTED HPLC APPLICATIONS

A. Biomedical/Biochemical Applications

By means of the RPLC methods described in the previous section, the levels of hypoxanthine, xanthine, and uric acid were determined in the physiological fluids of patients with gout [50], renal failure [54,59], xanthine oxidase deficiency (xanthinuria) [64], enzyme deficiencies [59], and Lesch–Nyhan syndrome [62]. Crawhall et al. [75] reported an isocratic RPLC method for the measurement of urine and plasma oxypurines in a patient with xanthinuria, and demonstrated its application to the measurement of erythrocyte HGPRT and adenine phosphoribosyltransferase (APRT, EC 2.4.2.7) in a disease similar to the Lesch–Nyhan syndrome.

RPLC methods were also applied to the analyses of nucleosides and bases, including oxypurines, in serum samples from patients with congestive heart failure [51] and with depressive illness [52]. Harmsen et al. [76] developed an isocratic RPLC system for the estimation of purine nucleosides and oxypurines in a perchloric acid (PCA) extract of blood, plasma, and serum from normal subjects and patients with ischemic heart disease. They concluded that the HPLC assay of blood hypoxanthine is a useful tool in the diagnosis of ischemic heart disease. Recently, Hallgren et al. [77] used an isocratic RPLC method to measure the concentrations of oxypurines in cerebral spinal fluid from patients with ischemic brain diseases. It was suggested that oxypurines accumulated rapidly in acute cerebral hypoxia but more gradually with cerebrovascular lesions (CVL); thus urate is a sensitive marker of dysfunction of the blood-brain barrier. Furthermore, Agren et al. [78], using an isocratic RPLC method [63], analyzed hypoxanthine and xanthine in the cerebral spinal fluid of patients with major depressive disorders. They found that both xanthines appeared to be linked with the expression of depressive symptomatology.

HPLC methods have also been applied to the determination of nucleotides, nucleosides, and bases, including oxypurines, in acid extracts of yeast or rat liver [47], and skin fibroblasts [48,49], via anion exchange, and of suspensions of rat cardiac myocytes [79] and bacterial cells [80], via reversed phase. The RPLC methods were also used for the analyses of purine nucleotides, nucleosides, and bases in cat spleen perfusates [56], of nucleosides and bases in rabbit kidney perfusates [81], and of UV-absorbing compounds in serum, urine, and dialyzate from dialyzed patients [82].

In addition, HPLC techniques were applied to the rapid determination of uric acid in physiological fluids. For example, for the specific determination of serum uric acid, Kiser et al. reported an isocratic RPLC method [83], and Pachla and Kissinger reported an anion-exchange method with electrochemical detection [84].

We examined the profile of UV-absorbing compounds in saliva [85], and in the PCA extract of gastrointestinal (GI) mucosa tissue [86], using an RPLC method with gradient elution. Compared with the serum [51,54] and saliva [85] profiles, the GI mucosa tissue extract profile showed relatively high levels of hypoxanthine, xanthine, and uracil, and a low level of uric acid, as shown in Fig. 6 [86].

B. Pharmacological Applications

HPLC methods have been used for the determination of methylated xanthine drugs and drugs that are used to influence xanthine metabolism [38]. An example of the latter is allopurinol, which is used as a therapeutic drug for gout, hyperuricemia, and cancer because it is a powerful inhibitor of xanthine oxidase and inhibits the accumulation of hypoxanthine. Allopurinol and its metabolite, oxipurinol, have been determined in physiological fluids with no interference from endogenous compounds, by means of several of the HPLC methods described in the previous section [57,58,60,70,71,81]. Specific HPLC methods for the determination of allopurinol and oxipurinol were also developed with anion exchange [87] and isocratic reversed-phase [88]. The chromatographic conditions for selected HPLC applications are listed in Table 4.

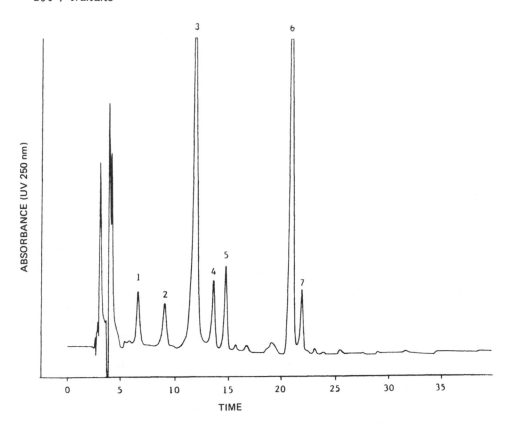

Figure 6 Chromatogram of a PCA extract of human gastric mucosa tissue homogenate. Column: Develosil ODS-5 (5 μm, 250 × 4.6 mm i.d., Nomura Chemical Co.). Eluents: A, 0.02 M KH_2PO_4, pH 5.0; B, CH_3OH-H_2O (3:2); gradient, linear from A to 40% B in 35 min; flowrate, 1.2 ml/min. Detector: UV at 250 nm (0.16 aufs). Temperature: ambient. Peaks: 1 = uracil, 2 = uric acid, 3 = hypoxanthine, 4 = xanthine, 5 = uridine, 6 = inosine, 7 = guanosine. (From Ref. 86.)

Table 4 Chromatographic Conditions for the Selected HPLC Applications

Compounds resolved[a]	Stationary phase[b]	Mobile phase	Flow rate (ml/min)	Temp. (°C)	Analysis time (min)	Ref.
UA, Hyp, Xan	RP: LiChrosorb RP-8 (10 µm, 0.46 × 25 cm, Merck)	0.05 M potassium hydrogen phosphate buffer (pH 2.85), isocratic	—	—	14	75
UA, Ura, Urd, Hyp, Xan, nucleosides	RP: µ-Bondapak C-18 (10 µm, 0.4 × 30 cm, Waters)	0.01 M $NH_4H_2PO_4/CH_3OH$ (10:1, v/v), pH 5.50, isocratic	1	—	20	76
UA, Creat, Xan, Hyp	RP: Nucleosil C-8 (5 µm, 0.4 × 25 cm, Macherey–Nagel)	0.2 M KH_2PO_4 (pH 7.2), isocratic	1.0	—	8	77
Nucleosides and bases including Hyp	RP: Partisil ODS-1 C-18 (10 µm, 0.46 × 25 cm, Whatman)	0.010 M potassium phosphate (pH 5.5)/methanol (90:10, v/v), isocratic	1.12	amb.	30	79
Purine and degradation products including Hyp, Xan	RP: LiChrosorb RP-18 (10 µm, 0.46 × 25 cm, E. Merck)	0.1 M potassium phosphate (pH 3.7), isocratic	1.5	amb.	20	80

Table 4 (continued)

Compounds resolved[a]	Stationary phase[b]	Mobile phase	Flow rate (ml/min)	Temp. (°C)	Analysis time (min)	Ref.
Nucleosides, bases including Hyp, allopurinol, oxipurinol	RP: μ-Bondapak C-18 (0.4 × 30 cm, Waters)	A: K_2HPO_4 (10 mg/liter, pH 4.5) B: 20% CH_3OH in A buffer, convex curve gradient from A to 50% B in 10 min	2	amb.	18	81
UV-absorbing compounds including Hyp and Xan	RP: μ-Bondapak C-18 (0.4 × 60 cm, Waters)	A: 0.025 M acetic acid buffer (pH 4.5) B: 0.1 M acetic acid in methanol, concave gradient from A to B in 90 min	1.0	—	80	82
UA, Ade	RP: μ-Bondapak C-18 (Waters)	35 ml/liter solution of acetonitrile in 0.010 M sodium acetate (pH 4.0), isocratic	2.5	—	4	83

UA	AE: SC-SAX (0.21 × 50 cm, Vydac)	0.10 M acetate buffer (pH 5.25), isocratic	0.39	40	5	84
Ura, UA, Hyp, Xan, Urd, Ino, Guo	RP: Develosil ODS-5 (5 μm, 0.46 × 25 cm, Nomura Chem.)	A: 0.02 M KH$_2$PO$_4$ (pH 5.0) B: CH$_3$OH/H$_2$O (3:2, v/v), linear gradient from A to 40% B in 35 min	1.2	amb.	35	86
Hyp, Xan, allopurinol and oxipurinol	AE: Aminex A-27 (12–15 μm, 0.45 × 7 cm, Bio-Rad)	1 M ammonium acetate (pH 8.7), isocratic	1	71	20	87
Allopurinol, oxipurinol and acetaminophen	RP: Spherisorb ODS (5 μm, 0.41 × 25 cm, Lab. Data Control)	0.05 M phosphate buffer (pH 6.0), isocratic	2	—	30	88

[a] Abbreviations used: Ura, uracil; Ade, adenine; Guo, guanosine. See Tables 1 and 2 for others.
[b] RP, reversed phase; AE, anion exchange.

C. Normal Oxypurine Levels in Physiological Samples

By means of the HPLC methods described in the previous section, the levels of oxypurines in physiological samples were determined. Table 5 shows the hypoxanthine and xanthine levels in physiological samples of normal subjects. Most values of oxypurines in serum vary widely. Recently, the effects of several anticoagulants such as heparin, acid-citrate-dextrose (ACD), and ethylenediaminetetraacetic acid (EDTA) on plasma RPLC chromatograms were investigated [89]. This study revealed that the presence of UV-absorbing compounds in ACD and EDTA solutions interfered with the determination of some plasma constituents. However, there were no interferences in the chromatograms of plasma samples if the anticoagulant was heparin.

Jorgensen and Poulsen [90] found a 1000-fold increase in hypoxanthine in serum left in contact with blood cells for 24 hr; thus, the timing of blood-sample preparation after venipuncture is very important. Zakaria and Brown also found changes in peak areas of several compounds in serum samples [54] and heparinized plasma samples [89] incubated at 25°C (in contact with blood cells). Drastic changes in adenosine, inosine, hypoxanthine, and xanthine levels were observed, indicating that the changes were due significantly to the by-products of nucleotide catabolism in blood cells (Fig. 7). A parallel increase in hypoxanthine levels over a period of time was found in erythrocytes [58,65,89] (for example, see Fig. 5), and platelets [58]. Even when whole blood was stored in an ice bath, before separation from the blood cells, increases were observed in hypoxanthine levels, although the increases were not as pronounced [65]. Therefore, these results emphasize that the blood must be centrifuged immediately after collection in order to achieve accurate and reproducible results in the determination of hypoxanthine levels.

In 1980, Sutton et al. [91] reported on the influence of exercise on purine metabolism in humans and suggested that there is accelerated degradation of purine nucleotides in skeletal muscle during vigorous exercise.

D. Enzyme Assays Related to Xanthine Metabolism

By means of HPLC techniques, the activity of several enzymes related to purine metabolism has been monitored in terms of the quantitation of substrates or products of the enzymatic reactions.

Hartwick and Brown [92] developed a rapid assay for adenosine deaminase in erythrocytes via an isocratic RPLC method. In this assay, the decrease of substrate adenosine, which had been added to an erythrocyte lysate, corresponded to the increase in both inosine and hypoxanthine. Chen and Hsu [93] reported a simple anion-exchange HPLC procedure for the automated quantitation of adenosine to determine the 5'-nucleotidase activity in human cervix homogenate.

Table 5   Normal Hypoxanthine (Hyp) and Xanthine (Xan) Levels in Physiological Samples[a]

Methods	Samples[b]	Hyp	Xan	Ref.
IEX	Plasma (M, 5)	$3.1 - 10.9^d$	$<1.6^d$	41
	Plasma (F, 3)	$1.8 - 2.5^d$	$<1.6^d$	
RPLC	Serum (M, 17; F, 14)	7.16 ± 2.81	2.62 ± 1.04	54
RPLC	Serum	$13.2 - 16.2^d$	$9.9 - 20.4^d$	57
	Urine	$0 - 1290^d$	$0 - 2040^d$	
RPLC	EDTA plasma[c]	0.46 ± 0.21	0.40 ± 0.27	58
RPLC	Heparin plasma (10)	1.03 ± 0.9	4.9 ± 1.5	59
RPLC	Heparin plasma (13)[c]	1.2 ± 0.72	—	60
RPLC	Plasma (4)[c]	1.5 ± 0.2	0.46 ± 0.09	61
	Erythrocytes (4)	11.0 ± 5.8	<0.1	
	Lymphocytes (4)	139 ± 22	0.2 ± 0.1	
	PMN (4)	375 ± 98	9.7 ± 5.5	
RPLC	Heparin plasma (6)[c]	3.6 ± 0.8	1.1 ± 0.7	76
	Serum (6)[c]	5.6 ± 1.9	6.6 ± 2.1	
	Blood (6)[c]	2.2 ± 1.3	0.2 ± 0.1	
RPLC	ACD plasma (M, 9)	$1.34 - 6.41$	—	89
	ACD plasma (F, 5)	$0.94 - 5.40$	—	
	EDTA plasma (M, 8)	$3.20 - 10.4$	—	
	EDTA plasma (F, 5)	$2.18 - 6.41$	—	
	Heparin plasma (M, 9)	$4.79 - 14.6$	—	
	Heparin plasma (F, 5)	$3.84 - 12.1$	—	
	Serum (M, 9)	$15.6 - 24.0$	—	
	Serum (F, 6)	$13.7 - 19.8$	—	
RPLC	Saliva (5)	1.09 ± 1.00	1.18 ± 0.81	85
RPLC	Serum (5)	6.42	2.02	67
	Urine (5)	116.13	89.92	
RPLC	CSF (M, 14; F, 12)	2.6 ± 0.5	1.7 ± 0.4	77
RPLC	Heparin plasma (11)[c]	2.5 ± 1.0	1.4 ± 0.7	65
	Erythrocytes (11)[c]	8.0 ± 6.2	n.d.	
	Urine (11)	48 ± 26^e	68 ± 42^e	

[a] Data represent mean values ± SD or range. Unit: micromolar concentration (μM).

[b] Abbreviation used: PMN, polymorphonuclear neutrophil leukocytes. Sex (M, male; F, female) and number of samples are shown in parentheses.

[c] Blood samples were centrifuged immediately after collection

[d] Unit was converted to micromolar concentration (μM).

[e] Unit is μM/24-hr urine.

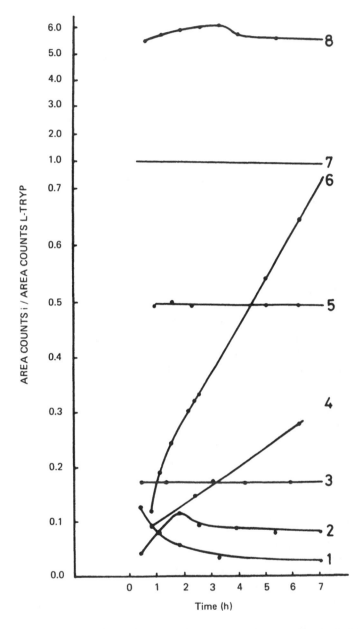

Figure 7 Changes in the normalized peak areas of a plasma sample incubated at 25°C for a period of 7 hr. The plasma was allowed to be in contact with the blood cells during this study. The compounds in the plasma are (1) adenosine, (2) inosine, (3) guanosine, (4) xanthine, (5) creatinine, (6) hypoxanthine, (7) L-tryptophan, and (8) uric acid. (From Ref. 89.)

Using an isocratic RPLC method, Halfpenny and Brown [94] were able to optimize an assay for purine nucleoside phosphorylase (PNP). The substrate, inosine, when incubated with the cell lysate, was converted into hypoxanthine. Xanthine oxidase was added to the reaction mixture to prevent the accumulation of hypoxanthine, which would inhibit the forward reaction of PNP. Uric acid, hypoxanthine, xanthine, and inosine were resolved for this assay with no interference from other compounds (Fig. 8). In addition, using HPLC, Halfpenny and Brown were able to determine simultaneously the activity of the enzymes HGPRT and PNP by monitoring the disappearance of inosine or guanosine and the increase in the concentration of IMP or GMP [95].

Vasquez and Bieber [96] developed an isocratic RPLC method for the assay of HGPRT purified from bovine brain. In this method, the product, inosine monophosphate (IMP) or guanosine monophosphate (GMP), was separated from the substrate, hypoxanthine or guanine.

Recently, Rylance et al. [97] reported a RPLC method with gradient elution for the assay of HGPRT in erythrocytes. The product of

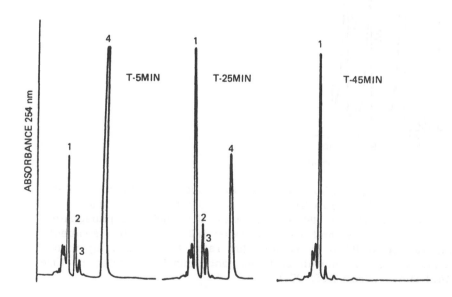

Figure 8 Reaction of purine nucleoside phosphorylase as a function of time. Chromatograms at various time intervals show the decrease of the substrate inosine (4) and the increase of the products uric acid (1), hypoxanthine (2) and xanthine (3). Chromatographic conditions: isocratic elution, 2 ml/min, 0.02 F KH_2PO_4 (pH 4.2) - 3% methanol. (From Ref. 94.)

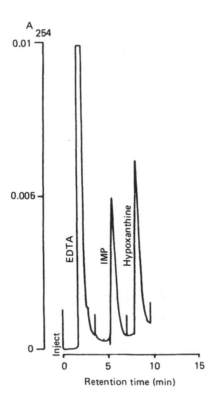

Figure 9 Typical HPLC chromatograph, showing separation of inosine monophosphate from hypoxanthine. Conditions: column; μ-Bondapak C-18 (3.9 × 300 mm); eluents, 0.02 mol/liter phosphate buffer (pH 5.6) (A) and methanol/water (60%, v/v) (B); gradient, linear 100% A to 40% B in 35 min; flow rate, 1.5 ml/min; injection volume, 40 μl. Inosine under these conditions has a retention time of 14.4 min. (From Ref. 97.)

Table 6 Chromatographic Conditions for the Enzyme Assays

Enzyme: compounds[a]	Stationary phase[b]	Mobile phase	Flow rate (ml/min)	Temp. (°C)	Analysis time (min)	Ref.
ADA: Hyp, Ino, Ado	RP: μ-Bondapak C-18 (10 μm, 0.4 × 30 cm, Waters)	0.01 F KH_2PO_4 (pH un-adjusted)/methanol (86:14, v/v), isocratic	2.0	amb.	6	92
5'-ND: Ado	AE: AG MP-1 (0.9 × 4 cm, Bio-Rad)	Linear gradient of HCl	2	—	10	93
PNP: UA, Hyp, Xan, Ino	RP: Partisil 5-ODS (5 μm, 0.4 × 25 cm, Whatman)	0.02 F KH_2PO_4 (pH 4.2)/methanol (97:3, v/v), isocratic	2.0	amb.	7	94
HGPRT: IMP, Hyp/GMP. Gua/Xan, Gua	RP: μ-Bondapak C-18 (0.4 × 30 cm, Waters)	0.125 M potassium phos-phate (pH 6.0), isocratic	2.5	—	7	96
HGPRT: EDTA, IMP, Hyp	RP: μ-Bondapak C-18 (10 μm, 0.39 × 30 cm, Waters)	A: 0.02 M phosphate buffer (pH 5.6) B: methanol/water (60:40), linear gradient from A to 40% B in 35 min	1.5	amb.	10	97

[a]For abbreviations, see the legend for Fig. 2.
[b]For abbreviations, see Table 4.

this enzymatic reaction, IMP, was separated from the substrate, hypoxanthine, and EDTA was added to the incubation medium to stop the reaction (Fig. 9). The chromatographic conditions for the enzyme assays are listed in Table 6.

ACKNOWLEDGMENTS

The author acknowledges Dr. T. Yasaka (PL Medical Data Center) for encouragement, and Patriarch T. Miki and "Perfect Liberty" Organization for a grant.

REFERENCES

1. H. M. Kalckar, J. Biol. Chem. *167*, 429 (1947).
2. J. R. Klinkenberg, S. Goldfinger, H. H. Bradley, and J. E. Seegmiller, Clin. Chem. *13*, 834 (1967).
3. G. G. Guibault, P. Brignac, Jr., and M. Zimmer, Anal. Chem. *40*, 190 (1968).
4. T. Sumi and Y. Umeda, Clin. Chim. Acta *95*, 291 (1979).
5. O. D. Saugstad, Pediatr. Res. *9*, 575 (1975).
6. H. A. Simmonds and J. D. Wilson, Clin. Chim. Acta *16*, 155 (1967).
7. T. T. Hayashi and B. Gilling, Anal. Biochem. *36*, 343 (1970).
8. P. J. Orsulak, W. Haab, and M. D. Appleton, Anal. Biochem. *23*, 156 (1968).
9. I. Akaoka, T. Nishizawa, and Y. Nishida, Biochem. Med. *14*, 285 (1975).
10. M. Krzyzanowska and W. Niemierdo, J. Chromatogr. *100*, 95 (1974).
11. W. Snedden and R. B. Parker, Anal. Chem. *43*, 1651 (1971).
12. V. Miller, V. Pacakova, and E. Smolkova, J. Chromatogr. *119*, 355 (1976).
13. J. L. Chabard, C. Lartique-Mattei, F. Vedrine, J. Petit, and J. A. Berger, J. Chromatogr. *221*, 9 (1980).
14. M. Lesch and W. L. Nyhan, Am. J. Med. *36*, 561 (1964).
15. J. E. Seegmiller, F. M. Rosenbloom, and W. N. Kelley, Science *155*, 1682 (1967).
16. C. E. Dent and G. R. Philpot, Lancet *1*, 182 (1954).
17. K. Engelman, R. W. E. Watts, J. R. Klinkenberg, A. Sjoerdsma, and J. E. Seegmiller, Am. J. Med. *37*, 839 (1964).
18. O. Sperling, U. A. Liberman, M. Frank, and A. De Vries, Am. J. Clin. Pathol. *55*, 351 (1971).
19. E. W. Holmes, D. H. Mason, L. T. Goldstein, R. E. Blant and W. N. Kelly, Clin. Chem. *21*, 76 (1974).

20. E. R. Giblett, J. E. Anderson, F. Cohen, B. Pollara, and H. J. Meuwissen, Lancet 2, 1067 (1972).
21. E. R. Giblett, A. J. Ammann, D. W. Wara, R. Sandman, and L. K. Diamond, Lancet 1, 1010 (1975).
22. S. M. Johnson, G. L. Asherson, R. W. E. Watts, M. E. North, J. Allsop, and A. D. B. Webster, Lancet 1, 168 (1977).
23. G. Kugler, Eur. J. Cardiol. 9, 227 (1979).
24. A. C. Fox, G. E. Reed, H. Meilman, and B. B. Silk, Am. J. Cardiol. 43, 52 (1979).
25. O. D. Saugstad, Pediatr. Res. 9, 158 (1975).
26. M. C. O'Connor, R. A. Harkness, R. J. Simmonds, and F. E. Hytten, Br. J. Obstet. Gynaecol. 88, 381 (1981).
27. H. Manzke and W. Staemmler, Neuropediatrics 12, 209 (1981).
28. R. F. Adams, F. L. Vandemark, and G. L. Schmidt, Clin. Chem. 22, 1903 (1976).
29. J. W. Nelson, A. L. Cordry, C. G. Aron, and R. A. Bartell, Clin. Chem. 23, 124 (1977).
30. R. K. Desiraju, E. T. Sugita, and R. L. Mayock, J. Chromatogr. Sci. 15, 563 (1977).
31. J. R. Miksic and B. Hodes, J. Pharm. Sci. 68, 1200 (1979).
32. P. M. Kabra and L. J. Marton, Clin. Chem. 28, 687 (1982).
33. C. D. Scott, J. E. Attrill, and N. G. Anderson, Proc. Soc. Exp. Biol. Med. 125, 181 (1967).
34. C. D. Scott, R. L. Jolley, W. W. Pitt, and W. F. Johnson, Am. J. Clin. Pathol. 53, 701 (1970).
35. J. E. Mrochek, W. C. Butts, W. T. Rainey, Jr., and C. A. Burtis, Clin. Chem. 17, 72 (1971).
36. M. Zakaria and P. R. Brown, J. Chromatogr. 226, 267 (1981).
37. K. Nakano, in *HPLC in Nucleic Acid Research* (P. R. Brown, ed.), Marcel Dekker, New York, p. 247, 1984.
38. K. Nakano, in *HPLC in Nucleic Acid Research* (P. R. Brown, ed.), Marcel Dekker, New York, p. 339, 1984.
39. P. R. Brown, Cancer Inv. 1, 439 (1983).
40. R. P. Singhal and W. E. Cohn, Biochemistry 12, 1532 (1973).
41. E. H. Pfadenhauer, J. Chromatogr. 81, 85 (1973).
42. P. R. Brown, S. Bobick, and F. L. Hanley, J. Chromatogr. 99, 587 (1974).
43. V. P. Demushkin and Yu. G. Plyashkevich, Anal. Biochem. 84, 12 (1978).
44. R. Eksteen, J. C. Kraak, and P. Linssen, J. Chromatogr. 148, 413 (1978).
45. R. D. Thompson, H. T. Nagasawa, and J. W. Jenne, J. Lab. Clin. Med. 84, 584 (1974).
46. M. Weinberger and C. Chidsey, Clin. Chem. 21, 834 (1975).

47. A. Floridi, C. A. Palmerini, and C. Fini, J. Chromatogr. *138*, 203 (1977).
48. B. Bakay, E. Nissinen, and L. Sweetman, Anal. Biochem. *86*, 65 (1978).
49. E. Nissinen, Anal. Biochem. *106*, 497 (1980).
50. R. A. Hartwick and P. R. Brown, J. Chromatogr. *126*, 679 (1976).
51. A. M. Krstulovic, P. R. Brown, and D. M. Rosie, Anal. Chem. *49*, 2237 (1977).
52. A. M. Krstulovic, R. A. Hartwick, P. R. Brown, and K. Lohse, J. Chromatogr. *158*, 365 (1978).
53. R. A. Hartwick, S. P. Assenza, and P. R. Brown, J. Chromatogr. *186*, 647 (1979).
54. R. A. Hartwick, A. M. Krstulovic, and P. R. Brown, J. Chromatogr. *186*, 659 (1979).
55. R. A. Hartwick, C. M. Grill and P. R. Brown, Anal. Chem. *51*, 34 (1979).
56. F. S. Anderson and R. C. Murhpy, J. Chromatogr. *121*, 251 (1976).
57. G. J. Putterman, B. Shaikh, M. R. Hallmark, C. G. Sawyer, C. V. Hixson, and F. Perini, Anal. Biochem. *98*, 18 (1979).
58. W. E. Wung and S. B. Howell, Clin. Chem. *26*, 1704 (1980).
59. A. McBurney and T. Gibson, Clin. Chim. Acta *102*, 19 (1980).
60. G. A. Taylor, P. J. Dady, and K. R. Harrap, J. Chromatogr. *183*, 421 (1980).
61. R. J. Simmonds and R. A. Harkness, J. Chromatogr. *226*, 369 (1981).
62. R. A. De Abreu, J. M. Van Baal, C. H. M. M. De Bruyn, J. A. J. M. Bakkeren, and E. D. A. M. Schretlen, J. Chromatogr. *229*, 67 (1982).
63. F. Niklasson, Clin. Chem. *29*, 1543 (1983).
64. R. Boulieu, C. Bory, P. Baltassat, and C. Gonnet, J. Chromatogr. *233*, 131 (1982).
65. R. Boulieu, C. Bory, P. Baltassat, and C. Gonnet, Anal. Biochem. *129*, 398 (1983).
66. M. Kito, R. Tawa, S. Takeshima, and S. Hirose, J. Chromatogr. *231*, 183 (1982).
67. M. Kito, R. Tawa, S. Takeshima, and S. Hirose, J. Chromatogr. *278*, 35 (1983).
68. M. Zakaria, P. R. Brown, and E. Grushka, Anal. Chem. *55*, 547 (1980).
69. S. P. Assenza and P. R. Brown, J. Chromatogr. *289*, 355 (1984).
70. N. D. Brown, J. A. Kintzios, and S. E. Koetitz, J. Chromatogr. *177*, 170 (1979).
71. W. Voelter, K. Zech, P. Arnold, and G. Ludwig, J. Chromatogr. *199*, 345 (1980).

72. J. C. Kraak, C. X. Ahn, and J. Fraanje, J. Chromatogr. *209*, 369 (1981).
73. S. P. Assenza and P. R. Brown, J. Chromatogr. *181*, 169 (1980).
74. E. L. Esmans, Y. Luyten, and F. C. Alderweireld, Biomed. Mass Spectrometry *10*, 347 (1983).
75. J. C. Crawhall, K. Itiaba, and S. Katz, Biochem. Med. *30*, 261 (1983).
76. E. Harmsen, J. W. de Jong, and P. W. Serruys, Clin. Chim. Acta *115*, 73 (1981).
77. R. Hallgren, F. Niklasson, A. Terent, A. Akerblom, and E. Widerlov, Stroke *14*, 382 (1983).
78. H. Agren, F. Niklasson, and R. Hallgren, Psychiatry Res. *9*, 179 (1983).
79. R. J. Henderson, Jr., and C. A. Griffin, J. Chromatogr. *226*, 202 (1981).
80. P. Durre and J. R. Andreesen, Anal. Biochem. *123*, 32 (1982).
81. K. M. Taylor, L. Chase, and M. Bewick, J. Liq. Chromatogr. *1*, 849 (1978).
82. E. J. Knudson, Y. C. Lau, H. Veening, and D. A. Dayton, Clin. Chem. *24*, 686 (1978).
83. E. J. Kiser, G. F. Johnson, and D. L. Witte, Clin. Chem. *24*, 536 (1978).
84. L. A. Pachla and P. T. Kissinger, Clin. Chem. *25*, 1847 (1979).
85. K. Nakano, S. P. Assenza, and P. R. Brown, J. Chromatogr. *233*, 51 (1982).
86. K. Nakano, K. Shindo, H. Yamamoto, and T. Yasaka, J. Chromatogr. *332*, 127 (1985).
87. M. Brown and A. Bye, J. Chromatogr. *143*, 195 (1977).
88. W. G. Kramer and S. Feldman, J. Chromatogr. *162*, 94 (1979).
89. M. Zakaria and P. R. Brown, Anal. Biochem. *120*, 25 (1982).
90. S. Jorgensen and H. Poulsen, Acta Pharmacol. Toxicol. *11*, 223 (1955).
91. J. R. Sutton, C. J. Toews, G. R. Ward, and I. H. Fox, Metabolism *29*, 254 (1980).
92. R. A. Hartwick and P. R. Brown, J. Chromatogr. *143*, 383 (1977).
93. S. S. Chen and D. S. Hsu, J. Chromatogr. *210*, 186 (1981).
94. A. P. Halfpenny and P. R. Brown, J. Chromatogr. *199*, 275 (1980).
95. A. P. Halfpenny and P. R. Brown, Proceedings of Pittsburgh Conference on Analytical Chemistry and Applied Spectroscopy, March 1983.
96. B. Vasquez and A. Bieber, Anal. Biochem. *79*, 52 (1977).
97. H. J. Rylance, R. C. Wallence, and G. Nuki, Clin. Chim. Acta *121*, 159 (1982).

7

Liquid Chromatography of Carbohydrates

Toshihiko Hanai   *Gasukuro Kogyo Inc., Iruma, Japan*

I. INTRODUCTION

Carbohydrates are an important energy source and are involved in the structure of proteins. The analysis of carbohydrates by liquid chromatography has been therefore undertaken by many researchers. The high-speed analyses available today are possible because of the development of new packings and detection systems. The analysis time unit has become the minute instead of the hour; furthermore, in the not too distant future it will be shortened to the second. Because there are published books and reviews [1—15] on the details of reports published before 1979, in this review the state of the art of liquid chromatography of carbohydrates will be summarized from the publications of the past five years. Over 700 reports related to liquid chromatography of carbohydrates have been published since 1979, and 321 of these publications from the past few years were used for this review.

II. PACKINGS

For the analysis of carbohydrates, the separation of optical isomers is often necessary. The development of new packings, the increased efficiency of columns, and the new derivatization reagents now make possible the separation of such isomers. For example, the enantiomers of aldoses have been separated as diastereoisomeric 1-(N-acetyl-α-methylbenzylamino)-1-deoxyalditol acetates [16,17]. Some of the packings used in the separations of carbohydrates are discussed in the following paragraphs.

A. Silica Gels with Chemically Bonded Amino Groups

Alkylamino-modified silica gels, especially those modified with the propylamino group, are now widely used for the separation of mono- and oligosaccharides. The retention mechanism of saccharides on these packings in acetonitrile/water mixtures is described as partition [18]. In the separation of isomers, the number and the steric position of hydroxyl groups are important [19]. In situ modifications of silica with amines have been described [20—22]. An example of a chromatogram of oligosaccharides is shown in Fig. 1 [23].

The selectivity of organic modifiers for the chromatography of saccharides on a propylamine-bonded silica gel [24] and the selectivity of different packings [25] have been discussed. It has been found that an acetone/ethylacetate/water mixture can be used as the mobile phase instead of acetonitrile/water mixture [26], which is more toxic.

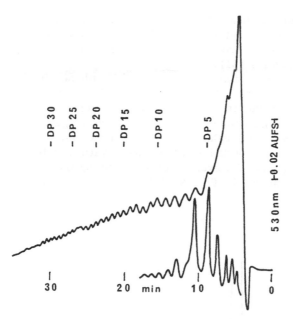

Figure 1 Chromatogram of oligosaccharides of exploded wood.
Column, 25 cm long, 4.3 mm i.d., packed with Chromosorb LC-9
(a propylamino-bonded silica gel); elution, a linear gradient from
acetonitrile:water (70:30) to acetonitrile:water (62.5:37.5) in 30 min;
flow rate, 1 ml/min; column temp, ambient; detection, tetrazolium
blue reduction reaction detection at 530 nm; DP, degree of polymerization, is indicated by numbers over the peaks.

B. Silica Gels Physically Modified with Amino Groups

Silica gel packings physically modified with amino groups are quite
stable and effective for the chromatography of mono- and oligosaccharides [27]. However, if n-alkylamine is present in the eluent, it
is possible to use octadecyl-bonded silica gel as a packing instead of
the virgin silica gel [28] that has been commonly used.

C. Anion-Exchange Resins

Although a combination of anion-exchange resins and borate buffer
gives the best separations of carbohydrates, the analysis time is
quite long and the maintenance of the system is not easy. However,
when this system was modernized, tri- to pentasaccharides were readily separated on an anion-exchange resin with the borate buffer [29].

For fluorometric detection of saccharides, ethylenediamine was added to the borate buffers, and the eluent was heated. This system has been applied to the analysis of environmental and natural products [30] and neutral monosaccharides [31].

The use of ethyl alcohol/water mixtures as the mobile phase simplifies the use of anion exchangers. For example, complex mixtures of sugars were separated on a sulfate form of an anion-exchange resin [32]. Because of the high capacity of many ion exchangers, it is easier to separate glucose and galactose by an ion exchanger than by means of a propylamine-bonded silica gel used with a mobile phase of acetonitrile/water [33]. An example of a chromatogram of reducing sugars in urine is shown in Fig. 2 [33].

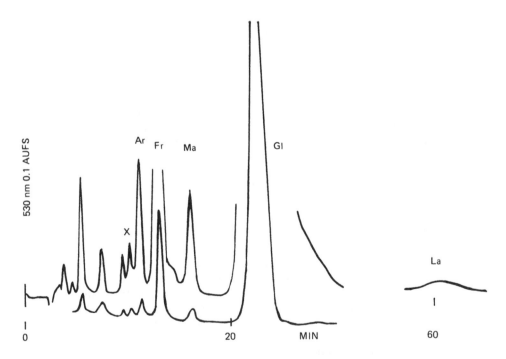

Figure 2   Chromatogram of saccharides in urine. Column, 15 cm long, 4.3 mm i.d. packed with Hitachi 3013N (phosphate form anion-exchange resin); eluent, acetonitrile:water (85:15); flow rate, 1 ml/min; column temp, 60°C; detection, tetrazolium blue reaction detection at 530 nm; X, xylose, Ar, arabinose; Fr, fructose; Ma, mannose; Gl, glucose; La, lactose.

D. Cation-Exchange Resins

Although cation-exchange resins that contain about 4% divinylbenzene are physically soft, the high performance of such a column is very effective for analyzing mixtures of oligo- and monosaccharides. This column has been particularly useful in the food industry. Preparative chromatography for acidic and neutral saccharides has been done on an H^+-form cation-exchange resin column with a mobile phase of 0.5% formic acid solution [34,35]. The calcium form of cation-exchange resins has been used for the analysis of sugarcane saccharides [36], polyols and their derivatives [37], 2- to 9-oligosaccharides [38], and cyclodextrins [39]. The addition to the eluent of triethylamine in water improves the column efficiency [40]. The silver form of a cation-exchange resin has also been used for the separation of oligosaccharides [41], and the sodium form with acetonitrile/water mixtures has been applied to the separation of mono- and disaccharides [42]. In addition, it has been found that the lanthanum form of these resins is effective in separation alditols [43].

E. Octadecyl-Bonded Silica Gel

Octadecyl-bonded silica gel, which is an especially effective packing for the chromatography of nonpolar compounds, can be used for the chromatography of some underivatized and derivatized saccharides. The compound 2-acetamido-2-deoxy-3-O-(4-deoxy-α-2-threo-hex-4-enopyranosyluronic acid)-D-galactose and its 4-sulfoxy and 6-sulfoxy derivatives have been separated on an octadecyl-bonded silica gel with a mobile phase of 0.035 M tetrabutylammonium phosphate, pH 7.54 [44].

Amadori compounds have been purified [45,248], and partially methylated sugars [46] and glucosides [48] have been separated on octadecyl-bonded silica gels. Peralkylated oligosaccharides have also been separated on these columns, and their structures characterized by GC-MS [47].

F. Packings for Size Exclusion

The development of new packings now makes it easy to separate oligosaccharides by size-exclusion chromatography [49]. Collagen gel [50], *N*-methylenechitosan gel [51], polysaccharide gel [52], diol-bonded silica gel [53], and hydrophilic-coated silica gel [54] can be prepared for the size-exclusion chromatography of polar compounds. Molecular-weight markers of polysaccharides can be made from hyaluronidase-digested hyaluronic acid; their molecular weights were found to be 4900, 3000, 1900, and 1150 [55]. Synthetic amylose is now used as a standard in the size-exclusion chromatography of starches [56], and size-exclusion chromatography with packings of cyclodextrin has been used to fractionate maltosaccharides [57,58], nonderivatized cellulose [59], and oligosaccharides from amylose [60]. Size exclusion was also utilized to determine the molecular weight of sea worm chlorocruorin [61], aorta glycosaminoglycans [62], starches [63], polysaccharides [64], heparins [65], dextran [66], and xanthan gum [67].

G. Other Packings

Affinity chromatography with ConA-Sepharose [68] has been used to purify homoglucan and heterogalactan. Liquid chromatography was also used to prepare and purify monospecific anti-*Salmonella* lipopolysaccharide antibodies [69]. Epoxy-activated agarose was synthesized as a packing for the affinity chromatography of various carbohydrates [70], and a charcoal column was used for the selective separation of cyclodextrins [71]. Ketoses and aldoses were selectively separated on a polyethyleneimine ion exchanger [72], and a study has been made of the hydrophobic interaction between heparin and phenyl-Sepharose CL4B [73].

III. DETECTION

A. Direct Detection

A refractive index detector is commonly used for the direct monitoring of carbohydrates; however, its sensitivity is not satisfactory for trace analysis [75,76]. The improved ultraviolet (UV) spectrophotometers now available can be used at low wavelength with high-purity solvents. The sensitivity of UV detection of saccharides, which is about 0.1 μM, is superior to that obtained with chromic acid, but inferior to that obtained with orcinol [74]. A light-scattering detector that is sensitive for higher-molecular-weight compounds has been applied to the analysis of polysaccharides [84] and dextrins [85], and a combined refractometer-polarimetric detector has been used in the analysis of methylated carbohydrates [83]. Other detectors that can be used for the analysis of carbohydrates include the electro-chemical detector [77,78], a pulsed amperometric detector [79], a triple-pulsed amperometric detector [80,81], and amperometric detection with copper bis(phenanthroline) as mediator [82].

B. Postcolumn Reaction Detection

Several approaches have been reported to improve the sensitivity of detection in both borate buffer and nonbuffer eluents. The sensitivity using 2,2'-bicinchroninate is 0.1—30 μg [86]. Other reactants include neocuproin (2,9-dimethyl 1,10-phenanthroline hydrochloride) [87], aminocarbonyl reaction (with the sensitivity of 10 μg) [88], and tetrazolium blue [19,23,89]. The tetrazolium blue reaction is useful in monitoring reducing saccharides in an effluent of acetonitrile/water mixtures, as is shown in Figs. 1 and 2.

Fluorescence is a powerful detection method in the separation of saccharides in a borate buffer. Fluorescent products that are formed by the reaction of reducing and nonreducing saccharides with ethylenediamine in borate media have been monitored with the sensitivity of less than 1 nmol [30], and saccharides reacting with ethanolamine [90—92], 2-aminopropionitrile-fumarate- and taurine-borate reagent have a sensitivity of less than 0.2 nmol [93,94], and with periodate less than 1 nmol [95,96]. A 2-cyanoacetamide borate complex [97—99], an arginine borate complex [100], and a cuprammonium borate complex [101] have also been used for fluorescence detection.

Aminoglycosides have been detected by means of an on-line post-column derivatization system with o-phthalaldehyde [102]. In addition, a photooxygenation-chemiluminescence detector has been developed for monitoring saccharides [103].

C. Precolumn Derivatization

Several precolumn derivatization methods have been applied to carbohydrates that are very polar and not sensitive to common optical detectors.

Acetylation can improve the chromatography of oligosaccharides, and up to 35 oligosaccharides have been well resolved in a reversed-phase mode of liquid chromatographic separation [104]. An example of a chromatogram of acetylated oligosaccharides is shown in Fig. 3.

By means of precolumn derivatization, benzoylated alditols [105], mono- and disaccharides [105], amino sugars [105], methylglucosides [105], intermediate anomers in the synthesis of glycosides [106] and oligosaccharides [107], and p-nitrobenzyloxime of monosaccharides [108] and glycolipids [109] have been chromatographed. The sensitivity of the nitrobenzyloxime derivatives is 0.5—5.0 ng. Phenyldimethylsilyl [111,112] and phenylisocyanate [110] have also been used as precolumn derivatization reagents with reversed-phase columns. Pyridylamino derivatization has been applied to glycoproteins [113] and oligosaccharides in deamination products of parcine and whale heparins [114]. Dansylhydrazones of mono- and oligosaccharides are synthesized for their high sensitivity in RPLC separations [115—118].

D. LC-MS for Saccharides

LC-MS techniques have been applied for the analysis of perbenzoyl, and O-benzyloximes of mono- and disaccharides, perbenzoates of alditols [119], and peralkylated oligosaccharide-alditols [120]. Chemical ionization with ammonia has also been used to determine the structure of glycosides, glucurosides, and sugars [121].

The difficulty of direct coupling of liquid chromatography-mass spectrometry (LC-MS) has been discussed [122—124]; however,

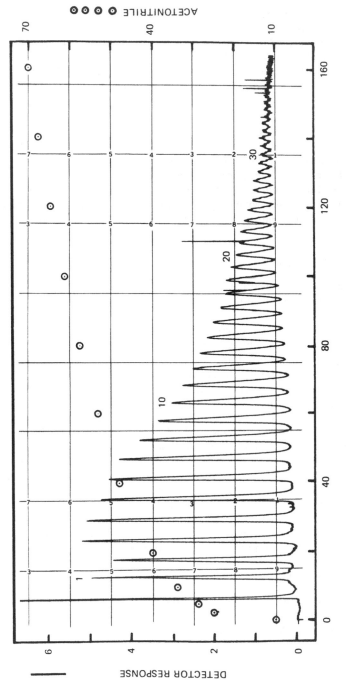

Figure 3 Chromatogram of acetylated oligosaccharides of partially hydrolyzed amylose. Column, 100 cm long, 3.2 mm i.d., packed with Vydac octadecyl-bonded silica gel (30—44 μm); elution, no. 4 gradient of a Waters model 660 programmer from acetonitrile:water (9:1) to acetonitrile:water (7:3) in 160 min; flow rate, 1 ml/min; column temp., 65°C; detector, Pye Unicum flame-ionization detector. (From Ref. 104.)

interfaces utilizing a moving belt [125] and nebulization by diaphrams [126] have been used in the HPLC analysis of saccharides.

A review of LC-MS systems for the analysis of polar compounds is available [9].

IV. APPLICATIONS

A. Biochemical

Biologically important compounds such as glycoproteins, glucosamino-glucans, proteoglycans, and cyclodextrins contain saccharides in their structure, and the development of liquid chromatographic analyses of oligosaccharides has made studies of their structure more feasible.

By means of liquid chromatography, oligosaccharides of human bronchial glycoprotein from a patient with chronic bronchitis have been analyzed [127], and oligosaccharides that are linked to asparagine have been purified [128–130]. In addition, comparative studies of asparagine-linked oligosaccharide structures of rat liver microsomal and lysosomal β-glucuronidases have been reported [128].

The carbohydrates of two membrane glycoproteins (HANA protein and F protein) contain fucose, mannose, galactose, and glucosamine [131]. Glycoprotein-derived oligosaccharides obtained by enzymatic release with endoglycosidases or by chemical release by hydrazinolysis have been separated by HPLC [132], as have been isomeric [133], acidic [134], and anionic [135] oligosaccharides in glycoprotein. Disaccharides in chondroitin sulfates have also been purified [136, 137] by liquid chromatography, and fucitol and mannitol were separated as perbenzoylated derivatives [138]. An example of a chromatogram of glycopeptides is shown in Fig. 4.

Because proteoglycans from tissue and cells can be isolated by liquid chromatography [139–143], the metabolism of proteoglycan and hyaluronate from human anticular cartilage can be studied in vitro [139]. In the past few years, glycosaminoglycans and their disaccharides were purified by LC [144–148]. Analyses of heparin sulfate of embryonic carcinoma and its degradation product [144] and reduced disaccharides derived from glycosaminoglycans [145] have also been achieved. In 1982, glycosaminoglycans in human spine [146] and the mitral valve of human heart [147] were isolated, and unsaturated disaccharides produced from chondroitin sulfates by chondroitinases were analyzed in a study of glycosaminoglucans [136]. In addition, chitooligosaccharides from *Pycnoporus cinnabarinus* β-N-acetylhexosaminidase have been isolated [149], the mutarotation of anomers was studied [150,151], and the time course of the human lysozyme-catalyzed reaction of chitopentanose was measured by gel filtration [152].

Isolation of a capsular polysaccharide from *Escherichia coli* O10: K5:H4 [153] and an oligosaccharide of the pneumococcal capsular

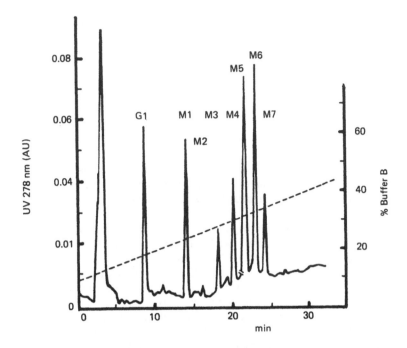

Figure 4 Separation of N-acetylated glycopeptides prepared from ovalbumin and containing Asn as their only amino acid. Column: Micropak AX-5; eluent, gradient 1% B/min (A: 3% acetic acid/20% water in acetonitrile, pH = 5.5 with trimethylamine; B: 3% acetic acid/water, pH = 5.5 with trimethylamine). G1 is the product of endo-β-N-acetylhexosaminidase H digestion of the hexamannosyl glycopeptide, and contains one N-acetylglucosamine residue. M1-M5 are derived from M6 by partial digestion with jack bean or *A. satoi* α-mannosidase. All mannosyl glycopeptides contain two residues of N-acetylglucosamine in their core. (From Ref. 135.)

polysaccharide type 18C have been reported [154]. Free sialyloligosaccharides or sialyloligosaccharides released from gangliosides were labeled by reduction with sodium borohydride and then chromatographed [155]; thereafter, the sialyloligosaccharide products were identified [156,157].

Investigations of the metabolism of xylitol and glucose in rats bearing hepatocellular carcinomas have been reported [158], and a chromatographic analysis of N-acetylglucosamine- and galactosamine-containing carbohydrates was performed on octadecyl-bonded silica gels [159]. Other interesting oligosaccharide analyses include even- and odd-numbered oligosaccharides of hyaluronate from human umbilical cord [160], oligosaccharides and glucopeptides accumulating in lysosomal storage disorders [161], oligosaccharides produced by bovine testes hyaluronidase [162], and oligosaccharides in hen ovomucoid [163].

Polysaccharides have been isolated from *Leptotricia buccalis* strain L11 [164], squid cranial cartilage [165], multivalent bacterial capsular polysaccharide vaccines [166], and an extracellular polysaccharide produced by *Zoogloca remigera* 115 [167]. Liquid chromatographic analyses of amino sugars [168] and nucleotide sugars [169] have also been reported. Oligosaccharides in mannosidosis urine have been analyzed [170—172], and several oligosaccharides were found in the urine [173—175] of GM1 gangliosidosis [173] of a patient with Morquio syndrome type G [174]. A chromatogram of oligosaccharides from mannosidosis urine is shown in Fig. 5.

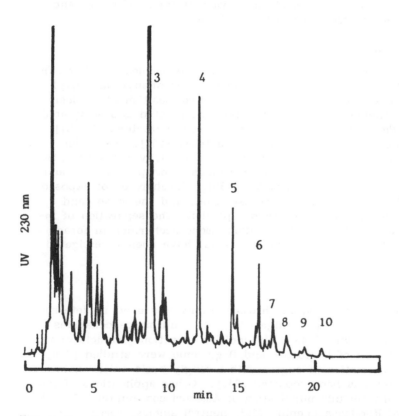

Figure 5 Separation of nonreduced benzoylated oligosaccharides from mannosidosis urine. Column, Ultrasphere octyl, 4.6 mm i.d., 25 cm long; eluent, 15-min linear gradient of 80% acetonitrile/water to 100% acetonitrile at 2 ml/min. The number beside each peak indicates its DP. (From Ref. 170.)

The structure of the phosphorylated oligosaccharides of the high-mannose type in human β-glucosidase has been identified [176]. Also reported have been oligogalacturonic acid analyses [177], identification of urinary isomeric chondroitin sulfates from patients with mucopolysaccharidoses [178], and synthesis of 2-deoxy-D-[1-^{11}C]-glucose in metabolic studies of regional cerebral glucose [179]. Chromatographic conditions of biochemically important oligo- and polysaccharides are listed in Table 1.

B. Food

The analysis of oligo- and monosaccharides is very important in the brewing industry [180—190] and in the sugar industry [191—198]. The analysis of saccharides in food is common, and the conditions for liquid chromatographic analysis of various types of mono- and oligosaccharides in foods are listed in Table 2.

C. Pulp and Wood

Via liquid chromatography, lignin-carbohydrate complexes isolated from *Pinus densiflora* Sieb. et Zucc. were analyzed on phenyl- and octyl-Sepharose CL-4B gels [229], as was sugar composition of hemicellulose from Brittlehart [230]. In addition, methylated and acetylated extracts of *Picea abies* (Norway spruce) were characterized [231], and the structure of xyloglucan was identified [232]. Exclusion chromatography was used to separate lignin carbohydrates [233,234], to study the chemical composition of cellulose from pulp [235], and to isolate allergen from birch pollen [236]. Saccharides of exposed wood can also be purified by size exclusion, and the mono- and oligosaccharides can be rechromatographed for the separation of individual species [23,213]. In addition, monosaccharides in forage grass [238—241] and tropical grasses [242] have been investigated via liquid chromatography.

D. Other Applications

In environmental work, liquid chromatography was used to determine polysaccharides in sewage sludge [243,244], and monosaccharides found in natural water [245]. In addition, reaction products of the catalytic oxidation of D-glucose and D-gluconic were studied [246], and the transformation of sugars with strongly basic anion-exchange resins in water has been reported [247]. Other applications of liquid chromatogrpahy include purification of Amadori compounds [45,248], glycosides of Rhodeasapogenin [249], neutral sugars from tobacco cell walls, and lime flower mucilages [250]. Hydroxyl exchange resins are being used as a cleanup method of biological samples in the chromatography of sugars [251].

For preparative work, medium- and high-pressure liquid chromatography has been used in oligosaccharide synthesis [252]. Other

Table 1 Experimental Condition of Biochemical Samples

Sample	Column[a]	Eluent	Detector	Ref.
Human bronchial glycoprotein	DAX4(30 × 0.9 cm)	0.2 M Na-borate buffer, pH 8.5	orcinol	127
Phosphorylated oligosaccharides in human β-glucosidase	DAX 8-11	0.5 M borate buffer, pH 8.6	radioactivity	176
Oligosaccharides linked to asparagine	Bio-Gel P-4	—	radioactivity	128
Carbohydrates of HANA protein and F protein	Bio-Gel P-4	—	radioactivity	131
Oligosaccharides in urine GM1-gangliosidosis	Bio-Gel P-4	—	phenol/H_2SO_4	173
Oligosaccharides of the pneumococcal capsular polysaccharide type 18-C	Bio-Gel P-2	water	anthrone/H_2SO_4	154
Oligosaccharides in mannosidosis urine	Bio-Gel P-2 (57 × 2.6 cm)	water	orcinol	171
Glycosaminoglycans	Fractogel TSK HW-55(S)	0.2 M NaCl	radioactivity and UV 210 nm	144
Anomers of chitooligosaccharides	TSK G2000 PW	water	UV 220 nm	151
Chitooligosaccharides	TSK G2000 PW	water	UV 220 nm	152
Multivalent bacterial capsular polysaccharide vaccines	Sepharose gel	—	—	166
Acidic oligosaccharides	Hitachi 2630	water-0.15 M NaCl	UV 210 nm	134

Table 1 (continued)

Sample	Column[a]	Eluent	Detector	Ref.
Sialyloligosaccharides	DEAE-cellulose	0.02 M pyridine-acetate buffer, pH 5.4	radioactivity	155
Proteoglycans from tissue and cells	DEAE-Sephadex	LiCl(0.3–2 M) in 0.05 M tris-HCl, pH 8.0	radioactivity	142
Anomers of chitooligosaccharides	Shodex Ionpak S-614	6% acetonitrile	UV 220 nm	150
Extra cellular polysaccharide produced by *Zoogloca renrigera* 115	Shodex S-614, Shodex S-801	acetone/water water	RI RI	167 167
Oligosaccharides produced by bovine testes hyaluronidase	silica gel	20 mM NaOAc, pH 4.0 containing 1.5 mg hyaluronic acid/liter	— —	166
Nucleotide sugar	(packing: Bandiraeosimplicifolia I isolectins immobilized on Sepharose)	—	—	169
Oligosaccharides in urine of Morquio syndrome G	Charcoal-Celite	water/ethanol	(pretreatment)	174
N-Acetylglucosamine- and galactosamine-containing carbohydrates	ODS-silica gel	water	UV 202 nm	159

Sample	Column	Mobile phase	Detection	Ref
Polysaccharides from squid cranial cartilage	ODS-silica gel	—	RI	165
Oligosaccharides in mannosidosis urine	C8-silica gel	acetonitrile/water	UV 230 nm	170
Glycoprotein-derived oligosaccharides	MicroPak AX-5	acetonitrile/water	radioactivity	132
Isomer oligosaccharides	NH_2	acetonitrile/water	UV 190 nm	133
Anionic aligosaccharides	NH_2	acetonitrile/water + 3% acetic acid/tri-ethylamine, pH 5.5	radioactivity	135
Disaccharides from glycosaminoglycans	NH_2	acetonitrile/water/ 0.5 M NH_4HCOO, pH 4.5 or 6.0	UV 244 nm	145
Chitooligosaccharides from *Pyenoporus cinnabaisnus* β-*N*-acetylhexosaminidase	NH_2	acetonitrile/water	UV 210 nm	149
Oligosaccharides of hyaluronidate from hyman umbilical cord	NH_2	0.1 M KH_2PO_4, pH 4.75	UV 206 nm	160
Oligosaccharides in lysosomal storage disorders	NH_2	acetonitrile/1% AcOH acetonitrile/0.1 M NaOAc, pH 5.0	RI	161
Oligosaccharides in hen ovomucoid	NH_2	acetonitrile/water	UV 200 nm	163
Oligosaccharides in mannosidosis urine	NH_2	acetonitrile/water 0.01% 1,4-diaminobutane	RI	172

Table 1 (continued)

Sample	Column[a]	Eluent	Detector	Ref.
Oligogalacturonic acids	NH_2	0.11 M acetate buffer, pH 7.5	UV 235 nm	177
Chondroitin sulfate from patient with mucopolysaccharidoses	NH_2	acetonitrile/water/ 0.5 M NH_4COOH, pH 6.8	UV 254 nm	178
2-Deoxy-D-(1-^{11}C)glucose	NH_2	acetonitrile/water	radioactivity	179
Oligosaccharides in chondroitin sulfates	Partisil 10-SAX	7.5 mM K-phosphate buffer, pH 6.5	UV 232 nm	136

[a]NH_2 = propylamino bonded silica gel.

Table 2 Liquid Chromatography of Food[a]

Sample	Compound	Column	Eluent	Detector	Ref.
Agricultural products	mono	NH$_2$	ACN/H$_2$O	RI	199
Cantaloupe juice	mono	SI	Amine/ACN/H$_2$O	RI	200
Cereal (breakfast)	mono	NH$_2$	ACN/H$_2$O		201
(Presweetened)	mono, di		ACN/H$_2$O	RI	202
Chewing gum	sorbitol, xylitol, mannitol	Aminex-HPX87	H$_2$O	RI	203
Chicory root	mono	SI	Amine/ACN/H$_2$O	TZB	189
Corn syrup	oligo				204
Dairy product	mono	NH$_2$	ACN/NH$_2$	RI	205
	mono	Aminex-HPX87	H$_2$O	RI	206
Dietary fiber	mono	Aminex-HPX87P	H$_2$O	RI	207
Drink	mono	NH$_2$	ACN/H$_2$O		208
	mono	NH$_2$	Me$_2$CO/EtOAc/H$_2$O	RI	209
Food	mono	NH$_2$	ACN/H$_2$O	RI	210
	mono	NH$_2$	ACN/H$_2$O	RI	211
	mono, di, sorbitol	NH$_2$	EtOH/ACN/H$_2$O	RI	212
	sugar/NaCl soln.	NH$_2$	ACN/H$_2$O	RI	213
Food and plant	mono	SI	Amine/ACN/H$_2$O	TZB	214
Fruit	mono	NH$_2$	ACN/N$_2$O	UV	215
	mono	NH$_2$	EtOH/ACN/H$_2$O	RI	216

Table 2 (continued)

Sample	Compound	Column	Eluent	Detector	Ref.
Guava	mono				217
Leguminous seeds	oligo	NH_2	ACN/H_2O	RI	218
Loquat	mono				219
Milk, ice cream	mono	NH_2	ACN/H_2O	RI	220
Neutral detergent fiber	mono, di	NH_2	ACN/H_2O		221
Liquid nutritional supplement (hospital)	mono	NH_2	ACN/H_2O	UV	222
Seeds oil	mono	NH_2	ACN/H_2O	RI	223
Onion powder	mono	NH_2	ACN/H_2O	RI	224
Soybeans	mono	NH_2	ACN/H_2O	RI	225
	pinitol, cyclitols	NH_2	ACN/H_2O	RI	226
Vegetable fiber	aldoses	ion-exchange	borate buffer		227
Wheat starch	oligo	NH_2	ACN/H_2O	RI	228

[a] Abbreviations: NH_2, propylamino-bonded silica gel or μ-Bondapak-CH; ACN, acetonitrile; H_2O, water; mono, monosaccharide; SI, silica gel; di, disaccharide; oligo, oligosaccharide; TZB, tetrazolium blue reaction detector; Me_2CO, acetone; EtOAc, ethyl acetate; EtOH, ethanol.

carbohydrates for which liquid chromatography has been used include unsubstituted glycosides, isopropylidene, benzylidene, and derivatives of glycosides [253]; the anticancer reagent, β-glucan [254], amylostatins [255]; the antibiotic cephalosporin, penicillin, aminoglycoside, anthracycline [256]; and the N-nitroso sugar amino acids [257]. In addition, thermodynamic functions of liquid chromatography of saccharides have been investigated [258].

V. SUMMARY

Liquid chromatography is used in the qualitative and quantitative analysis of monosaccharides, both in the food industry and in biochemical research. However, better methods for the isolation and purification of poly- and oligosaccharides are needed, especially in the biomedical field. It is hoped that in the near future, the improvements in both analytical and preparative HPLC techniques that will become available will be helpful in the structural determination of oligosaccharides which are important compounds in the study of the biological mechanisms of organs.

ACKNOWLEDGMENTS

The author gratefully acknowledges the support of Professor J. Hubert and the assistance of Mme C. Haumont of the Université de Montreal for the library search. The author thanks Mr. A. Zyukurogi for technical assistance, and Miss H. Ohtsubo of Gasukuro Kogyo, Inc. for typing the manuscript. The assistance of Ms. Y. Ueno is gratefully acknowledged for the correction of the English in this manuscript. Professor P. R. Brown's comments, suggestions, and patience are greatly appreciated.

REFERENCES

1. R. Schwarzenback, Carbohydrates analysis—a challenge to HPLC, Chromatogr. Sci. *12*, 193−222 (1979).
2. M. H. Simatupang, M. Sinner, and H. H. Dietrichs, Hochdruck-flussigkeits-chromatographische Trennung von Kohlenhydraten in der Lebensmittelchemie, Auto Anal., Innovationen: Problemloesungen Med. *2*, 43−54 (1979).
3. R. D. Wilson, The analysis of carbohydrates by high-pressure liquid chromatography, Rep. N. Z., Dept. Sci. Ind. Res., Chem. Div., C. D. 2280 (1979).

4. A Hegrand and M. Rinaudo, Liquid chromatography applied to oligosaccharide fractionation, J. Liq. Chromatogr. 4, 175—293 (1981).

5. K. M. Brobst and H. D. Scobell, Modern chromatographic methods for the analysis of carbohydrate mixtures, Starch 34, 117—121 (1982).

6. S. C. Churms, Carbohydrates, J. Chromatogr. Libr. 22B, 223—286 (1983).

7. Vor E. Truscheit, W. Frommer, B. Tunge, L. Miiller, D. D. Schmit, and W. Wingender, Angero, Chem. 93, 738—755 (1981).

8. T. K. Palmer, Sugars, sugar alcohols, and saccharin, Chromatogr. Sci. 9, 1317—1340 (1979).

9. D. E. Games, Applications of combined high-performance liquid chromatography-mass spectrometry, Anal. Proc. (London) 17, 322—326 (1980).

10. R. L. Whistler, Gel permeation chromatography, Methods Carbohydr. Chem. 8, 45—53 (1980).

11. G. D. McGinnis and P. Fang, High-performance liquid chromatography, Methods Carbohydr. Chem. 8, 33—43 (1980).

12. D. T. Folkes and P. W. Taylor, Determination of carbohydrates, HPLC Food Anal. (R. Macrae, ed.), Academic Press, London, 149—66 (1982).

13. V. Vockove, Survey of high-performance liquid chromatography techniques for the separation of sugars, Radioisotopy 23, 215—245 (1982).

14. P. L. Dubin, Aqueous exclusion chromatography, Sep. Purific. Methods 10, 287—313 (1981).

15. M. Gurkin and V. Patel, Aqueous gel filtration chromatography of enzymes, proteins, oligosaccharides and nucleic acids, Am. Lab. (Fairfield Conn.) 14, 64, 66, 68—70, 72—73 (1982).

16. R. Oshima, Y. Yamauchi, and T. Kumanotani, Carbohydr. Res. 107, 169 (1982).

17. R. Oshima and T. Kumanotani, Chem. Lett. 7, 943 (1981).

18. L. A. Th. Verhaar, B. F. Kustes, J. Chromatogr. 234, 57 (1982).

19. M. D'Amboise, T. Hanai, and D. Noel, Carbohydr. Res. 79, 1 (1980).

20. B. B. Wheals and P. C. White, J. Chromatogr. 176, 421 (1979).

21. B. Prosch, J. Chromatogr. 253, 49 (1982).

22. T. G. Baust, R. E. Lee, Jr., R. R. Rojas, D. L. Hendrix, D. Griday, and H. James, J. Chromatogr. 261, 65 (1983).

23. D. Noel, T. Hanai, and M. D'Amboise, J. Liq. Chromatogr. 2, 1325 (1979).

24. Y. Kurihara, T. Sato, and M. Umino, Toyo Soda Kenkyu Hokoku 24, 115 (1979).

25. P. E. Show and C. W. Wilson, III, J. Chromatogr. Sci. 20, 209 (1982).

26. V. H. Muller and V. Siepe, Deutsche Lebensmittel-Rundschan *76*, 156 (1980).

27. K. Aitzetmuller, J. Chromatogr. *156*, 354 (1978).

28. C. H. Lochmuller and W. B. Hill, Jr., J. Chromatogr. *264*, 215 (1983).

29. M. Torii, B. P. Alberto, S. Tanaka, Y. Tsumuraya, A. Misaki, and T. Sawai, J. Biochem. (Tokyo) *85*, 883 (1979).

30. K. Mopper, R. Dauson, G. Liebezeit, and H. P. Hanson, Anal. Chem. *52*, 2018 (1980).

31. M. H. Simatupang, J. Chromatogr. *178*, 588 (1979).

32. K. Mopper, Anal. Biochem. *85*, 528 (1978).

33. M. D'Amboise, T. Hanai, and D. Noel, Clin. Chem. *26*, 1348 (1980).

34. T. Kumanotani, R. Oshima, Y. Yamauchi, N. Takai, and Y. Kurose, J. Chromatogr. *176*, 462 (1979).

35. R. Oshima, Y. Kurose, and T. Kumanotani, J. Chromatogr. *179*, 376 (1979).

36. J. Wong-Chong and F. A. Martin, J. Agric. Food Chem. *27*, 929 (1979).

37. S. J. Angyal, G. S. Bethell, and R. J. Beveridge, Carbohydr. Res. *73*, 9 (1979).

38. J. Schmidt, M. John, and C. Wandrey, J. Chromatogr. *213*, 151 (1981).

39. H. Hokse, J. Chromatogr. *189*, 99 (1980).

40. L. A. T. Verhaar and B. F. M. Kuster, J. Chromatogr. *210*, 279 (1981).

41. H. D. Scobell and K. M. Brobst, J. Chromatogr. *212*, 51 (1981).

42. T. Kawamoto and E. Okada, J. Chromatogr. *258*, 284 (1983).

43. L. Petrus, V. Bilik, L. Kuniak, and L. Stankovic, Chem. Zuesti *34*, 530 (1980).

44. N. Ototani, N. Sato, and Z. Yoshizawa, J. Biochem. *85*, 1383 (1979).

45. N. Moll and B. Gross, J. Chromatogr. *206*, 186 (1981).

46. N. W. H. Cheetham and P. Sirimanne, J. Chromatogr. *196*, 171 (1980).

47. B. S. Valent, A. G. Daroill, M. McNeil, B. K. Robertsen, K. Boerce, and P. Albersheim, Carbohydr. Res. *79*, 165 (1980).

48. A. Roy and N. Roy, Curr. Sci. *50*, 983 (1981).

49. K. Tanaka, T. Kitamura, T. Matsuda, H. Yamasaki, and H. Sasaki, Toyo Soda Kenkyu Hokoku *25*, 81 (1981).

50. M. Shaw and A. Schy, J. Chromatogr. *170*, 449 (1979).

51. S. Hirano, N. Matsuda, O. Miura, and T. Tanaka, Carbohydr. Res. *71*, 344 (1979).

52. Y. A. Eltekov, N. M. Strakhova, T. Kalal, T. Peska, and T. Stamberg, J. Polym. Sci. Polym. Symp. *68*, 247 (1981).

53. D. P. Herman, L. R. Field, and S. Abbott, J. Chromatogr. Sci. *19*, 470 (1981).

54. H. G. Barth, J. Liq. Chromatogr. *3*, 1481 (1980).
55. H. Nakagawa, N. Enomoto, and K. Yamaguchi, Agric. Bull. Saga Univ. *49*, 65 (1979).
56. W. Praznik and R. Ebermann, Starch/Staerke *3*, 288 (1979).
57. T. Nakakuki and K. Kainuma, J. Jpn. Soc. Starch Sci. *29*, 27 (1982).
58. H. Kondo, H. Nakatani, and K. Hiromi, Agric. Biol. Chem. *45*, 2369 (1981).
59. Y. T. Bao, A. Base, M. R. Ladisch, and G. T. Tsao, J. Applied Polymer Sci. *25*, 263 (1980).
60. H. Kondo, H. Nakatani, R. Matsuno, and K. Hiromi, J. Biochem. *87*, 1053 (1980).
61. M. E. Himmel and D. G. Squire, J. Chromatogr. *210*, 443 (1981).
62. B. Radhakrishnamurthy, E. R. Dalferes, Jr., P. Vijayagopal, and G. S. Berenson, J. Chromatogr. *192*, 307 (1980).
63. F. Meuser, R. W. Klingler, and E. A. Niediek, Getreide, Mehl. Brot. *33*, 295 (1979).
64. L. V. Kosenko, Mikrobiol. Zh. (Kieo) *45*, 82 (1983).
65. T. Harenberg and J. X. De Vries, J. Chromatogr. *261*, 287 (1983).
66. A. A. Soetemen and F. A. H. Peeters, Bull. Soc. Chim. Belg. *91*, 875 (1982).
67. F. Lambert, M. Milas, and M. Rinando, Polym. Bull. (Berlin) *7*, 185 (1982).
68. S. Mizumo, Kagaku to Seibutsu, *20*, 420 (1982).
69. R. Girard and T. Goichot, Ann. Immunol. *132C*, 211 (1981).
70. I. Matsumoto, H. Kitagaki, Y. Akai, Y. Ito, and N. Seno, Anal. Biochem. *116*, 103 (1981).
71. S. Yoshikawa, S. Kitahata, and S. Okada, Kagaku to Kogyo, *57*, 67 (1983).
72. V. Bilik, L. Petrus, and L. Kuniak. Chem. Zvesti *33*, 118 (1979).
73. A. Ogamo, K. Matsuzaki, H. Uchiyama, and K. Nagasawa, J. Chromatogr. *213*, 439 (1981).
74. L. A. Verhaar and J. M. H. Dirkx, Carbohydr. Res. *62*, 197 (1978).
75. S. I. M. Johncock and P. J. Wagstaffe, Analyst *105*, 581 (1980).
76. H. Binder, J. Chromatogr. *189*, 414 (1980).
77. W. Bushberger, K. Winsauer, and C. Breitwieser, Fresenius Z. Anal. Chem. *315*, 518 (1983).
78. W. Buchberger, F. Winsauer, and C. Breitwieser, Fresenius Z. Anal. Chem. *311*, 517 (1982).
79. R. D. Rocklin and C. A. Pohl, J. Liq. Chromatogr. *6*, 1577 (1983).
80. S. Hughes and D. C. Johnson, Anal. Chim. Acta *149*, 1 (1983).
81. S. Hughes and D. C. Johnson, J. Agric. Food Chem. *30*, 712 (1982).

82. N. Watanabe and M. Inoue, Anal. Chem. *55*, 1016 (1983).
83. A. Heyraud and P. Salemis, Carbohydr. Res. *107*, 123 (1982).
84. N. Suzuki, K. Wada, and K. Suzuki Carbohydr. Res. *109*, 249 (1982).
85. C. J. Kim, A. E. Hamielec, and A. Benedeh, J. Liq. Chromatogr. *5*, 425 (1982).
86. M. Sinner and J. Puls, J. Chromatogr. *156*, 197 (1978).
87. M. H. Simatupang and H. H. Dietrichs, Chromatographia *11*, 89 (1978).
88. Y. Iijima, J. Jpn. Soc. Starch Sci. *26*, 30 (1979).
89. S. K. Nuor and J. Vialle, J. -L. Rocca, Analysis 7, 381 (1979).
90. T. Kato and T. Kinoshita, Anal. Biochem. *106*, 238 (1980).
91. H. Mikami and Y. Ishida, Shimadzu Hyoron, *37*, 23 (1980).
92. S. Honda, Y. Matsuda, M. Takahashi, K. Kakehi, A. Honda, S. Ganno, and M. Ito, J. Pharmacobio-Dyn. *3*, S-11, S-31 (1980).
93. T. Kato and T. Kinoshita, Bunseki Kagaku *31*, 615 (1982).
94. T. Kato, F. Iijima, and T. Kinoshita, Nippon Kagaku Kaishi 1603 (1982).
95. P. Nordin, Anal. Biochem. *131*, 492 (1983).
96. S. Honda, M. Takahashi, S. Shimada, K. Kakehi, and S. Ganno, Anal. Biochem. *128*, 429 (1983).
97. S. Honda, Y. Matsuda, M. Takahashi, K. Kakehi, and S. Ganno, Anal. Chem. *52*, 1079 (1980).
98. S. Honda, M. Takahashi, K. Kakehi, and S. Ganno, Anal. Biochem. *113*, 130 (1981).
99. S. Honda, M. Takahashi, Y. Nishimura, K. Kakehi, and S. Ganno, Anal. Biochem. *118*, 162 (1981).
100. H. Mikami and Y. Ishida, Bunseki Kagaku *32*, E207 (1983).
101. G. K. Grimble, H. M. Barker, and R. H. Taylor, Anal. Biochem. *128*, 422 (1983).
102. G. Lachatre, G. Nicot, C. Gonnet, J. Tronchet, L. Merle, J. -P. Valette, and Y. Nousille, Analusis *11*, 168 (1983).
103. M. S. Gandelman and J. W. Birks, J. Chromatogr. *242*, 21 (1982).
104. G. B. Wells and R. L. Lester, Anal. Biochem. *97*, 184 (1979).
105. C. A. White, J. F. Kennedy, and B. T. Golding, Carbohydr. Res. *76*, 1 (1979).
106. M. Dreux, M. Lafosse, P. H. A. Zollo, J. R. Pougny, and P. Sinay, J. Chromatogr. *204*, 207 (1981).
107. P. E. Daniel, I. T. Lott, and R. H. McCluer, Chromatogr. Sci. *18*, 363 (1981).
108. M. H. Lawson and G. F. Russell, J. Food Sci. *45*, 1256 (1980).
109. T. Yamazaki, A. Suzuki, S. Honda, and T. Yamakawa, J. Biochem. (Tokyo) *86*, 803 (1979).
110. B. Bjorkqvist, J. Chromatogr. *218*, 65 (1981).

111. C. A. White, S. W. Vass, J. F. Kennedy, and D. G. Large, J. Chromatogr. *264*, 99 (1983).

112. C. A. White, S. W. Vass, J. F. Kennedy, and D. G. Large, Carbohydr. Res. *119*, 241 (1983).

113. S. Hase, T. Ikenaka, and Y. Matsushima, J. Biochem. (Tokyo) *90*, 407 (1981).

114. M. Kosakai and Z. Yoshizawa, J. Biochem. (Tokyo) *92*, 295 (1982).

115. G. Dutot, D. Biou, G. Durand, M. Pays, Feuill. Biol. *22*, 101 (1981).

116. W. F. Alpenfels, Anal. Biochem. *114*, 153 (1981).

117. K. Mopper and L. Johnson, J. Chromatogr. *256*, 27 (1983).

118. M. Takeda, M. Maeda, and A. Tsuji, J. Chromatogr. *244*, 347 (1982).

119. R. M. Thomson and D. A. Cory, Biomed. Mass Spectrom *6*, 117 (1979).

120. P. Aman, M. McNeil, L. E. Franzen, A. G. Darvill, and P. Albersheim, Carbohydr. Res. *95*, 263 (1981).

121. D. E. Games and E. Lewis, Biomed. Mass. Spectrom 7, 433 (1980).

122. D. J. Dixon, Analysis *10*, 343 (1982).

123. C. Eckers, D. E. Games, M. L. Games, W. Kuhnz, E. Lewis, N. C. A. Weerasinghe, and S. A. Westwood, Anal. Chem. Symp. Ser. 7, 169 (1981).

124. D. E. Games, Biomed. Mass Spectrum. *8*, 454 (1981).

125. D. E. Games, P. Hirter, W. Kuhnz, E. Lewis, N. C. A. Weerasinghe, and S. A. Westwood, J. Chromatogr. *203*, 131 (1981).

126. P. J. Arpino, P. Krien, S. Vajta, and G. Devant, J. Chromatogr. *203*, 117 (1981).

127. G. Lamblin, M. Lhermitte, A. Boereman, P. Roussel, and V. Reinhold, J. Biol. Chem. *255*, 4595 (1980).

128. T. Mizuochi, Y. Nishimura, K. Kato, and A. Kobata, Arch. Biochem. Biophys. *209*, 298 (1981).

129. G. B. Wells, S. T. Turco, B. A. Hanson, and R. L. Lester, Anal. Biochem. *110*, 397 (1981).

130. P. I. Clark, S. Narasimhan, J. M. Williams, and J. P. Clamp, Carbohydr. Res. *118*, 147 (1983).

131. H. Yoshima, M. Nakanishi, Y. Okada, and A. Kobata, J. Biol. Chem. *256*, 5355 (1981).

132. S. J. Mellis and J. U. Baenzinger, Anal. Biochem. *114*, 276 (1981).

133. M. L. Bergh, P. L. Koppen, D. H. Van den Eijnden, J. Arnarp, and J. Loenngren, Carbohydr. Res. *117*, 275 (1983).

134. T. Tsuji, K. Yamamoto, K. Konami, T. Irinuma, and T. Osawa, Carbohydr. Res. *109*, 259 (1982).

135. S. J. Mellis and J. U. Baenzinger, Anal. Biochem. *134*, 442 (1983).

136. A. L. Fluharty, J. A. Glick, N. M. Matusewicz, and H. Kihara, Biochem. Med. *27*, 352 (1982).

137. A. Hjerpe, C. A. Antonopoulos, B. Engfeldt, and M. Nurminen, J. Chromatogr. *242*, 193 (1982).

138. P. F. Daniel, J. Chromatogr. *176*, 260 (1979).

139. P. Gysen, G. Heynen, and P. Franchimont, C. R. Séances Soc. Biol. Ses. Fil. *174*, 867 (1980).

140. E. R. Schwartz, J. Stevens, and D. E. Schmidt, Jr., Anal. Biochem. *112*, 170 (1981).

141. B. Radhakrishnamurthy, N. Jeansonne, and G. S. Berenson, J. Chromatogr. *256*, 341 (1983).

142. G. Verbruggen and E. M. Veys, Int. Cong. Ser. Excerpta Med. *573*, 113 (1982).

143. R. V. Iozzo, R. Marroguin, and T. N. Wight, Anal. Biochem. *126*, 190 (1982).

144. T. Irimura, M. Nakajima, N. Di Ferrante, and G. L. Nicolson, Anal. Biochem. *130*, 461 (1983).

145. G. J. -L. Lee, D. -W. Liu, J. W. Pao, and H. Tiechelmann, J. Chromatogr. *212*, 65 (1981).

146. E. Gurr, R. Schubert, A. Delbrueck, and W. Koeller, J. Clin. Chem. Clin. Biochem. *20*, 723 (1982).

147. H. Masuda and S. Shichijo, Saishin Igaku *37*, 423 (1982).

148. E. C. Lau and J. V. Ruch, Anal. Biochem. *130*, 237 (1983).

149. A. Ohtakara, M. Mitsutomi, and E. Nakamae, Agric. Biol. Chem. *46*, 293 (1982).

150. T. Fukamizo and K. Hayashi, J. Biochem. *91*, 619 (1982).

151. T. Fukamizo, T. Torikawa, T. Nagayama, T. Minematsu, and K. Hayashi, J. Biochem. *94*, 115 (1983).

152. T. Fukamizo, T. Torikawa, S. Kuhara, and K. Hayashi, J. Biochem. *92*, 709 (1982).

153. W. F. Vann, M. A. Schmidt, B. Jann, and K. Jann, Eur. J. Biochem. *116*, 359 (1981).

154. L. R. Phillips, O. Nishimura, and B. A. Fraser, Carbohydr. Res. *121*, 243 (1983).

155. D. F. Smith, J. L. Magnari, and V. Ginsburg, Arch. Biochem. Biophys. *209*, 52 (1981).

156. M. L. E. Bergh, P. L. Koppen, and D. H. Van den Eijnden, Biochem. J. *201*, 411 (1982).

157. G. Lamblin, A. Klein, A. Boersma, Nasir-Ud-Din., and P. Roussel, Carbohydr. Res. *118*, C1 (1983).

158. J. Sato, Y.-M. Nang, and J. Van Eys, Cancer Res. *41*, 3192 (1981).

159. K. Blumberg, F. Liniere, L. Pustilnik, and C. A. Bush, Anal. Biochem. *119*, 407 (1982).

160. P. Nebinger, M. Koel, A. Franz, and E. Werries, J. Chromatogr. *265*, 19 (1983).

161. N. M. K. Ng Ying Kin and L. S. Wolfe, Anal. Biochem. *102*, 213 (1980).

162. P. J. Knudsen, P. B. Eriksen, M. Flenger, and K. Florenty, J. Chromatogr. *187*, 373 (1980).

163. J. P. Parente, G. Strecker, Y. Leroy, J. Montreil, and B. Fournet, J. Chromatogr. *249*, 199 (1982).

164. N. K. Birkeland and T. Hofstad, Acta Pathol. Microbiol. Immunol. Scand. [B], *90B*, 435 (1982).

165. A. Hjerpe, B. Engfeldt, T. Tsegenidis, and C. A. Antonopoulos, J. Chromatogr. *259*, 334 (1983).

166. M. Porro, S. Fabbiani, I. Marsili, S. Viti, and M. Saletti, J. Biol. Stand. *11*, 65 (1983).

167. F. Ikeda, H. Shuto, T. Saito, T. Fukui, and K. Tomita, Eur. J. Biochem. *123*, 437 (1982).

168. D. A. Johnson, Anal. Biochem. *130*, 475 (1983).

169. D. A. Blake and I. J. Goldstein, Anal. Biochem. *102*, 103 (1980).

170. P. F. Daniel, D. F. De Feudis, I. T. Lott, and R. H. McCluer, Carbohydr. Res. *97*, 161 (1981).

171. P. F. Daniel, D. F. De Feudis, and I. T. Lott, Eur. J. Biochem. *114*, 235 (1981).

172. C. D. Warren, A. S. Schmit, and R. W. Jeanloz, Carbohydr. Res. *116*, 171 (1983).

173. K. Yamashita, T. Ohkura, S. Okada, H. Yabuushi, and A. Kobata, J. Biol. Chem. *256*, 4789 (1981).

174. J.-C. Michalski, G. Strecker, H. Van Halbeck, L. Dorland, and J. F. G. Vliegenthart, Carbohydr. Res. *100*, 351 (1982).

175. T. G. Warner, A. D. Robertson, and J. S. O'Brien, Clin. Chim. Acta *127*, 313 (1983).

176. M. Natowicz, Y. U. Baenziger, and W. S. Sly, J. Biol. Chem. *257*, 4412 (1982).

177. A. G. J. Voragen, H. A. Schols, J. A. DeVries, and W. Pilnik, J. Chromatogr. *244*, 327 (1982).

178. G. J. L. Lee, J. E. Evans, H. Tieckelmann, J. T. Dulaney. and E. W. Naylor, Clin. Chim. Acta *104*, 65 (1980).

179. M. M. Vora, T. E. Boothe, R. D. Finn, P. M. Smith, and A. T. Gilson, J. Lab. Compounds Radiopharmaceut. *20*, 417 (1983).

180. M. Dadic and G. Belleau, J. Am. Soc. Brew. Chem. *40*, 141 (1982).

181. A. J. Ritson, Proc. Conv. Inst. Brew. *16*, 121 (1980).

182. A. Rapp and A. Ziegler, Dtsch. Lebensm.-Rundsch *75*, 396 (1979).

183. H. Sedova and M. Kahler, Kvasny *28*, 193 (1982).

184. G. K. Buckee and D. E. Long, J. Am. Soc. Brew. Chem. *40*, 137 (1982).

185. Y. Shimadzu, M. Uehara, and M. Watanabe, J. Brew. Soc. Japan *75*, 327 (1980).

186. C. Otoguro, S. Ogino, and M. Watanabe, Nippon Jozo Kyokai Zassi *78*, 220 (1983).

187. J. Schmidt, M. John, and H. Niefind, Brauwissenschaft *34*, 114 (1981).

188. M. Toth, E. Laszlo, and J. Morvai, Soripar *28*, 127 (1981).

189. E. C. Samarco and E. S. Parente, J. Assoc. Off. Anal. Chem. *65*, 76 (1982).

190. W. Flah, Mitt. Klosterneuburg, *31*, 204 (1981).

191. D. F. Charles, Int. Sugar J. *83*, 169 (1981).

192. H. Heikkila, Chem. Eng. *90*, 50 (1983).

193. W. P. P. Abeydeera, Proc. Conf. Aust. Soc. Sugar Cane Technol. 171 (1983).

194. J. Wong-Chong and F. A. Martin, J. Agric. Food Chem. *27*, 929 (1979).

195. J. Wong-Chong and F. A. Martin, Sugar Azucar *75*, 64 (1980).

196. M. A. Clarke, Proc. Tech. Sess. Cane Sugar Refin. Res. 138 (1981).

197. A. R. Lara, B. Slutzky, and I. Reynaldo, Cell. Chem. Technol. *17*, 197 (1983).

198. P. C. Ivin, P. C. Atkins, and R. Russ, Proc. Conf. Aust. Soc. Sugar Cane Technol. *187* (1983).

199. M. Deki, Kanzei Chuo Bunsekishoho *23*, 15 (1983).

200. N. L. Wade and S. C. Morris, J. Chromatogr. *240*, 257 (1982).

201. G. P. Jones, D. R. Briggs, and H. Toet, Food Technol. Aust. *35*, 281 (1983).

202. L. C. Zygmunt, J. Assoc. Off. Anal. Chem. *65*, 256 (1982).

203. M. L. Richmond, D. L. Barfuss, B. R. Harte, J. I. Gray, and C. M. J. Stine, J. Dairy Sci. *65*, 1394 (1982).

204. A. W. Wight and P. J. Van Niekert, J. Agric. Food Chem. *31*, 282 (1983).

205. L. E. Fitt, W. Hassler, and D. E. Just, J. Chromatogr. *187*, 381 (1980).

206. T. Yasui, T. Furukawa, and S. Hase, J. Food Ind. *27*, 358 (1980).

207. M. J. Neilson and J. A. Marlett, J. Agric. Food Chem. *31*, 1342 (1983).

208. F. W. Scott and K. D. Trick, Food Chem. *5*, 237 (1980).

209. R. Gaub, Monatsscher. Brau. *36*, 125 (1983).

210. W. J. Hurst, R. A. Martin, and B. L. Zoumas, J. Food Sci. *44*, 892 (1979).

211. J. L. Iverson and M. P. Bueno, J. Assoc. Off. Anal. Chem. *64*, 139 (1981).

212. S. C. C. Brandao, M. L. Richmond, J. I. Gray, I. D. Morton, and C. M. Stine, J. Food Sci. *45*, 1492 (1980).
213. R. B. H. Wills, R. A. Francke, and B. P. Walker, J. Agric. Food Chem. *30*, 1242 (1982).
214. G. G. Birch and L. F. Green, Food Chem. *10*, 211 (1983).
215. P. E. Shaw and C. W. Wilson, III, J. Sci. Food Agric. *34*, 109 (1983).
216. M. L. Richmond, S. C. C. Brandao, J. I. Gray, P. Markakis, and C. M. Stine, Agric. Food Chem. *29*, 4 (1981).
217. C. W. Wilson, P. E. Shaw, and C. W. Campbell, J. Sci. Food Agric. *33*, 777 (1982).
218. B. Quemener and C. Mercier, Lebensm.-Wiss. U. -Technol. *13*, 7 (1980).
219. P. E. Shaw and C. W. Wilson, J. Sci. Food Agric. *32*, 1242 (1981).
220. J. J. Warthesen and P. L. Kramer, J. Food Sci. *44*, 626 (1979).
221. J. L. Slavin and J. A. Marlett, J. Food Chem. *31*, 467 (1983).
222. W. Messerschmidt, Die Krankenhaus-Apotheke *29*, 4ʒ (1979).
223. R. Yamauchi, K. Kato, and Y. Ueno, Nippon Nogei Kagaku Kaishi *56*, 123 (1982).
224. N. Gorin and F. T. Heidema, Agric. Food Chem. *28*, 1340 (1980).
225. L. T. Black and J. D. Glover, J. Am. Oil Chem. *57*, 143 (1980).
226. M. Ghias-Ud-Din, A. E. Smith, and D. V. Phillips, J. Chromatogr. *211*, 295 (1981).
227. R. R. Selvendran, J. F. March, and S. G. Ring, Anal. Biochem. *96*, 282 (1979).
228. M. Wootton and M. A. Chaudhy, Starch *33*, 200 (1981).
229. N. Takahashi, J. Azuma, and T. Koshijima, Carbohydr. Res. *109*, 161 (1982).
230. T. Fukuda, T. Nimura, and N. Terashima, Mokuzai Gakkaishi *25*, 55 (1979).
231. S. Pulkkinen and S. Vaisanen, Pap. Puu. *64*, 72 (1982).
232. T. Hayashi and K. Matsuda, J. Biol. Chem. *256*, 11117 (1981).
233. P. P. Nefedov, A. E. Rusakov, M. A. Ivanov, L. D. Scherbakova, M. A. Lazareva, and V. I. Zakharov, Khim. Drev. *4*, 66 (1981).
234. P. Kristersson, K. Lundquist, R. Simonson, and K. Tongsvik, Holzforschung *37*, 51 (1983).
235. I. M. Popa and D. Gavrilescu, Rev. Padurilor-Ind. Lemnului. Celul. Hirtie *31*, 184 (1982).
236. H. Vik, S. Elsayed, J. Apold, and B. S. Paulsen, Int. Arch. Allergy Immunol. *68*, 70 (1982).
237. R. H. Marchessault, S. Coulomb, T. Hanai, and H. Morikawa, Transactions *TR52* (1980).

238. S. L. Fales, D. A. Holt, V. L. Lechtenberg, K. Johnson, M. R. Ladisch, and A. Anderson, Agron. J. *74*, 1074 (1982).

239. F. E. Barton, II, W. R. Windham, and D. S. Himmelsbach, J. Agric. Food Chem. *30*, 1119 (1982).

240. S. Yamamoto, T. Kondo, and Y. Mino, J. Jpn. Grassl. Sci. *26*, 305 (1980).

241. S. Yamamoto and Y. Mino, J. Jpn. Grassl. Sci. *28*, 8 (1982).

242. C. W. Ford and L. Howse, Trop. Agron. Tech. Memor. *31*, 1 (1982).

243. S. Hashimoto, H. Watanabe, K. Nishimura, and W. Kawakami, Hakkokogaku *61*, 69 (1983).

244. S. Hashimoto, H. Watanabe, K. Nishimura, and W. Kawakami, Hakkokogaku *61*, 77 (1983).

245. K. Mopper, R. Dawson, G. Liebegeit, and U. Ittekkot, Marine Chem. *10*, 55 (1980).

246. J. M. H. Dirkx and L. A. T. Verhaar, Carbohydr. Res. *73*, 287 (1979).

247. Y. Okada and K. Koizumi, Yakugakuzassi *103*, 524 (1983).

248. N. Moll, B. Gross, T. Vinh, and M. Moll, J. Agric. Food Chem. *30*, 782 (1982).

249. K. Miyahara, K. Kudo, and T. Kawasaki, Chem. Pharm. Bull. *31*, 348 (1983).

250. W. Blaschek, J. Chromatogr. *256*, 157 (1983).

251. H. James, J. Liq. Chromatogr. *5*, 767 (1982).

252. D. R. Bundle, T. Iversen, and S. Jasephson, Am. Lab. *12*, 93 (1980).

253. G. D. McGinnis and P. Fang, J. Chromatogr. *153*, 109 (1978).

254. S. Mizuno, Kagaku to Seibutsu *21*, 474 (1983).

255. K. Fukuhara, H. Murai, and S. Murao, Agric. Biol. Chem. *46*, 1941 (1982).

256. E. R. White and J. E. Zarembo, J. Antibiotics *34*, 836 (1981).

257. H. Roper, S. Roper, and K. Heyns, IARC Sci. Pub. *41*, 87 (1982).

258. K. Fukuhara, A. Nakamura, and M. Tsuboi, Chem. Farm. Bull. *30*, 1121 (1982).

8

HPLC of Glycosphingolipids and Phospholipids

Robert H. McCluer   *Eunice Kennedy Shriver Center, Waltham, Massachusetts*

M. David Ullman   *Edith Nourse Rogers Memorial Veterans Hospital, Bedford, Massachusetts*

Firoze B. Jungalwala   *Eunice Kennedy Shriver Center, Waltham,, Massachusetts*

I. INTRODUCTION

A. General Definition and Biological Importance of Glycosphingolipids and Phospholipids

Cellular lipids, which make up cell membrane bimolecular layers, are classified as simple and complex. The simple lipids are primarily triglycerides and cholesterol, and the complex lipids consist primarily of the glycosphingolipids and phospholipids. This chapter reviews HPLC methodology developed for the analysis of the glycosphingo- lipids and phospholipids, which are ubiquitous components of bio- logical membranes. The isolation and measurement of glycosphingo- lipids and phospholipids are essential for many studies on complex biological systems. The preparative or analytical separation of these lipids is generally a tedious and time-consuming process. Because of the difficulties inherent in separations by older chromatographic techniques, the use of HPLC in the purification and analysis of these lipids is proving to be a welcome addition to the research tools for membrane biology.

Glycosphingolipids are glycosides of N-acylsphingosine (ceramide). The variation in fatty acids, sphingosine, and carbohydrate moieties results in a large number of distinct molecular species. The classi- fication of glycosphingolipids has been based primarily on their carbo- hydrate structures rather than on the ceramide residues [1], and several classification schemes are in use. They are classified as neutral glycosphingolipids, and as acidic glycosphingolipids which in- clude sulfatides and gangliosides. They are also classified according to the number of carbohydrate residues present. Glycosylceramides with a large number of carbohydrate residues (20−50) have been detected. The most useful classification depends on the structure of the carbohydrate moieties, i.e., the globo, lacto, muco, ganglio, and gala series. The globo series contains globotriaosylceramide, Gal- [α(1→4) or α(1→3)]Gal[β(1→4)]Glc[β(1→1)]Cer; the lacto series

contains lactotriaosylceramide, GlcNAc[β(1→3) or β(1→4)]Gal[β(1→4)]-
GlcCer; the muco series mucotriaosylceramide, Gal[β(1→4) or β(1→3)]-
Gal[β(1→4)]GlcCer; the ganglio series gangliotriaosylceramide, GalNAc-
[β(1→4)]Gal[β(1→4)]GlcCer; and the gala series galabiosylceramide,
Gal[α(1→3) or α(1→4)]GalCer as the core structures. Details on the
structural variations, nomenclature, and distribution of glycosphingolipids have been summarized by Kanfer and Hakomori [2].

Phospholipids are the major lipid components of all biological membranes. In the mammalian system, the phospholipid composition varies in different types of membranes. The major phospholipids in the
mammalian membranes are phosphatidylcholine (PC), phosphatidylethanolamine (PE), phosphatidylserine (PS), phosphatidylinositol (PI),
sphingomyelin (SPh), phosphatidic acid (PA) and some polyphosphoinositides, phosphatidylinositol phosphate (PIP), and phosphatidylinositol bisphosphate (PIP$_2$).

B. Extraction and Isolation of Lipid Classes

Glycosphingolipids and phospholipids can be coextracted with other
lipids and some nonlipid materials by homogenizing wet tissues with
20 vol of chloroform-methanol (2:1) [3]. Quantitative extraction of
polysialogangliosides, highly complex neutral glycosphingolipids, and
polyphosphoinositides requires specialized extraction procedures [4].
Gangliosides and polyglycosylceramides, along with nonlipid contaminants in the extract, can be removed from the bulk of the lipids
by the solvent phases formed from chloroform-methanol-water (8:4:3),
originally described by Folch et al. [3]. The general scheme involving Folch's procedures for the fractionation of tissue lipids is presented diagrammatically in Fig. 1. Partition of the gangliosides and
polyglycosylceramides into upper aqueous-methanol phase is not absolute. The presence of salts (0.73% NaCl or 0.88% KCl) prevents the
partition of acidic phospholipids and sulfatides into the upper phase
[5]. Repeated extraction of the lower phase can lead to recovery of
about 90% of GM3 in the upper phase, whereas about 90% of GM4 remains in the lower phase. The gangliosides and polyglycosylceramides
(more than five carbohydrate residues) in the upper phase can be
separated from the nonlipid contaminants by dialysis [3], gel filtration [6], or by reversed-phase chromatography [7]. The gangliosides and neutral polyglycosylceramides can be separated by DEAE-
Sephadex chromatography. The lipids in the lower chloroform phase
can be fractioned by a variety of procedures [6], but one widely
used method employs chromatography on silicic acid columns [8].
Lower-phase lipids are dissolved in chloroform and applied to a silicic
acid column; nonpolar neutral lipids are eluted with chloroform, the
glycosphingolipids are eluted with acetone-methanol (9:1), and phospholipids are eluted with methanol. Sulfatides in the acetone-methanol

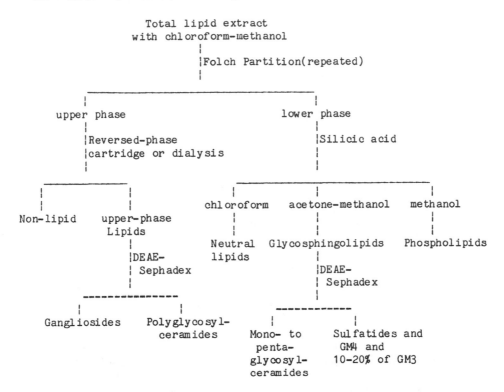

Figure 1 General scheme for extraction and fractionation of total cellular lipids based on extraction and partition with chloroform-methanol-water.

fraction can be separated from the neutral glycosphingolipids by DEAE-Sephadex chromatography [2].

The partitioning of lipids between the solvent phases and their recovery from the chromatographic steps depends to some extent on the lipid composition of the tissue source and concentrations of ions in the extract so that the procedures for extraction and fractionation of lipids need to be optimized for both the tissue and the quantity of material to be processed. Procedures applicable to individual lipids are discussed in the following sections.

C. Conventional Analytical Techniques

Neutral glycosphingolipids and ganglioside fractions can be resolved into individual components by TLC. A classical method for the analysis of plasma and urine neutral glycosphingolipids involves separation by TLC, elution from the plate, methanolysis, and subsequent analysis of the methylglycosides by GLC as their silylated derivatives [8]. Analogous methods for the analysis of gangliosides have also been described [9].

The use of high-performance thin-layer chromatography plates (HPTLC), which are prepared with uniform superfine silica gel, has provided improved resolution so that 10–15 gangliosides from brain can be distinguished on one-dimensional development. Methods for the quantitation of glycosphingolipids separated by TLC involve detection with specific spray reagents such as orcinol for neutral glycosphingolipids or resorcinol for gangliosides and subsequent scanning with commercially available TLC scanners [10]. A modification of these methods has been reported that allows quantification of gangliosides in the picomolar range [11]. Such methods require extreme care in controlling color development and spot geometry on the plates; the degree of color reaction occasionally fluctuates from band to band and from plate to plate, requiring that known amounts of standards be run close to the test samples. However, the procedure is highly sensitive and can provide accurate data. Elution of components from the TLC plates and subsequent color development in solution is probably more reliable but is less sensitive. TLC-separated gangliosides have also been analyzed by visualization with iodine or resorcinol, scraping of the silica gel, and utilization of the silica directly for the determination of ganglioside sialic acid by the resorcinol method [1].

D. Development of HPLC Technology and Its Potential Importance

Modern HPLC technology, which implies the use of injection ports for sample application, reusable columns packed with microparticulate (down to 3 µm) chromatographic adsorbants, and pumps for uniform solvent flow and on-line sample detection, provides the potential for rapid and highly sensitive methods for the separation and analysis of lipids. Numerous books [12–15], reviews, and courses that present the theoretical and practical aspects of HPLC are now available. Classical methods for the separation and analysis of the complex glycosphingolipids and phospholipids have primarily depended on class separation. The degree of heterogeneity in terms of molecular species has generally been established by analysis of the products of hydrolysis such as fatty acids, long-chain bases, and sugar residues. HPLC techniques have potential to separate and analyze the hundreds of individual molecular species of lipids. The technique provides the means not only to separate and analyze the lipids by class but also to resolve molecular species of individual classes of lipids via subsequent chromatography in a different mode.

No adequate HPLC procedures have yet been reported for the fractionation of crude lipid extracts. Currently, glycosphingolipid and phospholipid fractions are obtained by solvent partition and chromatographic procedures such as outlined above. Individual classes of glycosphingolipids or phospholipids are then separated by normal phase according to their polarity. Separation of molecular species can subsequently be obtained by reversed-phase HPLC, which separates them according to their nonpolar moieties. Argentation chromatography can further resolve components by their degree of unsaturation.

HPLC has proved to be a powerful tool for the isolation and quantitation of lipids. In this chapter, we shall first describe, in general terms, the HPLC techniques and instrumentation used for the preparative isolation and quantitative analysis of glycosphingolipids and phospholipids. We shall then describe, in greater detail, selected methods for specific separation and/or analysis. Finally, we shall present the current status of liquid chromatography-mass spectrometric (LC-MS) methods for the analysis of glycosphingolipids and phospholipids.

II. GENERAL TECHNIQUES

A. Introduction

The purpose of this section is to familiarize the reader with the general aspects of HPLC that are applicable to glycosphingolipid and phospholipid analysis. The general features of stationary and mobile phases, and detection methods that are pertinent to the isolation and analysis of glycosphingolipids and phospholipids are discussed.

B. Stationary and Mobile Phases

Most chromatography of glycosphingolipids and phospholipids involves the adsorption (normal-phase) mode with silica as the stationary phase or reversed-phase mode for molecular species separation. There are two types of silica that are predominantly used: totally porous and pellicular.

Totally porous silicas have mean particle diameters of 3, 5, or 10 μm, and have either a spherical or irregular shape. This form of silica has large linear (loading) capacities, high efficiencies, and some selectivity for the fatty acid chain length. The large linear capacities are particularly useful in preparative HPLC. The high efficiencies provide excellent resolution of glycosphingolipids and phospholipids, but the selectivity for the fatty acid chain length of glycosphingolipids can cause multiple-peak configurations (e.g., "doublets") or peak broadening, based on the population of medium- and long-chain fatty acids.

Pellicular packings have a thin porous layer (pellicle) of stationary phase coated on solid ($\sim$25-μm mean particle diameter) glass beads. They have lower linear capacities than totally porous packings, and their efficiencies are only slightly lower than the totally porous silica of 5-μm particle size. Further, two features make pellicular columns especially attractive for the analysis of glycosphingolipids: economy and selectivity. Pellicular columns can be inexpensively dry-packed in less than 20 min, and they routinely provide very high resolution for long periods of time (many months of extensive usage). The pellicular material also has less selectivity for the fatty acid portion of the complex lipids as compared to that observed for the totally porous silica. This lowered selectivity avoids peak splitting or broadening and thus actually makes it more advantageous for the quantitative analysis of neutral glycosphingolipid.

The mobile phases for the preparative isolation and quantification of glycosphingolipids and phospholipids are mixtures of relatively apolar solvents, such as hexane, with reasonably polar solvents, such as dioxane, isopropanol, or isopropanol and water. The mobile phase is commonly presented in a gradient of increasing or decreasing concentration of the polar solvent during adsorption or reversed-phase HPLC, respectively. The choice of mobile phases is limited by the mixing characteristics of the different solvents and by the method of detection.

Baseline instability can be created by poor mixing characteristics of the solvents that comprise the mobile phase. Some HPLC mixers work only by laminar flow (static mixer). However, laminar flow is usually inadequate for thorough mixing of solvents of diverse viscosities. A dynamic (mechanical) mixer, in our experience, is better. Sometimes two such mixers, or one with a dual mixing chamber is required.

Some of the polar solvents have residual UV absorption at the wavelength (203–230 nm) used for detection of glycosphingolipids, phospholipids, or their derivatives. This causes a positive baseline drift during gradient elutions. In fact, the drift can be so great that it either precludes the use of a given mobile phase, or decreases the sensitivity of the procedure. The drifting baseline is eliminated by directing the mobile phase through the preinjector flow-through reference cell (Fig. 2). This generates a horizontal baseline (because the *rate* of change in the polar solvent concentration is the same in the two detector cells when a linear gradient is generated) during the major part of the gradient run, with a negative and positive deflection at the beginning and end of the gradient, respectively. Such a flow path can also be useful for systems that have only a minor amount of absorption (e.g., isopropanol/hexane), but where high sensitivity is sought.

C. Detection Methods

A variety of detection methods have been employed to detect the separated native or derivatized complex lipids. These include refractive index (RI), ultraviolet (UV) or visible light absorption,

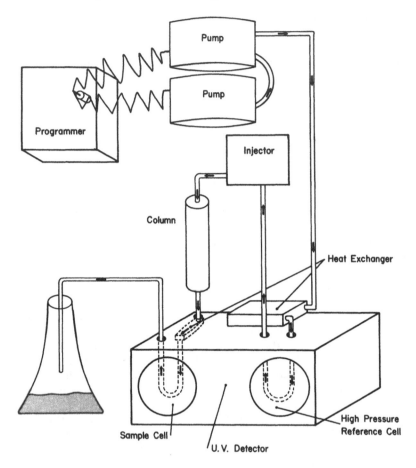

Figure 2 Diagram of chromatographic system, showing flow of solvent first through reference cell, followed by flow through injector, column, and sample cell, respectively. (From Ref. 21.)

fluorescence, and moving wire detectors. None of the detectors is universally applicable because RI detectors must be used with iso-cratic mobile phases; UV detectors must be used at low wavelengths where they are extremely sensitive to refractive index changes and to small amounts of UV absorbing contaminants, and moving wire detectors are difficult to stabilize and are no longer commercially available. Mass spectrometers are expensive; in addition, LC interfacing methods need further development. Many laboratories, therefore, have resorted to either "off-line" TLC detection of fraction aliquots [16] or "on-line" detection of lipid derivatives [17–20].

The "off-line" detection of complex lipids by TLC has been very useful over the years, but it has also been laborious and time consuming. In general, fractions are collected during the course of the gradient run, aliquots of the fractions are spotted on TLC plates, and the plates are developed in a suitable mobile phase. The developed plates are then sprayed with a specific spray reagent (e.g., orcinol for neutral glycosphingolipids) and treated (usually with heat) until color development is complete. The fractions that contain the separated neutral glycosphingolipid classes are thus identified, and the appropriate ones are combined.

The "on-line" detection of complex lipids entails either the low-wavelength detection of native lipids or the formation of UV-abosrbing derivatives prior to their injection into the HPLC. The derivatives are separated by adsorption chromatography, and in some cases the native material can be regained by mild alkaline methanolysis of the isolated derivatives.

UV detectors may be of either the *fixed*-wavelength or *variable*-wavelength type. Generally, fixed-wavelength detectors have a better signal-to-noise ratio (higher sensitivity) than variable-wavelength detectors, but the variable-wavelength detectors provide greater versatility (different absorption maxima can be accommodated), and they are more economical in the long run. For example, low-wavelength detection of phospholipids and neutral glycosphingolipids can be performed at 205 nm, perbenzoylated neutral glycosphingolipids have their absorption maxima at 230 nm, and the biphenylcarbonyl derivatives of sphingoid bases have their absorption maxima at 280 nm. Fixed-wavelength detectors for each of these derivatives would provide greater sensitivities, but the total cost of the several detectors would, or course, be very high.

It is an advantage to have a UV detector equipped with a high-pressure flow-through reference cell [21]. This type of cell allows elimination of the rising baseline that occurs during gradient elution when the polar solvent has some residual abosrption of UV light as discussed above.

UV detectors are vulnerable to ambient-temperature fluctuations at the tubing that leads from the column outlet to the detector inlet. Inadequate insulation around this tubing will lead to unstable baselines; therefore, it should be well insulated. We use a piece of rubber tubing slit from end to end and slipped over the tubing to be insulated. It is also helpful if there is a heat exchanger in the detector to stabilize the mobile-phase temperature [21]. Some manufacturers have incorporated such an exchanger into their detectors. An easy method to detect ambient temperature problems is simply to place your finger on the tubing in question for a few seconds. If there is a baseline disturbance, it is an indication that there is an ambient temperature problem.

III. GLYCOSPHINGOLIPIDS

A. Introduction

This section describes specific techniques used for the isolation of native glycosphingolipids and for the quantification of glycosphingolipids as various derivatives. The analysis of glycosphingolipids as their perbenzoylated derivatives is dealt with in detail because these derivatives have been successful in overcoming the lack of a characteristic chromophore and because the authors have had extensive experience with them. Descriptions for the analysis of ceramides, mono-, di-, tri-, tetra-, and polyglycosylceramides and gangliosides are presented. A descritpion of the available techniques for the separation of molecular species of these lipids as their perbenzoylated derivatives is also given.

B. Separation of Native Glycosphingolipids

A method has been described [22] for the purification of native neutral glycosphingolipids that utilizes columns packed with porous spherical silica particles (Iatrobeads) and a solvent gradient of increasing water content and decreasing hexane content. Specifically, lower- and upper-phase neutral glycosphingolipid fractions from human type O erythrocyte membranes were utilized. A stainless-steel column (4 mm × 50 cm) was slurry packed with 10-µm porous silica spheres (Iatrobead 6RS-8010, Iatron Chemical Co. 1-11-4 Higashi-Kanda, Chiyoda, Tokyo 101, Japan) in a mixture of tetrabromethane-tetrachloroethylene 60:40 at 400 kg/cm^2. The column was washed with isopropanol-hexane 55:45 and equilibrated with starting solvent (isopropanol-hexane-water, 55:44:1). The glycosphingolipids were injected and eluted with a linear gradient of isopropanol-hexane-water in proportions varied according to the glycosphingolipids to be separated (Fig. 3). Fractions were collected, and each fraction was analyzed by HPTLC. Glycosphingolipids containing mono- to tetrakaidecasaccharides were separated within 60 min. The method was utilized for the preparative separation of highly complex glycosphingolipids with blood group activity [23—26]. Similar procedures for the resolution of closely related neutral glycosphingolipids and gangliosides from umbilical cord erythrocyte membranes and other sources have been used extensively [27].

C. HPLC Analysis of Perbenzoylated Derivatives

1. General Comments on Perbenzoylation

The analysis of glycosphingolipids becomes increasingly sensitive and practical if derivatives with large extinction coefficients are prepared. It was initially demonstrated that this could be accomplished by derivatizing brain galactosylceramides with 10% benzoyl chloride in

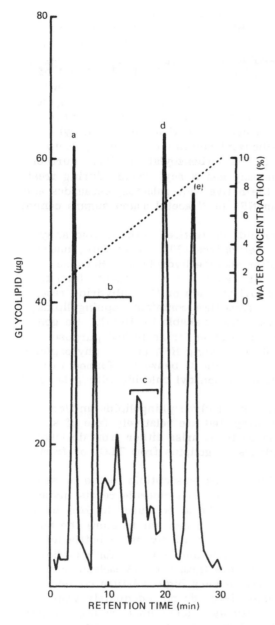

Figure 3 Preparative HPLC of glycosphingolipids without derivatization through an Iatrobead (10 μm) column (8 mm × 50 cm). Elution was with a linear gradient of isopropanol-hexane-water (55:44:1 to 55:35:10) at a flow rate of 2 ml/min. The eluates were collected every 0.5 min by a fraction collector. Peak a, GlcCer; b, LacCer; c, GbOse$_3$Cer; d, GbOse$_4$Cer; and e, GbOse$_5$Cer. (From Ref. 22.)

pyridine at 60°C for 1 hr [28]. Interestingly, the derivatives were separated into two components by HPLC on pellicular silica. The two components were shown to be the derivatives of hydroxy fatty acid (HFA) and nonhydroxy fatty acid (NFA) galactosylceramides. Attempts to recover the parent galactosylceramides by treatment of the perbenzoylated derivatives with mild alkali were successful with the HFA-galactosylceramides, but the NFA-galactosylceramides gave rise to both the native NFA-galactosylceramides and benzoyl psychosine [28]. Subsequently, it was determined that the N-benzoyl (forming a diacylamine) and O-benzoyl derivatives of NFA-galactosylceramides were formed during the perbenzoylation [29]. Further, the diacylamine derivatives randomly lost N-acyl groups during treatment with mild alkali. Thus, the native NFA-galactosylceramides and other sphingolipids that contain NFA or N-acetyl amino sugars cannot be recovered in high yields.

NFA- and HFA-glycosphingolipids derivatized with benzoic anhydride yield only the O-benzoyl derivatives [30]. Therefore, native glycolipids can be recovered after alkaline hydrolysis. Because sphingolipids that contain only HFA as N-acyl substituents form the same derivative with either benzoyl chloride or the anhydride reaction, they can easily be distinguished from HFA-containing sphingolipids, which form different derivatives distinguishable by HPLC. We generally utilize the benzoyl chloride reaction for analytical purposes, because resolution of components containing HFA and/or phytosphingosine is superior to that obtained with the O-benzoates formed with benzoic anhydride. The reactions of benzoyl chloride with galactosylceramides are shown in Fig. 4.

It is important that isolated neutral glycosphingolipids and/or ganglioside samples to be perbenzoylated are relatively free of silica or silicic acid particles. These particles arise either from dissolved silica after extraction of the glycosphingolipids from TLC, or from *fines* that avoid filtration.

The perbenzoylation procedure is also susceptible to the presence of moisture in the reaction medium. Benzoyl chloride reacts instantaneously with water (especially atmospheric moisture) to form benzoic acid and hydrochloric acid. Thus, it is important to store the bottle of benzoyl chloride in a dry place. We routinely use a bottle for only three to six months. The pyridine and toluene (for ganglioside reactions) must also be dry. We store them over 4A molecular sieves. Some batches of pyridine may have a very high moisture content, and it is likely that the molecular sieves do not adequately trap all of the water. This situation is usually evident upon mixing the pyridine with the benzoyl chloride, when a crystalline precipitate comes out of the solution almost immediately. There are other, more subtle, indications that moisture has gotten into the reaction medium also. For example, the *solvent front* of the chromatogram may be broader

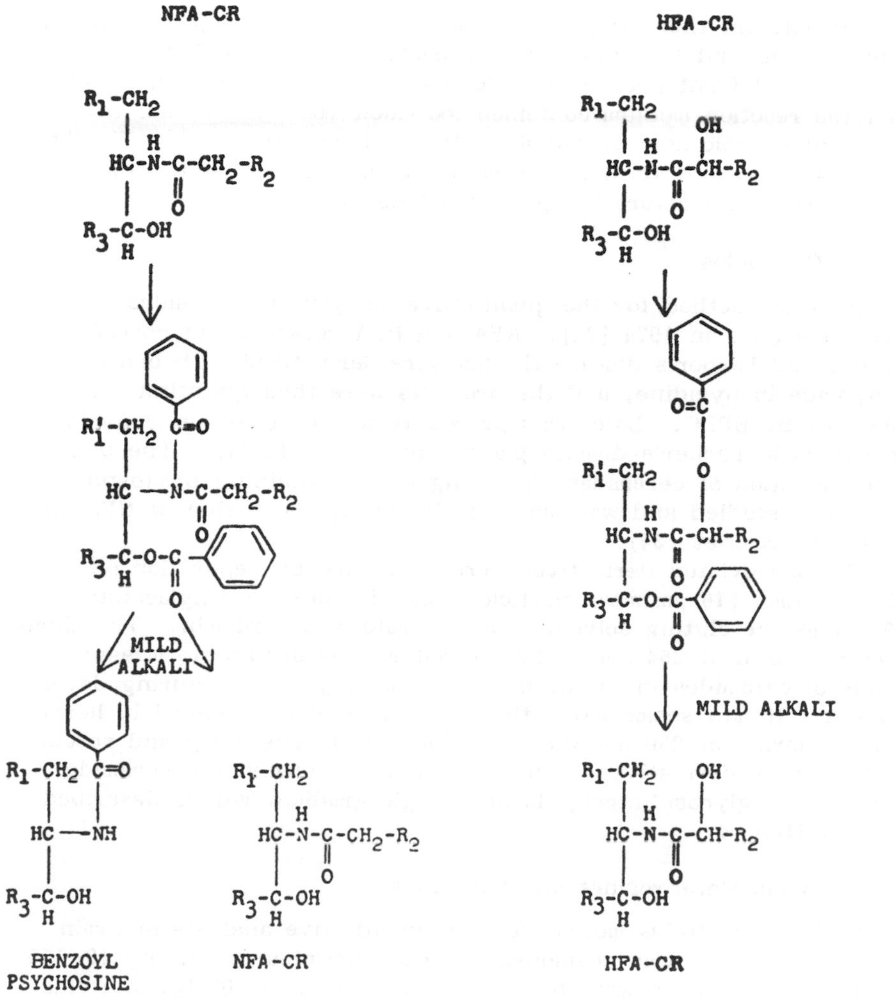

R_1 = GALACTOSYL-O-

R_1' = TETRABENZOYLGALACTOSYL-O-

R_2 = $-(CH_2)_{20}-CH_3$

R_3 = $-CH{=}CH-(CH_2)_{12}-CH_3$

Figure 4 The formation of benzoylated cerebroside derivatives by reactions with benzoyl chloride and their degradation with mild alkali.

than usual; or, there may be a broad peak that elutes just after the solvent front and that tails into the peaks to be quantified. The large solvent-front peak is benzoic acid, and it generally indicates that the reaction medium contained too much moisture so that the resultant benzoic acid overwhelmed the isolation procedure. Finally, it is also important to store the sample over phosphorus pentoxide for at least 1 hr before the perbenzoylation is performed.

2. Ceramides

An HPLC method for the quantitative analysis of ceramides was first developed in 1974 [31]. NFA and HFA ceramides in normal tissues and Farber's disease tissues were derivatized with benzoic anhydride in pyridine, and the products were then quantitatively analyzed by HPLC. Later this procedure was utilized for analysis of ceramides in Farber's disease plasma and urine [32,33]. The benzoylation reaction of ceramides, including those containing phytosphingosine, was studied and was shown to be analogous to that of NFA and HFA cerebrosides [34].

The benzoylated derivatives were quantified by separation on a silica column (10-μm mean particle size) with hexane-ethylacetate (97:3) as the eluting solvent at a flow rate of 1.5 ml/min. The eluate was monitored at 254 nm. This procedure was utilized to measure levels of ceramides in brain, liver, and kidney of rats during development. It now seems likely that a gradient of isopropanol in hexane with monoring at 230 nm would provide better sensitivity and resolution of these derivatives. The separation of benzoylated ceramides and neutral glycosphingolipids in a single gradient run is described in Sec. III.C.4.

3. Brain Cerabrosides and Sulfatides

An isocratic HPLC method for the quantitative analysis of brain cerebrosides [29] was developed in 1976. Samples containing 10–150 nmol of cerebrosides were benzoylated with 1 ml of 10% benzoyl chloride in pyridine at 60°C for 1 hr. Excess reagents were removed by solvent partition, and benzoylated NFA and HFA cerebrosides were separated on a 2.1 mm × 50 cm pellicular (Zipax) column. The cerebrosides were eluted isocratically with 7% ethyl acetate in hexane with detection at 280 nm. The procedure was utilized to measure cerebrosides by direct benzolyation of chloroform-methanol extracts of adult brain, and to analyze more purified cerebroside samples. Sulfatides did not interfere with the cerebroside analysis. Furthermore, the sulfatides could be measured as cerebrosides after desulfation with mild acid hydrolysis. Total lipid extracts of adult brain could be analyzed directly for cerebrosides without interference from other lipids. However, tissue or tissue fractions that contained low concentrations of cerebrosides required the use of larger aliquots of total lipids, and consequently inadequate chromatographic separations were obtained. Later, analysis of samples with low concentrations of cerebrosides was performed with gradient elution and 230-nm detection of the perbenzoyl derivatives [35].

Gluco- and galactosylceramides can be separated and analyzed as their perbenzoylated derivatives on a 5- or 10-μm porous silica column with isopropanol/hexane mobile phase [36]. However, the ratio of glucocerebroside to galactocerebroside must be roughly 1; otherwise the resolution degenerates as the ratio moves away from 1. A consistent separation of perbenzoylated glucocerebroside from perbenzoylated galactocerebroside, at all ratios, is obtained with HPLC of the perbenzoylated derivatives on a pellicular silica column (Zipax, 2.1 mm × 50 cm) maintained at 60°C and a 10-min linear gradient of 1—20% dioxane in hexane at a flow rate of 2 ml/min [37]. The derivatives are detected by their UV absorption at 230 nm. The system is particularly useful for the analysis of brain cerebrosides because the ratio of galactocerebroside to glucocerebroside is usually very high. This system was utilized to measure gluco- and galactosylceramide levels in different regions of Gaucher's and control human brains.

Sphingolipids containing 4-sphingenine have also been determined by HPLC after they are converted to their corresponding 3-keto derivatives by oxidation with 2,3-dichloro-5,6-dicyanobenzoquinone [33]. The 3-keto derivatives have high extinction coefficients at 230 nm, and as little as 10 nmol of ceramide, cerebrosides, sulfatides, or sphingomyelin can be quantified.

The benzoylation procedure for cerebrosides also allows the benzoylation of sulfatides, if present in the glycolipid fraction. However, during the solvent partitioning employed to remove reagents, the benzoylated sulfatides do not distribute into the hexane phase and thus do not interfere with the analysis of cerebrosides. The analysis of tissue and urinary sulfatides involves the isolation of a glycosphingolipid fraction by silicic acid chromatography and subsequent separation of sulfatides from neutral glycosphingolipids by chromatography on a small DEAE-Sephadex column [38]. The sulfatide fraction, eluted with sodium acetate, is then desulfated in anhydrous methanolic-HCl, and the resulting glycosphingolipids are recovered by reversed-phase chromatography on a reversed-phase cartridge (Sep Pak or Bond Elut) and are analyzed by HPLC. A method for the analysis of perbenzoylated sulfatides has also been developed [39]. Glycolipid fractions containing sulfatide (1—100 nmol) are perbenzoylated with 100 μl of 0.5% (w/v) benzoic anhydride and 50% saturated 4-dimethylamino-pyridine in tetrahydrofuran for 6 hr at 37°C. The sulfatide products are purified on a silicic acid column (0.4 × 30 mm) by elution with chloroform-methanol (19:1). Alternatively, they can be purified on a Bond Elut reversed-phase cartridge [40]. The perbenzoylated sulfatides are analyzed by HPLC on a Micropak-Si-10 column with hexane-2-propanol-propionic acid (100:35:2) at a flow rate of 1 ml/min.

HFA-sulfatide is eluted before NFA-sulfatide in the system. Reversed-phase HPLC of HFA- and NFA-sulfatide is performed on a Fatty Acid Analysis column from Water's Associates [39]. The individual molecular species were eluted isocratically with tetrahydrofuran-acetonitrile-water (15:25:40).

The simultaneous analysis of picomole quantities of benzoylated cerebrosides, sulfatides, and galactosyl diglycerides has been described [36]. The procedure involves benzoylation of total lipid extracts from brain and desulfation with mild acid. In this manner, the cerebrosides derived from sulfatides have one free hydroxyl group. Subsequent chromatography is on a 5-μm porous silica gel column, and a gradient of isopropanol in hexane results in the separation of the totally benzoylated cerebrosides and the cerebrosides from sulfatides that have one free hydroxyl group.

4. Mono-, Di-, Tri-, and Tetraglycosylceramides

Three different derivatization procedures have been developed for the quantitative analysis of mono-, di-, tri-, and tetraglycosyceramides: O-,N-benzoyl, O-benzoyl, and O-acetyl-N-p-nitrobenzoyl. These procedures entail the formation of the perbenzoylated or p-nitroperbenzoylated esters and/or diacylamines. The UV absorption for the derivatives is several thousand times greater than for the native lipids.

The history of perbenzoylated neutral glycosphingolipid analysis by HPLC began with the derivatization and UV detection at 280 nm [41], progressed through the development of conditions for the derivatization of neutral glycosphingolipid mixtures from plasma (mono-, di-, tri-, and tetraglycosylceramides) [17], and is now performed on neutral glycosphingolipid from a variety of tissue sources with detection at 230 nm, the absorption maximum of the derivatives [21].

O-, N-benzoyl derivatives of neutral glycosphingolipids are formed by their reaction with 10% benzoyl chloride in pyridine at 37°C for 16 hr. The reaction is terminated by drying the reaction mixture under nitrogen, and adding 3 ml of hexane to the reaction vial (13 × 100 mm screw-capped culture tube with a Teflon-lined cap). The hexane layer is washed four times with 1.8 ml of methanol saturated with sodium carbonate. The lower phases are withdrawn and discarded. Each sample is then washed once with 1.8 ml of methanol-water (80:20), the lower phase is withdrawn and discarded, and the hexane is evaporated with a stream of nitrogen. The derivatives are dissolved in carbon tetrachloride, and an aliquot is injected into the HPLC column. Alternatively, the derivatives can be isolated from the reagents and reaction by-products with a C-18 reversed-phase sample preparation cartridge by a procedure that was developed for perbenzoylated gangliosides (see Sec. III.C.6). The procedure is fast, and the recoveries are comparable to the partition work-up

[40]. However, N-acetylpsychosine cannot be used as an internal standard with this method because its partition characteristics are substantially different from those for the native neutral glycosphingolipids. The benzoylated products are stable for months, providing that they are completely free of alkali.

The perbenzoylated neutral glycosphingolipids have been separated on a pellicular (Zipax, E. I. DuPont de Nemours, Inc., Wilmington, DE) column (2.1 mm × 500 mm i.d.) with a 10-min linear gradient of 2—17% water-saturated ethyl acetate in hexane with a flow rate of 2 ml/min. The derivatives are detected by their UV absorption at 280 nm. The minimum level of detection (twice baseline noise) by this procedure is approximately 70 pmol of each neutral glycosphingolipid. The sensitivity of the procedure can be increased by the use of a mobile phase that is transparent at the absorption maximum of the derivatives, 230 nm.

The increase in sensitivity is accomplished with a 13-min linear gradient of 1—7% dioxane in hexane and a flow rate of 2 ml/min. However, it is necessary to direct the mobile phase through a preinjector flow-through reference cell to cancel the residual absorption of the dioxane (Fig. 2). Otherwise, the dioxane produces a rapid off-scale drift during the gradient run. Several UV detectors have flow-through reference cells that are pressure-rated high enough to be utilized in this system. The maximum pressure rating required is about 1000 psi. A modification of this procedure (15-min linear gradient of 0.23 to 20% dioxane in hexane) can be used to separate and quantify ceramides (derivatized by the same procedure) and the neutral glycosphingolipid in a single chromatographic run [42].

Alternatively, mobile phases of isopropanol in hexane have been used. They do not have to be directed through a preinjector flow-through reference cell because they are nearly UV transparent at 230 nm [36,43]. These systems are usually used with a totally porous 5- or 10-μm silica column, but they can be used with the pellicular silica columns as well. The separation of glucocerebroside, (with or without an internal standard of monogalactosyldiglyceride) can be carried out on a totally porous silica column with an isocratic mobile phase of 0.85% isopropanol in hexane and a flow rate of 1.5 ml/min. Alternatively, mono-, di-, tri-, and tetraglycosylceramides can be separated on the same column with a 6-min isocratic run of 0.85% isopropanol in hexane, followed by a 10-min linear gradient of 0.85 to 15% isopropanol in hexane and a flow rate of 1.5 ml/min.

Both systems—dioxane/hexane and isopropanol/hexane—have specific advantages and disadvantages. The dioxane/hexane system provides consistent resolution of the major glycosphingolipids, and in our laboratories seems to provide better resolution of the minor plasma and tissue neutral glycosphingolipids. Further, the selectivity and efficiency of this system are such that the neutral glycosphingolipid derivatives yield only one peak for each neutral glycosphingolipid rather than further partial separation of each neutral glycosphingolipid on the basis of its fatty acid composition. This system provides more consistent integration of the peaks and yields a less confusing chromatogram. The disadvantage of the system is that the mobile phase, in a gradient run but not in an isocratic run, must pass through a preinjector flow-through reference cell. Many detectors will accommodate the rather modest pressure requirements, and the hexane/dioxane system is the one of choice in our laboratories.

The isopropanol/hexane mobile pahse used with a porous silica column produces adequate resolution of the major plasma glycosphingolipids. It also shows some selectivity for the fatty acid portion of derivatized neutral glycosphingolipids, and the chromatographic peaks are often broadened or split into doublets on the basis of the population of medium- and long-chain fatty acids. It has been our experience that this contributes to a slightly lesser degree of reproducibility in the quantitation of the neutral glycosphingolipid derivatives.

The isopropanol/hexane mobile phase can be used with a pellicular silica column, but the resolution of the neutral glycosphingolipid derivatives is not as great as that obtained with either the hexane/dioxane with a pellicular column or the isopropanol/hexane system with a totally porous silica column. The isopropanol/hexane mobile phase with a pellicular column also provides resolution of some hydroxy fatty acids (HFA) or phytosphingosine-containing neutral glycosphingolipids, which are not resolved in the dioxane system [44] (Fig. 5). Obviously, the system of choice in any laboratory is dependent upon the specific objectives. Sometimes combinations of systems can be used. For example, if one system is not adequately resolving peaks of interest, the peaks can be collected and separated in the other system [45] (Fig. 5, inset).

O-benzoyl derivatives of neutral glycosphingolipids are produced via their reaction with 10% benzoic anhydride in pyridine with 5% N-dimethylaminopyridine (DMAP) as a catalyst. The reaction is run at 37°C for 4 hr [30]. Excess reagents are removed by solvent partition between hexane and aqueous alkaline methanol. The derivatives are then separated by HPLC on either pellicular or totally porous silica columns, as described for the O-,N-benzoyl derivatives (Fig. 6). The O-benzoyl derivatives can be used to distinguish between HFA- and NFA-glycosylceramides by comparing their elution times to the N-,O-perbenzoylated derivatives (see Sec. III.C.1). The O-benzoyl derivatives have also been used for preparative purposes because the native neutral glycosphingolipids are regained by mild alkaline methanolysis of the isolated derivatives. Even though the absorption maximum of O-perbenzoylated neutral glycosphingolipids is at 230 nm, the UV detector utilized for their preparative isolation can be 254-nm or a 280-nm fixed-wavelength detector because the derivatives do have some absorption at these wavelengths and large amounts of material can be chromatographed.

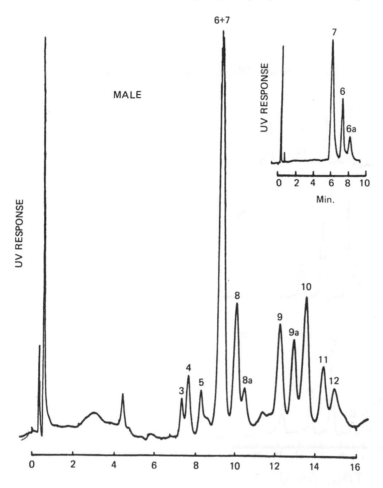

Figure 5 HPLC of male (C57BL/6J) mouse kidney perbenzoylated glycosphingolipids on a Zipax column with detection at 230 nm. A 13-min linear gradient of 1—20% dioxane in hexane at a flow rate of 2 ml/min was used as the mobile phase. Inset shows the separation of the collected peaks 6 and 7, when reinjected on the same column but with a 14-min linear gradient of 0.25—1% isopropanol in hexane as the solvent. Peak 3, Glc-Sph-Nfa; 4, Gal-Sph-Nfa; 5, Glc-Phyto-Nfa; 6, Glc-Sph-Hfa + Glc-Phyto-Hfa; 6a, Gal-Sph-Hfa + Gal-Phyto-Hfa; 7, GaOse$_2$-Sph-Nfa; 8, GaOse$_2$-Sph-Hfa + GaOse$_2$-Phyto-Nfa; 8a, Lac-Sph-Nfa; 9, GbOse$_3$-Sph-Nfa; 9a, GbOse$_3$-Phyto-Nfa; 10, GbOse$_3$-Sph-Hfa; 11, GbOse$_4$-Sph-Nfa and 12, GbOse$_4$-Phyto-Nfa + GbOse$_4$-Sph-Hfa. Sph refers to C$_{18}$-sphingosine; Phyto, to C$_{18}$-phytosphingosine; Nfa, to nonhydroxy fatty acid; Hfa, to hydroxy fatty acid. (From Ref. 45.)

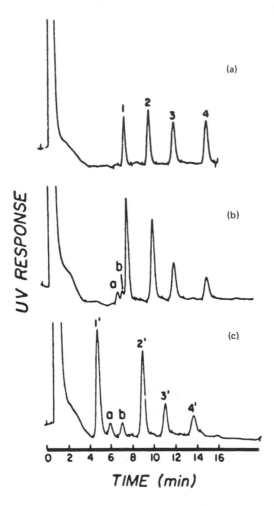

Figure 6   HPLC of perbenzoylated standard and plasma glycosphingolipids.   The derivatized glycosphingolipids were injected into a Zipax
column (2.1 mm × 50 cm) and eluted with a 13-min linear gradient of
2.5—25% dioxane in hexane with detection at 230 nm.   (a) Standard
glycosphingolipids (GSL) per-O-benzoylated with benzoic anhydride
and 4-dimethylaminopyridine (DMAP).   (b) Plasma GSL per-O-benzoylated with benzoic anhydride and DMAP.   (c) Plasma GSL perbenzoylated with benzoyl chloride.   Glycosphingolipid peaks are identified
as (1) glucosylceramide, (2) lactosylceramide, (3) galactosyllactosyl-
ceramide, (4) N-acetylgalactosaminylgalactosyllactosylceramide.   Peak
A is unidentified, and peak B is hydroxy fatty acid containing
galactosylceramide.   (From Ref. 30.)

The *p*-nitrobenzoyl derivatives are formed from O-acetylated derivatives that can be utilized for the isolation of glycosphingolipid fractions [46], by reaction with *p*-nitrobenzoylchloride in pyridine at 60°C for 6 hr. The O-acetylated glycosphingolipids (1—2 mg) are *N-p*-nitrobenzoylated with 10 mg *p*-nitrobenzoyl chloride in 0.3 ml of pyridine at 60°C for 6 hr. Samples that contain less than 0.1 mg of O-acetylated neutral glycosphingolipid are derivatized with 4 mg *p*-nitrobenzoyl chloride. The mixture is cooled, and 2 ml of 3% aqueous sodium bicarbonate solution and 2 ml of chloroform are added. The mixture is centrifuged at 600 × *g* for 3 min. Then the lower layer is washed twice with sodium bicarbonate solution and three times with 2 ml of water. The layer is then dried with a stream of nitrogen. Excess reagents and by-products are removed with a Sephadex LH-20 column. The residue is dissolved in 50 μl chloroform-methanol (1:1) (v/v) and placed on the column (10 mm × 57 cm). The column is eluted with chloroform-methanol (1:1) in 1.1-ml fractions. Neutral glycosphingolipid derivatives are isolated in fractions 11—13. The derivatives are separated on a silica column (Zorbax SIL, E. I. duPont de Nemours, Inc., Wilmington, DE) with a 40-min linear gradient of 0.5—7% isopropanol in hexane-chloroform and a subsequent hold for an additional 10 min. The flow rate is 0.5 ml/min and the derivatives are detected by their UV adsorption at 254 nm. Alternatively, the derivatives can be separated on a phenyl column with 3% isopropanol-3% chloroform in hexane and a flow rate of 1 ml/min.

In our laboratories, the formation of *O-,N*-benzoyl derivatives is used most frequently for the quantitative analysis of neutral glycosphingolipid mixtures. This procedure provides excellent resolution of the derivatives and provides slightly better resolution of the HFA- and NFA-glycosphingolipids. However, the *O*-benzoyl derivatives are also used analytically, especially as an adjunct procedure for the tentative identification of HFA-glycosphingolipids.

It is important to know the absolute recovery that one obtains for analytical procedures. There are several instances in the literature that report neutral glycosphingolipid determinations but do not account for the losses that occur during their isolation. It is well documented that neutral glycosphingolipid recoveries from silicic acid columns decrease with the length of the neutral glycosphingolipid carbohydrate chains [4]. It is therefore important to add, if possible, a radiolabeled (or other) standard of high specific activity that can be perbenzoylated and used to determine the total recovery, and, consequently, the absolute lipid concentrations of the tissue source. *N*-Acetylpsychosine has been used as an internal standard [21] for neutral glycosphingolipid determinations. However, some difficulty is encountered because the short-chain *N*-acetyl group has a slightly different solubility during partitioning than that of long-chain fatty acid glycolipids (unpublished observation).

The O-acetyl-N-p-nitrobenzoyl derivatives appear to be slightly more difficult to work with because they require (a) the formation of or the isolation as the acetylated derivatives, (b) both solvent partition and column chromatography to isolate the derivatives, and (c) long gradient times to separate them on the columns used. However, their absorption maximum is at 254 nm, which may make these derivatives more convenient for some laboratories to use. Further, conditions have been established for the separation of the ceramide molecular species of the derivatives [19,47,49] (see Sec. III.D.2).

5. Polyglycosylceramides

Polyglycosylceramides with up to ten sugar residues have been analyzed by the benzoyl chloride perbenzoylation technique [48]. The derivatives are separated on a commercially available pellicular column (Zipax, E. I. DuPont de Nemours, Inc.) and a dioxane/hexane solvent system with up to 25% dioxane at the end of the gradient run. This procedure gives promise that even higher polyglycosylceramides can be quantitifed by HPLC of their perbenzoylated derivatives.

6. Gangliosides

Gangliosides can be quantified by the analysis of their N-,O-perbenzoylated derivatives. The ganglioside carboxylic acids create two populations of derivatives, the protonated and unprotonated, as they migrate through the column. These two populations migrate through the column as two fused peaks (doublet) that never attain baseline resolution (because they are in near equilibrium). In the past, acetic acid or phosphoric acid has been added to the mobile phase to suppress this ionization. This technique has met with limited success because acetic acid absorbs UV light below 280 nm and is difficult to work with, and because phosphoric acid has solubility limitations [49,50]. Further, both acids tend to decrease column life because silica dissolves in solvents with extremes of pH.

Monosialogangliosides have been quantified by perbenzoylation with 10% benzoyl chloride in pyridine at 60°C for 1 hr [50]. These derivatives are separated on a custom-packed silica Si-4000 column with an 18-min linear gradient of 7—23% dioxane in hexane.

The monosialogangliosides and the disialoganglioside GD1a have also been perbenzoylated in 10% benzoyl chloride in pyridine at 37°C for 12—24 hr [48]. The derivatives are isolated by the same procedure as the one for the neutral glycosphingolipid derivatives formed from benzoyl chloride [21]. They are then separated on a commercially available pellicular silica column (Zipax) with a 33-min linear gradient of 7—31% dioxane in hexane. Our results showed that the procedure was not satisfactory for the perbenzoylation of GT1b because it produced a second product that comigrated with GD1a on the

HPLC columns tested. Further, both the pellicular and totally po-rous silica columns of 5 or 10 µm (mean particle size) did not sepa-rate all of the major brain ganglioside derivatives, i.e. GM1, GD1a, GD1b, GT1b, GQ1b.

Mono- and polysialogangliosides from brain can be quantified by HPLC of their perbenzoylated derivatives, which are formed in a re-action medium consisting of 5% benzoyl chloride in 25% toluene in pyridine (v/v/v) [51] (Fig. 7). Gangliosides (at least up to 20 nmol) are perbenzoylated in a 1-ml reaction vial with 100 µl of the reaction medium. The reaction is run at 45°C for 16 hr. It is ter-minated by drying the mixture under nitrogen and adding 0.8 ml of methanol. The methanolic extract is then transferred to a prewashed (3 ml of methanol) sample preparation cartridge (Bond Elut, Analyti-chem International, Inc., Harbor City, CA). The solvent is suctioned through the cartridge with a slight vacuum and saved. The reaction vial is rinsed with an additional 0.8 ml of methanol. The methanol rinse is added to the cartridge, suctioned through as before, and combined with the initial methanol eluate. The combined methanol fractions are reapplied to the cartridge. The cartridge is then washed with 4 ml of methanol, and the perbenzoylated derivatives are eluted with 3 ml of methanol-benzene 8:2. The isolated derivatives are dis-solved in carbon tetrachloride, and an aliquot is injected into the HPLC column. They are separated on a silica (3-µm mean particle size) column (4.6 mm i.d. × 150 mmL), which is maintained at 90°C. The derivatives are eluted from the HPLC column with a 15-min linear gradient of 1.8 to 12% isopropanol in hexane with a flow rate of 2 ml/min. They are detected by their UV absorption at 230 nm and are quantified by comparing the response with the external standard.

The reaction mixture contains toluene for several reasons. First, it may act as a π base and suppress the activity of the HCl generated during the reaction. This virtually eliminates the formation of multi-ple peaks for GT1b. Second, it slows the reaction rate so that a convenient (16 hr) reaction time is provided. Finally, it also facili-tates the evaporation of the HCl that is generated during the reaction.

The carboxylic acid functional group of gangliosides is not deriva-tized under these perbenzoylation conditions. If the derivatives are chromatographed at ambient temperature, the polysialoganglioside peaks become broadened and/or form doublets. Therefore, the derivatives are chromatographed on a column that is maintained at 90°C. The preparative isolation of gangliosides necessitates some additional con-siderations because their sialic acids can interact with the silanol groups of adsorption columns in either the protonated or unprotonated form. Thus, the peaks are broad and have a tendency to tail. Fur-ther, the sialic acid groups may also interact with metal ions that often contaminate HPLC columns.

The monosialogangliosides and GD1a can also be quantitated as the 2,4-dinitrophenylhydrazine derivatives of their carboxylic acids [52].

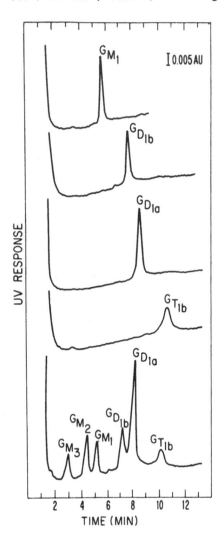

Figure 7 HPLC separation of brain gangliosides as their perbenzoylated derivatives on a silica (3 μm) column (4.6 mm i.d. × 150 mmL) maintained at 90°C. A 15-min linear gradient of 1.8–12% isopropanol in hexane with a flow rate of 2 ml/min was used as the mobile phase. Detection was at 230 nm. (From Ref. 51.)

D. Molecular Species Separation by Reversed-Phase HPLC

1. Native Gangliosides

The analysis of ganglioside fatty acids and long-chain bases has customarily been performed after acid hydrolysis. However, this type of data does not reveal the exact combination of the fatty acid with the sphingoid base in an individual ganglioside molecular species. Reversed-phase HPLC of underivatized gangliosides has been used to isolate and provide information about individual molecular species [53,54].

Human brain monosialogangliosides GM4, GM3, GM2, and GM1 were separated into molecular species by reversed-phase HPLC on a 5-μm mean-particle-size ODS column (4.6 mm × 250 mm, Altex) [54]. The ganglioside molecular species were eluted isocratically with methanol-water mixtures, with detection at 205 nm. The method permits good separation of as much as 400 μg or as little as 25 μg of monosialogangliosides. Improved resolution was obtained with slightly elevated temperature and less methanol in the mobile phase. A semipreparative reversed-phase HPLC procedure for the isolation of molecular species of GM1 and GD1a gangliosides containing a single long-chain base has also been described [55,56].

2. Derivatized Glycosphingolipids

a. Neutral Glycosphingolipids. Molecular species of neutral glycosphingolipid can be obtained by analysis of their benzoyl or O-acetyl-N-p-nitrobenzoyl derivatives [19,46,57—59] by reversed-phase chromatography. The neutral glycosphingolipid species are O-acetylated and derivatized with p-nitrobenzoyl chloride and separated by normal-phase chromatography as described above in Sec. III.C.4. The neutral glycosphingolipid derivatives are collected, as the classes elute from the column. The respective classes are then separated into their ceramide molecular species by reversed-phase HPLC. The derivatives are injected into a C-18 column (μ-Bondapak, Waters Associates, Milford, MA) and separated with acetonitrile as the mobile phase with a flow rate of 1 ml/min. The derivatives are detected by their UV absorption at 254 nm.

b. Gangliosides. The molecular species of perbenzoylated gangliosides can also be separated by reversed-phase HPLC on a C-18 column with methanol-acetonitrile-methylene chloride. Derivatives of bovine brain and mouse cerebellum have been separated, and they are currently being characterized [60]. The separation of molecular species of perbenzoylated GM1, LM1, and GM3 on C-18 reversed-phase, Accupak (3 μm) column with methanol/methylene chloride (4:1) as the mobile phase has recently been reported [61].

E. Analysis of Long-Chain Bases

Gas chromatographic methods for the analysis of sphingoid bases, produced by aqueous acid methanolysis, have been very useful [62–64]. However, they do not provide resolution of all sphingoid isomers and products. More recently, an isocratic reversed-phase HPLC method for the analysis of sphingoid bases as their biphenyl-carbonyl derivatives has been developed. These derivatives have a high extinction coefficient at 280 nm, which allows analysis of as little as 0.1 μg (0.33 nmol) of the long-chain bases [65].

Unique hydrolysis conditions are required to provide maximum yields of sphingoid bases from different sphingolipids. For maximum yields, sphingomyelins (500 μg or less) are hydrolyzed with 1 ml of 1 N HCl in water-methanol (67:33, by volume) at 70°C for 16 hr, and cerebrosides (500 μg or less) with 0.5 ml of 3 N HCl in water-methanol (1:1) at 60°C for 1.5 hr, whereas gangliosides require more vigorous conditions. To avoid formation of O-methyl ethers, ganglio-sides (10–200 μg) are hydrolyzed with 0.3 ml of aqueous acetonitrile-HCl (0.5 N HCl and 4 M H_2O in acetonitrile at 72°C for 2 hr [66].

After hydrolysis of sphingolipids, the sample is evaporated with nitrogen and the long-chain bases are partitioned between 1 ml of alkaline upper-phase methanol-saline-chloroform (48:47:3) containing 0.05 N NaOH and 5 ml of a lower phase, consisting of chloroform-methanol-water (86:14:1). The lower phase is washed once with alkaline upper phase and then three times with neutral upper phase. The lower phase is evaporated, and the residue is treated with 50 μl of 1% biphenylcarbonyl chloride in tetrahydrofuran and 100 μl of 50% aqueous sodium acetate. The biphasic mixture is agitated for 2 hr at room temperature. Five milliliters of upper-phase solvent is then added, and the mixture is shaken and centrifuged and the upper phase is removed. The lower phase is washed twice with 1 ml of methanol-0.9% saline-chloroform-15 M NH_4OH (96:92:2:3), followed by 3 washes with upper phase without alkali. The lower phase is evaporated to dryness and dissolved in methanol for HPLC analysis. The biphenylcarbonyl derivatives are separated and quantified by reversed-phase chromatography on a column (4.6 mm × 250 mm) packed with 5-μm ODS particles (Ultrasphere-ODS, Altex). Two solvent systems, tetrahydrofuran-methanol-water (25:40:20) or methanol-water (94:6), have been used to separate the long-chain bases from gangliosides [66].

IV. PHOSPHOLIPIDS

A. Class Separations of Phospholipids by Normal-Phase HPLC

Several HPLC methods for the class separation of phospholipids have been reported. The phospholipid classes can be separated on a silica

HPLC column with UV light detection at about 200 nm [67]. The absorption of low UV light for phospholipids is mainly due to double bonds in the fatty acid side-chains as well as functional groups such as carbonyl, carboxyl, phosphate, and amino groups. The response of the detector varies with the number of functional groups, primarily the number of double bonds. Thus, direct quantitation of phospholipids with an unknown degree of unsaturation is not possible with the UV detection mode. However, separated phospholipids can be collected and quantified by independent micromethods.

An HPLC separation with ultraviolet detection of commonly occurring phospholipids (Fig. 8) has been reported [68]. The analysis of brain and other tissue phospholipids in the lipid extracts is

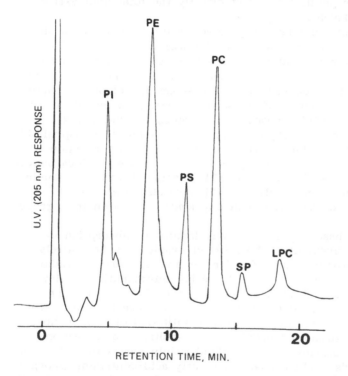

Figure 8   HPLC separation of pig liver phosphatidylinositol (PI), bovine brain phosphatidylethanolamine (PE), bovine brain phosphatidylserine (PS), bovine brain sphingomyelin (SP), and lysophosphatidylcholine (LPC) on a Micropak-SI 5-μm column. The elution was with a 20-min linear gradient of acetonitrile/methanol/water/ammonium hydroxide (15 M) from 95:3:2:0.05 to 65:21:14:0.35, pumped at 2 ml/min. The rise in the base-line at 205 nm was corrected by using a Schoffel memory module with the monitor. (From Ref. 68.)

usually performed after isolation of the phospholipid fraction from a
silicic acid column chromatography [8]. Approximately 1 nmol of a
phospholipid containing at least one double bond per molecule can be
detected. This method has been used for the separation of phospholipids from tissues and cells. Almost baseline resolution is obtained
for all phospholipids in natural mixtures. However, the separation
of PE and PS at times is variable if the amount of PS is relatively
large. The individual phospholipids have been collected from several
chromatographic runs, and the amount of each has been quantified
after solvent removal and phosphorus determination. Generally, we
have performed HPLC after separating the phospholipid fraction from
other neutral lipids and glycolipids, as discussed previously. However, if the relative amounts of these lipids are small compared to
phospholipids, it is possible to inject directly the total lipid extract
partitioned in the Folch lower phase.

Phospholipids have also been separated by HPLC on a 10-μm silica
(LiChrosorb Si-60) [69] or a silica radial compression column [70],
with n-hexane-2-propanol-water as the solvent system and with detection at 206 nm. Several other methods that utilize low-wavelength
UV detection have been developed for the separation of phospholipid
classes. Some of them successfully separate most phospholipid classes
but require complex mobile phases [71] or multiple columns [72].
Other systems utilize a variety of techniques to separate phospholipid
classes, but result in either incomplete separation or degradation of
some phospholipids. These techniques include the use of silica bonded
with benzene-sulfonate [73] or amino [74] functional groups. Mobile
phases that contain strong acids [75—77] have also been used to
separate phospholipid classes, but strong acids are known to degrade
plasmalogens.

HPLC detection based upon short-wavelength UV absorption described above is convenient. The injected samples can be entirely
recovered after chromatography for further analysis such as radio-
activity, and phosphate or fatty acid species determinations. However, there are certain drawbacks to this mode of detection. The
choice of solvents is limited because solvents that are UV-transparent
at around 200 nm are required. Direct quantitation of phospholipids
is not possible, as discussed previously. Sensitivity of detection
for the same molar amount of phospholipid mainly varies with the number of double bonds. Thus disaturated fatty acid-containing phospholipids, such as dipalmitoylglycerylphosphorylcholine, are detected with
rather low sensitivity. Small amounts of UV-absorbing impurities
could interfere with the analysis. For these reasons, other modes
of detection have been tried for the HPLC analysis of phospholipids.
An automatic phosphorus analyzer for quantitative determination of
phospholipids after HPLC has been described [78]. The eluate from
the column is directed into a rotating tabel containing 40 silica cups.
At 30-sec intervals the aliquots are automatically oxidized with nitric

acid, dried, and ashed to inorganic phosphorus. Molybdenum blue color reagents are automatically added to each cup, and the colored solution is then pumped through a 25-μl flow cell and read at 820 nm. Although reasonable linearity and reproducibility for samples in the range of 1—100 nmol has been reported, the detector is not widely used.

An alternate approach to quantitative analysis of phospholipids is to form ultraviolet-absorbing or fluorescent derivatives and analyze these by HPLC. Most of the phospholipids have functional groups available for such derivatization except for PC. The inability to derivatize PC precludes simultaneous analysis of all the phospholipids after derivatization. However, many phospholipids have been derivatized and quantitatively analyzed in the picomole range by HPLC. HPLC of biphenylcarbonyl derivatives of amino group-containing phospholipids, such as PE, PS, lysoPE, and lysoPS with UV detection at 280 nm has been reported [68,79]. The amino phospholipids containing vinylether bonds (plasmalogens) can be determined separately from diacyl and alkylacyl phospholipids by HPLC before and after treatment of the samples with HCl as done with the TLC method [80]. The lower limit of detection by HPLC of these lipids is about 10—15 pmol or 0.3—0.4 ng of phospholipid phosphorus.

A similar method for the analysis of amino group-containing phospholipids after formation of their fluorescent dimethylaminonaphthalene-5-sulfonyl derivatives has been described [81]. The sensitivity of the fluorescent method is similar to that reported for the biphenylcarbonyl group UV absorbance method. Sphingomyelin, PG, and PI can also be derivatized with 10% benzoic anhydride in tetrahydrofuran or with dimethylaminopyridine to form benzoyl derivatives of the phospholipids [68,82]. The derivatized phospholipids are then quantitatively analyzed by HPLC. Thus derivatization of phospholipids permits sensitive quantitative HPLC analysis of some phospholipid classes; however, a universal method using this approach has not emerged.

B. Molecular Species of Phospholipids by Reversed-Phase HPLC

Recently, great interest in the analysis of the molecular species of individual phospholipids has been evident. Reversed-phase HPLC has been generally used to achieve molecular species separation. The important feature of the reversed-phase chromatographic process is the magnitude of the hydrophobic interaction, which is determined by the nonpolar contact area between the solute and the ligand of the stationary phase. The actual molecular geometry involved in the binding of the solute to ligand is unknown because of scant knowledge of the topography of the stationary-phase surface and the arrangement of the hydrocarbonaceous ligands. Nevertheless, in practical chromatographic systems, it has been determined that the free-energy change is minimized when the contact area between the solute and the ligand and the net energy of interaction with the solvent are maximal [83].

Jungalwala et al. [84] reported the separation of molecular species of sphingomyelin by reversed-phase HPLC. Sphingomyelin species from bovine brain, sheep, and pig erythrocytes were resolved into 10—12 separate peaks on a μ-Bondapak C-18 or Nucleosil-5 C-18 column with methanol-5 mM potassium phosphate buffer, pH 7.4 (9:1, v/v) as the isocratic solvent and detection at 203—205 nm. The separation was based on the number of carbon atoms and degree of unsaturation in the fatty acid and sphingoid side chains of the various sphingomyelin species.

The HPLC analysis of sphingomyelin species was made quantitative and more sensitive by separating the benzoylated sphingomyelins on a Fatty Acid Analysis column from Waters Associates [82]. About 5 μg of benzoylated sphingomyelin was separated into various molecular species in 30 min with detection at 230 nm. LeBaron et al. [85] used this method to determine the turnover rates of molecular species of sphingomyelin in rat brain subcellular fractions. Although good separation was achieved on the reversed-phase column, sphingo-myelins having a double bond in the acyl side-chain (e.g., 24:1) were not resolved from the lipid having a saturated fatty acid with two less carbon atoms (e.g., 22:0). This difficulty can be overcome by the combined use of argentation HPLC and reversed-phase HPLC as reported by Smith et al. [82]. First, separation based on the degree of unsaturation was achieved with a commercially prepared silica-bonded silver column on which 3-O-benzoylated sphingomyelin was resolved into two peaks. One contained the lipid with only saturated fatty acids, whereas the other contained sphingomyelin with only the monounsaturated fatty acids. The two peaks separated on the silver column were collected and individually reinjected on the reversed-phase column to resolve all the possible molecular species of sphingomyelin.

Separation of molecular species of other phospholipids has been also achieved by reversed-phase and argentation-HPLC. Smith and Jungalwala [86] have reported the reversed-phase HPLC separation of PC from various sources. It is also shown that from the chromatographic behavior of each PC species, one can determine the relative hydrophobicity of various molecular species. All molecular species are not resolved on the reversed-phase column. However, separation based upon the degree of unsaturation can be achieved on a silver-coated silica gel HPLC column.

Patton et al. [71] also have described similar separation of molecular species of PC, PE, PI, and PS by reversed-phase HPLC with detection at 205 nm. Hsieh et al. [87] reported preparation of phosphatidic acid dimethyl esters from PC after enzymatic hydrolysis of PC followed by esterification of the phosphatidic acid with diazomethane.

The dimethyl esters of PA were then chromatographed on a reversed-phase column. Improved resolution using two Partisil-10 ODS columns in tandem was later reported for these derivatives [88]. Batley et al. [89] reported separation of diacylglycerol *p*-nitrobenzoates by reversed-phase HPLC. Phospholipids were first degraded by the action of phospholipase C to diacylglycerols, which were then derivatized and quantitatively analyzed. The latter method is based upon partial degradation of phospholipids, and may not be suitable if phospholipids labeled with radioisotopes in the base or in the phosphorus moiety are to be analyzed.

HPLC methods for the separation of alkenylacyl, alkylacyl, and diacyl acetylglycerols derived from ethanolamine glycerophospholipids and for separation of the individual molecular species from each of the separated classes have been described by Nakagawa and Horrocks [90].

V. LIQUID CHROMATOGRAPHY-MASS SPECTROMETRIC (LC-MS) METHODS

A. General Comments on LC-MS Methodology

The online combination of liquid chromatography with mass spectrometry (LC-MS) has attracted attention recently because of the success of the related technique of GC-MS. It is recognized that the combination of the versatility of liquid chromatography with the specificity and sensitivity of mass spectrometry will provide a procedure of enormous power. We have utilized the Finnigan polyiimide moving-belt interface for the LC-MS analysis of long-chain bases, glyco-sphingolipids, and phospholipids [91,92]. With this interface, as shown in Fig. 9, the HPLC eluate is applied directly onto a polyimide moving belt, which proceeds through vacuum chambers where most of the chromatographic solvent is removed. The residue on the belt is transported into an evaporator chamber adjacent to the ion source, where it is rapidly thermally desorbed.

A common problem for all LC-MS methods utilizing a moving belt is the maintenance of chromatographic resolution during transport of the sample into the ion chamber of the mass spectrometer. Karger and co-workers [93] have developed a hot gas nebulizer that provides uniform application of the solvent on the belt and maintains good chromatographic resolution. We have utilized a fine liquid stream, generated with a small orifice, to apply the chromatographic eluate as an even coating and to maintain resolution. This form of application performs well until the water composition of the mobile phase exceeds about 50%. At this water concentration, beads begin to form on the imide belt and resolution is degraded. The chromatographic effluent was formed into the stream by inserting a 10-cm section of polyimide-coated fused silica capillary GC column (0.25 mm i.d.),

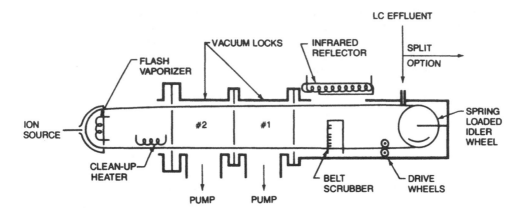

Figure 9   Simplified drawing of LC-MS interface. (Reproduced from Finnigan Catalog.)

pulled to a fine orifice into the bore of the 0.009 in. i.d. × 1/16 in. Teflon tube from the HPLC column. A 1/16-in. Swagelok fitting was then tightened over the sleeved portion of the tubing to provide a compression seal. A flow rate of 0.2 ml/min usually results in streaming, depending on solvent composition and exact size of the fine orifice. The capillary orifice is positioned about 1—2 cm above the moving belt so that the stream impacts the middle of the belt.

B. Long-Chain Bases

A reversed-phase HPLC method for the analysis of long-chain bases as their biphenylcarbonyl derivatives has been devised [65,66] and is described in Sec. III.E. During the hydrolysis of sphingolipids, several unidentified side products are formed, particularly from gangliosides, which require more vigorous conditions for quantitative liberation of the long-chain bases. These side products are not completely resolved after HPLC analysis, which leads to uncertainty of peak identification. An LC-MS method for the analysis of long-chain bases that utilizes chemical ionization-mass spectrometry (CI-MS) has recently been developed [92].

The biphenylcarbonyl derivatives of standard long-chain bases and bases liberated from gangliosides are separated. The effluent is directed through a UV detector (280 nm) and then applied to the polyiimide belt through the small orifice (described above) and flash evaporated in the ion chamber. The release of ganglioside long-chain bases with acetonitrile-HCl avoids formation of O-methyl ether derivatives. However, the allylic migration of the double bond and the formation of steroisomers and dehydration products are not avoided by this procedure. The multiple products observed by UV detection

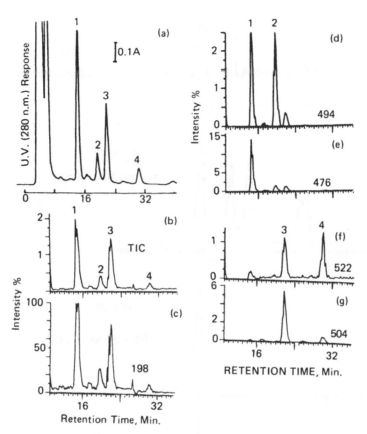

Figure 10 HPLC-MS of biphenylcarbonyl derivatives of an *O*-methyl-sphingoid fraction obtained from brain gangliosides. Ten micrograms of the sphingoid derivatives were analyzed by HPLC-MS in ammonia CI mode. Inset: (a) UV chromatogram; (b) reconstructed plot of total ion current (TIC). Reconstructed plot of specific ions monitored at (c) m/z 198, (d) m/z 494, (e) m/z 476, (f) m/z 522, and (g) m/z 504. In inset (a), peak 1, 5-*O*-methyl-C_{18}-sphingenine; peak 2, 3-*O*-methyl-C_{18}-sphingenine; peak 3, 5-*O*-methyl-C_{20}-sphingenine, and peak 4, 3-*O*-methyl-C_{20}-sphingenine as their biphenylcarbonyl derivatives. Other minor peaks were unidentified. (From Ref. 92.)

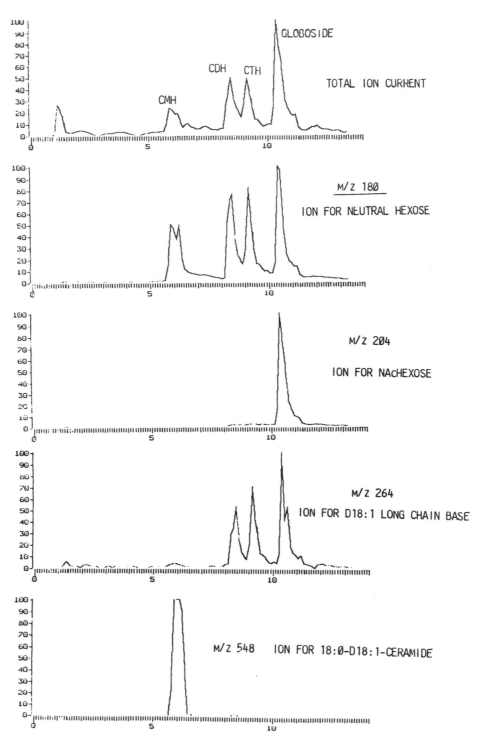

were identified directly by their CI mass spectra as they were eluted from the HPLC column (Fig. 10). The data show that by specific ion monitoring and analysis of mass spectra of HPLC resolved peaks, it is possible to directly characterize components of complex mixtures.

C. Glycosphingolipids

Informative spectra can be obtained with underivatized glycosphingo-lipids, containing up to five sugar residues, with the use of the polyiimide moving-belt LC-MS interface and CI-MS [94,95]. Chroma-tographic conditions for the separation of neutral glycosphingolipids which allow on-line CI-MS detection have been devised. Two columns (2 mm × 3 cm) packed with 5-μm spherical silica were used. A linear gradient with a ternary solvent system (methylene chloride-methanol-water) at a flow rate of 0.5 ml/min provided baseline resolution of all four glycosphingolipids within 12 min.

The total ion current and several selected ion chromatograms were obtained by LC-MS of a mixture of 5 μg each of GlcCer (synthetic *N*-stearoylglucosyl-D,L-sphinganine), LacCer (from blood plasma), GbOse$_3$Cer (from Fabry's kidney), and GbOse$_4$Cer (globoside from erythrocytes). The specific ion plots for neutral hexose (m/z 180), hexosamine (m/z 204), d18:1 long-chain base (m/z 264), and the ceramide ion for 18:0-d18:1 (m/z 548) illustrate the specificity and power of the LC-MS technique (Fig. 11).

The spectra obtained for the neutral glycosphingolipids demon-strate that LC-MS with a moving polyimide belt transport interface is useful for separation and structural analysis of complex glyco-sphingolipids. The inertness of the polyiimide belt and the relatively rapid thermal desorption of the sample from the belt make it possible for useful CI spectra to be obtained for these labile compounds. Considerable information concerning the carbohydrate composition and sequence for each glycosphingolipid is probably lost because of the mass range limitation of our instrument. However, as the number of carbohydrate residues increases, a corresponding increase in low mass fragments and a decrease in the yield of high mass ion is seen. The decrease in high mass information as the carbohydrate residue increases suggests that this approach may not be useful with glyco-sphingolipids with more than five sugar residues.

Figure 11   LC-ammonia CI-MS of mixed neutral ceramide hexosides: CMH, synthetic stearoyl-d18:0 glucosylceramide (3 μg); CDH, 16:0-d18:1 lactosylceramide, (5 μg); CTH, ceramide trihexoside isolated from Fabry's kidney, (5 μg); globoside was from red blood cells (5 μg). HPLC was on a 2 mm × 15 cm column of spherical silica (3 μm) with a 5-min linear gradient of methylene chloride-methanol-water from 93:6.5:0.5 to 65:31:4 with a 10-min hold at the final con-ditions and a flow rate of 0.4 ml/min. (From Ref. 94.)

The procedure has proven useful for the analysis of microgram quantities of complex glycosphingolipid fractions isolated from tissues. The separation and detection of various glycosphingolipid classes, as defined by carbohydrate composition, can be achieved in a short period of time. Mass spectrometric data concerning ceramide composition and data concerning the hexose composition of the carbohydrate moiety, i.e., neutral hexose and/or N-acetylhexosamine, is obtained. The method also has the potential for providing quantitative measurement of all of these components by analysis of the areas from specific ion plots. Further studies on the quantitative aspects of the LC-MS analysis of underivatized glycosphingolipids are required and will be the subject of future studies.

D. Phospholipids

As discussed previously, although significant improvement in the separation of individual phospholipids and molecular species of phospholipids by HPLC have been achieved, facile quantitative detection of these lipids is not quite satisfactory. HPLC-MS of a standard phospholipid mixture was performed with a variety of solvents and columns. The MS monitoring was generally done with ammonia as a reagent gas in the positive ion mode. Initially a silica gel column (Accupak 3 μm, 4.6 mm i.d. $\times$ 10 cm) with a gradient solvent mixture of dichloromethane-methanol-water-acetic acid was used. Although excellent separation of all the phospholipid standards was achieved, the separation deteriorated after repeated injections and variable chromatographic resolutions were obtained. The separation of acidic phospholipids PS and PI from each other and from PE was difficult under these conditions, possibly due to formation of various ionic forms of these phospholipids. This difficulty was resolved by performing HPLC with an ammonia-containing solvent and by equilibration of the sample in an ammonia-containing solvent prior to injection.

A reconstructed plot of the total ion current after HPLC-MS analysis of rat brain phospholipids (80 μg) is shown in Fig. 12a. The separation of phospholipids was achieved on a Brownlee silica gel (5 μm) cartridge column (2 $\times$ 60 mm). The column was developed with a linear gradient of solvent A, dichloromethane-methanol-water (93:6.5:0.5 v/v/v) and B, dichloromethane-methanol-water-15 M NH$_4$OH (65:31:4:0.2 v/v/v/v), starting at 88% A + 12% B to 55% A + 45% B in 10 min and programming to 100% B in 2 min at a flow rate of 0.8 ml/min. Under these conditions, fairly good reproducibility was achieved. All the phospholipids were well resolved except PS, which tailed to some extent into the lyso-PI peak. The specific ion plots for m/z 105, 141, and 142 are given in Fig. 12 (b—h). These ions are fairly specific for the individual phospholipid bases. Thus m/z 105, specific for PS, is not found to be associated with other phospholipids. Similarly, m/z 141 is relatively specific for ethanol-amine-containing phospholipids, whereas m/z 142 is specific for choline-

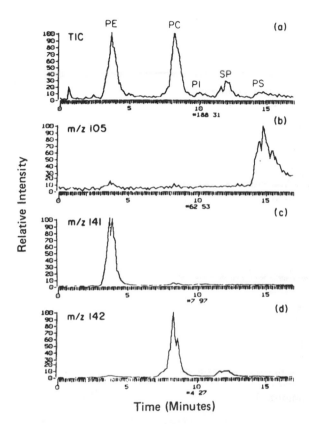

Figure 12 Reconstructed plots of total ion current (a) and various specific ions (b–h) monitored after HPLC-CIMS of rat brain phospholipids (80 µg) on a Brownlee silica gel (5 µm) cartridge HPLC column and eluted as described in the text. The eluate was applied to the moving belt of a Finnigan HPLC-MS interface. The solvent was removed by heating the belt at 330°C under vacuum, and the phospholipid was introduced into the ion source (150°C) of the mass spectrometer. Positive-ion mass spectra were continuously collected in the chemical ionization mode with ammonia as the reagent gas from m/z 100–900, every 7 sec, under the control of Teknivent model 56K MS data system. (From Ref. 91.)

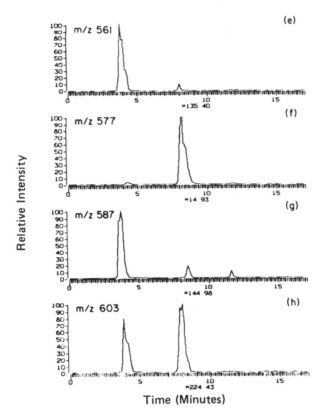

Figure 12   (continued)

containing phospholipids; m/z 198 is found to be associated only with inositol-containing phospholipids (not shown). HPLC-MS of phospholipids also shows that individual phospholipids were chromatographically resolved to some extent on the basis of molecular species. The ethanolamine-containing phospholipids were resolved into two separate peaks (Fig. 12c). The mass spectral analysis of the earlier peak showed that it contained mostly alkenylacyl GPE, m/z 561 (Fig. 12e), 587 (Fig. 12g), and 589 corresponding to plasmalogens with 16:1-18:1 (33%), 18:1—18:2 (36%), and 18:2—18:2 (31%), respectively. The front peak, however, also contained some diacyl GPE, m/z 623, 627, and 651 corresponding to 16:0—22:6 (20%), 18:0—20:4 (28%), and 18:0—22:6 (19%), respectively. The later-eluting peak contained mostly diacyl GPE with 18:0—18:1 (m/z 605, 15%) and 18:0—18:2 (m/z 603, 19%, Fig. 12h). The ratio of alkenylacyl GPE to diacyl GPE was 58:42. These results agree well with the previously published analysis [79,96].

Rat brain PC peak also split into two major peaks; the earlier peak contained mostly long-chain fatty acid species, whereas the latter contained short-chain fatty acid species. From the relative intensity of the diglyceride ions of PC, it was calculated that of the total rat brain PC 5.4% was ether-containing PC, mostly 16:1–18:0 (m/z 563, 27%) and 18:0–20:4 (m/z 613, 25%). The major diacyl-containing PC were 16:0–18:1 (m/z 577, 43%, Fig. 12f), 16:0–16:0 (m/z 551, 19%); 18:0–18:1 (m/z 605, 11%), 16:0–18:0 (m/z 579, 9%), 16:0–16:1 (m/z 549, 6%), and 14:0–16:0 (m/z 523, 6%). Eight other minor species were also recognized in the mass spectrum. It is surprising to note that rat brain PC contained several bis-saturated fatty acid-containing species in large amounts. Saturated species represented 35% of the total diacyl PC. Freysz and van don Bosch [97] also reported, in a preliminary communication, high amounts of saturated fatty acid-containing PC in rat brain. The percentage composition of individual fatty acids in PC was calculated and found to be in good agreement with the previously reported amounts.

The amount of PI in rat brain phospholipids is very small (about 4%). The specific ion m/z 198, however, clearly identified the peak as PI. The relative intensity of the $[M + 35]^+$-259 ions indicated that the major species of rat brain PI were 18:0–20:4 (m/z 662 and 627, 90%), 18:0–20:3 (m/z 664, 5%), and 18:0–22:5 (m/z 688, 5%).

The rat brain SP was resolved into two major peaks: The front peak was associated mostly with SP along with C_{18}-sphingenine and 22:0 (m/z 586 and 604, 4%), 24:1 (m/z 612 and 630, 12%), and 24:0 (m/z 614 and 632, 6%). The later peak contained SP with C_{18}-sphingenine and 16:0 (m/z 502 and 520, 5%), 18:0 9m/z 530 and 548, 54%), 20:0 (m/z 558 and 576, 5%), and C_{18}-sphinganine with 16:0 (m/z 522, 1%) and 18:0 (m/z 532 and 550, 10%). Small amounts (<3%) of other minor species of SP were also identified.

The last peak in the chromatogram was due to PS (m/z 105) The major molecular species of rat brain PS was identified as 18:0–18:1 fatty acid-containing PS.

Coupling of MS with HPLC offers an important advance in analytical methodology. The mass spectrometer not only functions as a highly sensitive universal detector, but also serves to provide valuable structural information not otherwise obtainable on small quantities of compounds. A transport type of interface has been found quite suitable for the removal of HPLC solvent and introduction of the residue on the belt into the mass spectrometer. Previously we and others had demonstrated that the moving-belt type of interface works well for such a purpose and that peak broadening is minimal [91,98]. Although theoretically any solvent could be used for HPLC analysis in ths system, practical considerations would require that solvents with high volatility be preferred to facilitate evaporation from the belt. Naturally occurring phospholipids are a complex mixture of compounds having similar polarity. Thus separation of phospholipids on a chromatographic column is difficult. Separation is also hampered by the existence of various salt forms of the same phospholipids or a mixture of salts and free bases. In our experience, the HPLC of phospholipids on silica columns is best achieved by the use of solvents containing ammonia. The same separation is consistently obtained, providing the solvent is well equilibrated with the column. Most of the commonly occurring phospholipids are well resolved; however the separation could be improved, especially for PS, which tails in this system, and to accommodate the resolution of other minor phospholipids in natural mixtures, such as PA, DPG, and lysophospholipids.

Quantitative analysis of individual phospholipids by HPLC-MS technique is yet to be developed. However, this should not be difficult because each phospholipid produces specific ions in the low mass range. We have also not explored the limits of sensitivity of this system; however, routinely we have injected about 5 µg of an individual phospholipid for complete HPLC-MS analysis. The sensitivity can be easily increased five fold to obtain reliable information on the molecular species of individual phospholipids. If one is just interested in knowing the amount of phospholipid in a mixture, specific ion monitoring in the low mass range for individual phospholipids should provide detection capability at the subnanogram level.

In summary, HPLC-CIMS overcomes the limitations of detection methods encountered in HPLC. Simultaneously the method provides extensive information on the molecular structure of each phospholipid and on the relative abundance of each molecular species of an individual class of phospholipids. This information is obtained in a few minutes. By conventional methods, such an analysis would require separation of large quantities of individual phospholipids by TLC or another chromatographic method, enzymatic and chemical degradation of the individual phospholipids, and gas chromatography.

ACKNOWLEDGMENTS

This work was supported by USPHS grants MS 16447, CA 16853, NS 10437, and HD 05515, and by the Veterans Administration. HPLC-MS results were obtained in collaboration with Mr. J. E. Evans.

REFERENCES

1. B. A. Macher and C. C. Sweeley, in *Methods in Enzymology*, Vol. 50, *Complex Carbohydrates*, Part C (V. Ginsberg, ed.), pp. 236–251 (1978).
2. J. N. Kanfer and S. Hakomori, *Sphingolipid Biochemistry*, Plenum Press, New York, 1983.

3. J. Folch, M. Lees, and G. H. Sloan Stanley, J. Biol. Chem. *226*, 497–509 (1957).

4. R. W. Ledeen and R. K. Yu, *Methods in Enzymology*, Vol. 83, *Complex Carbohydrates*, Part D (V. Ginsberg, ed.), Academic Press, New York, 1982, pp. 139–191.

5. K. Suzuki, Life Sci. *3*, 1227 (1964).

6. W. W. Christie, *Lipid Analysis*, Pergamon Press, New York, 1983.

7. M. A. Williams and R. H. McCluer, J. Neurochem. *35*, 266–269 (1980).

8. R. E. Vance and C. C. Sweeley, J. Lipid Res. *8*, 621–630 (1967).

9. G. Dawson, Ann. Clin. Lab. Sci. *2*, 274–284 (1972).

10. S. Ando, N. Chang, and R. K. Yu, Anal. Biochem. *89*, 437–450 (1978).

11. V. H. MacMillan and J. R. Wherret, J. Neurochem. *16*, 162 (1969).

12. L. R. Snyder and J. J. Kirkland, *Introduction to Modern Liquid Chromatography*, 2nd ed., Wiley-Interscience, New York, 1979.

13. E. L. Johnson and R. Stevenson, *Basic Liquid Chromatography*, Varian Associates, Palo Alto, Calif., 1978.

14. J. Q. Walker, M. T. Jackson, Jr., and J. B. Maynard, *Chromatographic Systems: Maintenance and Troubleshooting*, Academic Press, New York, 1977.

15. R. W. Yost, L. S. Ettre, and R. D. Conlon, *Prqctical Liquid Chromatography: An Introduction*, Perkin-Elmer Corp., Norwalk, Conn., 1980.

16. U. R. Tjaden, H. Karol, R. Van Hoeven, E. P. M. Oomen-Meulemans, and P. Emmelot, J. Chromatogr. *136*, 233–243 (1977).

17. M. D. Ullman and R. H. McCluer, J. Lipid Res. *18*, 371–377 (1977).

18. R. H. McCluer and M. D. Ullman, in *Cell Surface Glycolipids* (C. C. Sweeley, ed.), American Chemical Society, Washington, D.C., 1980, pp. 1–13.

19. A. Suzuki, S. Handa, and T. Yamakawa, J. Biochem. (Tokyo) *82*, 1185–1187 (1977).

20. A. Suzuki, S. K. Kundu, and D. M. Marcus, J. Lipid Res. *21*, 473–477 (1980).

21. M. D. Ullman and R. H. McCluer, J. Lipid Res. *19*, 910–913 (1978).

22. K. Watanabe and Y. Arao, J. Lipid Res. *22*, 1020–1024 (1981).

23. M. N. Fukada and S. Hakomori, J. Biol. Chem. *257*, 446–453 (1982).

24. R. Kannagi, E. Nudelman, S. B. Levery, and S. Hakomori, J. Biol. Chem. *257*, 1486 (1982).

25. S. Hakomori, E. Nudelman, S. B. Levery, D. Sotter, and B. B. Knowles, Biochem. Biophys. Res. Commun. *100*, 1578–1586 (1981).

26. S. Hakomori, E. Nudelman, R. Kannagi, and S. B. Levery, Biochem. Biophys. Res. Commun. *108*, 36 (1982).

27. Y. Okada, R. Kannagi, S. Levery, and S. Hakomori, J. Immunol. *133*, 835–842 (1984).

28. R. H. McCluer and J. E. Evans, J. Lipid Res. *14*, 611–617 (1973).

29. R. H. McCluer and J. E. Evans, J. Lipid Res. *17*, 412–418 (1976).

30. S. K. Gross and R. H. McCluer, Anal. Biochem. *102*, 429–433 (1980).

31. M. Sugita, M. Iwamori, J. Evans, R. H. McCluer, H. W. Dulaney, and J. T. Dulaney, J. Lipid Res. *15*, 233 (1974).

32. M. R. Iwamori and H. W. Moser, Clin. Chem. *21*, 725 (1975).

33. M. R. Iwamori, H. W. Moser, R. H. McCluer, and Y. Kishimoto, Biochim. Biophys. Acta *380*, 308–319 (1975).

34. M. Iwamori, C. Costello, and H. W. Moser, J. Lipid Res. *20*, 86 (1979).

35. F. B. Jungalwala, L. Hayes, and R. H. McCluer, J. Lipid Res. *18*, 285–292 (1977).

36. G. Nonaka and Y. Kishimoto, Biochem. Biophys. Acta *572*, 423 (1978).

37. E. M. Kaye and M. D. Ullman, Anal. Biochem. *138*, 380–385 (1984).

38. S. S. Raghavan, E. A. Finch, and E. H. Kolodny, Trans. Am. Soc. Neurochem. *15*, 170 (1984).

39. F. B. Jungalwala, H. Schea, S. K. Gross, M. D. Ullman, and R. H. McCluer, Abs. of Lectures and Posters for IV International Symposium on Column Liquid Chromatography, 1979, pp. 68–71.

40. D. A. Figlewicz, C. Nolan, I. N. Singh, and F. B. Jungalwala, J. Lipid Res. *26*, 140–144 (1985).

41. J. E. Evans and R. H. McCluer, Biochim. Biophys. Acta *270*, 565–569 (1972).

42. K-H. Chou and F. B. Jungalwala, J. Neurochem. *36*, 394–401 (1981).

43. P. M. Strasberg, I. Warren, M. A. Skomorowski, and J. A. Lowden, Clin. Chim. Acta *132*, 29–41 (1983).

44. R. H. McCluer, M. A. Williams, S. K. Gross, and M. H. Meisler, J. Biol. Chem. *256*, 13112–13120 (1981).

45. R. H. McCluer and S. K. Gross, J. Lipid Res. *26* (1985), in press.

46. S. Hakomori and K. Watanabe, in *Glycolipid Methodology* (L. A. Witting, ed.), American Oil Chemists Society, Champaign, Ill., 1976, pp. 13–47.

48. W. M. P. Lee, M. A. Westrick, and B. A. Macher, Biochim. Biophys. Acta *718*, 493–504 (1982).

49. R. H. McCluer and F. B. Jungalwala, in *Current Trends of Sphingolipidoses and Allied Disorders* (B. W. Volk and L. Schneck, eds.), Plenum, New York, 1976, p. 533.

50. E. G. Bremer, S. K. Gross, and R. H. McCluer, J. Lipid Res. *20*, 1028–1035 (1979).

51. M. D. Ullman and R. H. McCluer, J. Lipid. Res. *26*, 501–506 (1985).

52. Y. Kishimoto, N. Okamura, and Y. C. Lee, Trans. Am. Soc. Neurochem. *16*, 230 (1985).

53. S. Handa and Y. Kushi, Adv. Exp. Med. Biol. *153*, 23–31 (1981).

54. H. Kadowaki, J. E. Evans, and R. H. McCluer, J. Lipid Res. *25*, 1132–1139 (1984).

55. S. Sonnino, R. Ghidoni, G. Gazzotti, G. Kirschner, G. Galli, and G. Tettamanti, J. Lipid Res. *25*, 620–629 (1984).

56. S. Sonnino, G. Kieschner, R. Ghidoni, D. Acquotti, and G. Tettamanti, J. Lipid Res. *26*, 248–257 (1985).

57. A. Suzuki, S. Handa, T. Yamakawa, J. Biochem. *80*, 1181–1183 (1976).

58. O. Koul and F. B. Jungalwala, Biochem. J. *194*, 633–637 (1981).

59. S. Yahara, H. W. Moser, E. H. Kolodny, and Y. Kishimoto, J. Neurochem. *34*, 694–699 (1980).

60. M. D. Ullman, C. E. Nolan, R. H. McCluer, and F. B. Jungalwala, Trans. Am. Soc. Neurochem. *16*, 230 (1985).

61. K. H. Chou, C. E. Nolan, and F. B. Jungalwala, J. Neurochem. *44* (1985), in press.

62. C. C. Sweeley and E. A. Moscatelli, J. Lipid Res. *1*, 40–47 (1959).

63. R. C. Gaver and C. C. Sweeley, J. Am. Oil Chem. Soc. *42*, 294–298 (1965).

64. H. E. Carter and R. C. Gaver, J. Lipid Res. *8*, 391–395 (1967).

65. F. B. Jungalwala, J. E. Evans, and R. H. McCluer, J. Lipid Res. *24*, 1380–1388 (1983).

66. H. Kadowaki, E. G. Bremer, J. E. Evans, F. B. Jungalwala, and R. H. McCluer, J. Lipid Res. *24*, 1389–1397 (1983).

67. F. B. Jungalwala, J. E. Evans, and R. H. McCluer, Biochem. J. *155*, 55–60 (1976).

68. F. B. Jungalwala, S. Sanyal, and F. LeBaron, in *Phospholipids in the Nervous System*, Vol. 1, *Metabolism* (L. Horrocks, G. B. Ansell, and G. Porcellati, eds.), Raven Press, New York, 1982, pp. 91–103.

69. W. S. M. Geurts van Kessel, W. M. A. Hax, R. A. Demal, and J. DeGier, Biochim. Biophys. Acta *486*, 524–530 (1977).

70. J. L. James, G. A. Clawson, C. H. Chan, and E. A. Smuckler, Lipids *16*, 541–545 (1981).

71. G. M. Patton, J. M. Fasulo, and S. J. Robins, J. Chromatogr. *23*, 190–196 (1982).
72. R. L. Briand, S. Harold, and K. G. Blass, J. Chromatogr. *223*, 277–284 (1981).
73. R. W. Gross and B. E. Sobel, J. Chromatogr. *197*, 79–85 (1980).
74. V. L. Hanson, J. Y. Park, T. W. Osborn, and R. M. Kiral, J. Chromatogr. *205*, 393–400 (1981).
75. J. R. Yandrasitz, G. Berry, and S. Segal, J. Chromatogr. *225*, 319–328 (1981).
76. S. S. Chen and A. Y. Kou, J. Chromatogr. *227*, 25–31 (1982).
77. T. L. Kaduce, K. C. Norton, and A. A. Spector, J. Lipid Res. *24*, 1398–1403 (1983).
78. J. K. Kaitaranta and S. P. Bessman, Anal. Chem. *53*, 1232–1235 (1981).
79. F. B. Jungalwala, R. J. Turel, J. E. Evans, and R. H. McCluer, Biochem. J. *145*, 517–526 (1975).
80. L. A. Horrocks, J. Lipid Res. *9*, 469–472 (1968).
81. S. S. Chen, A. Y. Kou, and H. Y. Chen, J. Chromatogr. *208*, 339–346 (1981).
82. M. Smith, P. Monchamp, and F. B. Jungalwala, J. Lipid Res. *22*, 714–719 (1981).
83. C. Horvath and W. Melander, Chromatographia *11*, 262–273 (1978).
84. F. B. Jungalwala, V. Hayssen, J. M. Pasquini, and R. H. McCluer, J. Lipid Res. *20*, 579–587 (1979).
85. F. N. Le Baron, S. Sanyal, and F. B. Jungalwala, Neurochem. Res. *6*, 1081–1089 (1981).
86. M. Smith and F. B. Jungalwala, J. Lipid Res. *22*, 697–704 (1981).
87. J. Y. K. Hsieh, D. K. Welch, and J. G. Turcotte, J. Chromatogr. *208*, 398–403 (1981).
88. J. Y. K. Hsieh, D. K. Welch, and J. G. Turcotte, Lipids *16*, 761–763 (1981).
89. M. Batley, N. H. Packer, and J. W. Redmond, J. Chromatogr. *198*, 520–525 (1980).
90. Y. Nakagawa and L. A. Horrocks, J. Lipid Res. *24*, 1268–1275 (1983).
91. F. B. Jungalwala, J. E. Evans, and R. H. McCluer, J. Lipid Res. *25*, 738–749 (1984).
92. F. B. Jungalwala, J. E. Evans, H. Kodawaki, and R. H. McCluer, J. Lipid Res. *25*, 209–216 (1984).
93. E. P. Lankmayr, M. J. Hayes, B. L. Karger, P. Vourus, and J. M. McGuire, Int. J. Mass Spectrum Ion Phys. *46*, 177 (1983).
94. J. E. Evans and R. H. McCluer, Am. Soc. Mass Spec. 32nd Ann. Conference, 1984, pp. 683–684.

95. J. E. Evans, R. H. McCluer, and H. Kadowaki, Am. Soc. Mass Spec. 31st Ann. Converence, 1983, pp. 793–794.

96. W. T. Norton and S. E. Poduslo, J. Neurochem. *21*, 759–773 (1973).

97. L. Freysz and H. Van der Bosch, Abst. 7th Meeting of Int. Soc. Neurochem. 1979, p. 334.

98. M. J. Hayes, E. P. Lankmayer, P. Vouros, B. L. Karger, and J. M. McGuire, Anal. Chem. *55*, 1745–1752 (1983).

Author Index

Numbers in parentheses are reference numbers and indicate that an author's work is referred to although the name may not be cited in text. Italic numbers give the page on which the complete reference is listed.

A

Abbott, S., 64(1), *103*, 283(53), *299*

Abbott, S. R., 12(103), 25(103), *58*

Abeydeera, W. P. P., 290(193), *305*

Acquotti, D., 333(56), *351*

Adams, R. F., 249(28), *275*

Agren, H., 263(78), 265(78), *277*

Ahn, C. X., 257(72), 261(72), *277*

Ahr, G., 144(13), 146(13), *216*

Aitzetmuller, K., 281(27), *299*

Akai, Y., 284(70), *300*

Akaoka, I., 246(9), *274*

Åkerblom, A., 262(77), 265(77), 269(77), *277*

Albersheim, P., 283(47), 285(120), *299, 302*

Albert, A., 40(157), *60*

Alberto, B. P., 281(29), *299*

Alderlieste, E. T., 6(38), *56*

Alderweireld, F. C., 262(74), *277*

Alessi, P., 50(189,190), 51(190), *62*

Allen, T. W., 12(146), *60*, 205(66), *218*

Allsop, J., 248(22), *275*

Alpenfels, W. F., 285(116), *302*

Aman, P., 285(120), *302*

Klineberg, J. R., 246(2), 248(17), *274*

Klingen, T. J., 12(123), 35(123), *59*

Klinger, R. W., 284(63), *300*

Klussmann, R., 12(87), 20(87), *58*

Knipe, J. O., 42(176), 45(176), *61*

Knowles, B. B., 318(25), *350*

Knox, J. H., 3(17), 5(17), *55*

Knudsen, P. J., 289(162), *304*

Knudson, E. J., 263(82), 266(82), *277*

Kobata, A., 287(128,131), 289(173), 291(128,131,173), *302, 304*

Kobayoshi, M., 220(2), 225(2), 228(2), *241*

Koel, M., 288(160), 293(160), *304*

Koeller, W., 287(146), *303*

Koetitz, S. E., 257(70), 261(70), 263(70), *276*

Koizumi, K., 290(247), *307*

Kolodny, E. H., 323(38), 333(59), *350, 351*

Kolthoff, J. M., 266(18), *242*

Konami, K., 287(134), 291(134), *302*

Kondo, H., 283(58), 284(60), *300*

Kondo, T., 290(240), *307*

Könemann, H., 12(59), 16(59), *57*

Kong, R. C., 42(169,171), *61*, 141(7), 198(58), *216, 218*

Koppen, P. L., 287(133), 288(156), 293(133), *302, 303*

Kopperman, H. L., 7(40), *56*

Kosakai, M., 285(114), *302*

Kosenko, L. V., 284(64), *300*

Koshijima, T., 290(229), *306*

Kou, A. Y., 336(76), 337(81), *352*

Kouchiyama, K., 226(23,24,25), 227(23), *242*

Koul, O., 333(58), *351*

Kovats, E., 146(25), *216*

Kraak, J. C., 12(85), 19(85), *58*, 250(44), 252(44), 257(72), 261(72), *275, 277*

Králik, J., 12(79), 18(79), *57*

Kramer, P. L., 296(220), *306*

Kramer, W. G., 263(88), 267(88), *277*

Krien, P., 287(126), *302*

Kristersson, P., 290(234), *306*

Kriz, J., 12(126), *59*

Kroeff, E. P., 41(162,163), *61*

Kroll, M. G. F., 226(22), *242*

Krstulović, A. M., 12(58,139), 28(58), *57, 60*, 251(51,52,54), 254(52), 258(51,52,54), 262(51,52,54), 263(51,54), 268(54), *276*

Krzyzanowska, M., 246(10), *274*

Kucera, P., 2(7), *55*

Kudo, K., 290(249), *307*

Kuehl, D. T., 110(14,15), 113(15), 115(14), 131(14), *138*

Kugler, G., 248(23), *275*

Kuhara, S., 287(152), 291(152), *303*

Kuhnz, W., 285(123), 287(125), *302*

Kumanotani, T., 280(16,17), 283(34,35), *298, 299*

Kundu, S. K., 316(20), *349*

Kuniak, L., 283(43), 284(72), *299, 300*

Kura, G., 231(42,44), 233(44), 236(42,55), *243*

Kurás, M., 12(126), *59*

Kurihara, Y., 280(24), *298*

Kurose, Y., 283(34,35), *299*

Kurz, W., 8(44), *56*

Kushi, Y., 333(53), *351*

Kuster, B. F. M., 283(40), *299*

Kustes, B. F., 280(18), *298*

Kuz'menkov, M. I., 220(8), 231(8,45,46), 233(45,46), *241, 243*

Subject Index

S

Printed and bound by CPI Group (UK) Ltd, Croydon, CR0 4YY

17/10/2024

01775696-0015